Biotechnology and Biology of Plants

Biotechnology and Biology of Plants

Dr. Kishore Kumar

RANDOM PUBLICATIONS
NEW DELHI (INDIA)

Biotechnology and Biology of Plants

ISBN 978-93-5111-323-2

Published in 2014 in India by

RANDOM PUBLICATIONS

4376-A/4B, Gali Murari Lal, Ansari Road
New Delhi-110 002
Phone : +9111-43580356, 011-23289044
e-mail: randompublications@hotmail.com

Reprinted 2025

Type Setting by : Friends Media, Delhi-110089
Digitally Printed at: Replika Press Pvt. Ltd.

Contents

Preface

Plant biotechnology may be defined as the art, science and application of knowledge obtained from the study of life sciences to create technological improvements and change the genetics of plants in order to produce desired characteristics in plant species. This can be accomplished through many different techniques ranging from simply selecting plants with desirable characteristics for propagation, to more complex molecular techniques.

Genetic engineering opens a totally new dimension for bioprospecting. The search for new genes and their application is the primary objective of the biotech industry. Gene technology now enable humans to integrate revolutionary new properties in to cultivated plants through inter-specific or inter-generic gene transfer which was not possible through classical approach of crop improvement. Genetic engineering is the direct introduction into a plant of an isolated or modified single gene using transformation techniques. Now it is possible to transform almost all of the plants cultivated by man. Where this has not happened, it is more to do with the insignificance of the plant in agriculture or forestry rather than a reflection of a stubborn resistance to all attempts to transform it. A couple of years ago, cereals were regarded as being recalcitrant species. This has new changed and transformation has been reported as being successful for nearly all cereal species.

Plants provide an essential foundation for life on earth, the food we eat and the beauty of the natural environment and as a result botany is considered to be an extremely important science! Botany is no longer confined to the study of how and why plants survive in the way they do. With the advancement of technology the study of botany means the study of how plants are used in biotechnology, the significant part that they play in ecosystems and of course how they can continue to provide food, medicines and even fuel to an ever-expanding global population. This book will be highly beneficial to student, teachers and research workers in the field of plant biotechnology, physiology, agriculture, and plant science.

Author

1

Biotechnology and Plant

BIOTECHNOLOGY IN PLANT CELL

Clonal propagation refers to the process of asexual reproduction by multiplication of genetically identical copies of individual plants. The vegetative propagation of plants is labour-intensive, low in productivity and seasonal. The tissue culture methods of plant propagation, known as 'micropropagation' utilizes the culture of apical shoots, axillary buds and meristems on suitable nutrient medium. The regeneration of plantlets in cultured tissue was described by Murashige in 1974. Fossard gave a detailed account of stages of micropropagation.

The micropropagation is rapid and has been adopted for commercialization of important plants such as banana, apple, pears, strawberry, cardamom, many ornamentals (*e.g.* Orchids) and other plants. The micropropagation techniques are preferred over the conventional asexual propagation methods because of the following reasons: (a) In the micropropagation method, only a small amount of tissue is required to regenerate millions of clonal plants in a year., (b) micropropagation is also used as a method to develop resistance in many species., (c) in vitro stock can be quickly proliferated as it is season independent,. (d) long term storage of valuable germplasm possible.

The steps in micropropagation method are:

- *Initiation of culture*: From an explant like shoot tip on a suitable nutrient medium,
- Multiple shoots formation from the cultured explant,
- Rooting of in vitro developed shoots and,
- Transplantation - transplantation to the field following acclimatization.

The factors that affect micropropagation are:

- Genotype and the physiological status of the plant *e.g.* plants with vigorous germination are more suitable for micropropagation.,
- The culture medium and the culture environment like light, temperature etc. For example an illumination of 16 hours a day and

8 hours night is satisfactory for shoot proliferation and a temperature of 25°C is optimal for the growth.

The benefits of micropropagation this method are:

- Rapid multiplication of superior clones can be carried out through out the year, irrespective of seasonal variations.
- Multiplication of disease free plants *e.g.* virus free plants of sweet potato (Ipomea batatus), cassava (Manihot esculenta)
- Multiplication of sexually derived sterile hybrids
- It is a cost effective process as it requires minimum growing space.

SOMACLONAL VARIATION

The genetic variations found in the in vitro cultured cells are collectively referred to as somaclonal variation and the plants derived from such cells are called as 'somaclones'. It has been observed that the long-term callus and cell suspension culture and plants regenerated from such cultures are often associated with chromosomal variations. It is this property of cultured cells that finds potential application in the crop improvement and in the production of mutants and variants (*e.g.* disease resistance in potato).

Larkin and Scowcroft working at the division of Plant Industry, C.S.I.R.O., Australia gave the term 'somaclones' for plant variants obtained from tissue cultures of somatic tissues. Similarly, if the tissue from which the variants have been obtained is having gametophytic origin such as pollen or egg cell, it is known as 'gametoclonal' variation.

They explained that it may be due to:

- Reflection of heterogeneity between the cells and explant tissue,
- A simple representation of spontaneous mutation rate, and
- Activation by culture environment of transposition of genetic materials.

Shepard *et al.* also contributed by screening about 100 somaclones produced from leaf protoplasts of Russet Burbank. They found that there was a significant amount of stable variation in compactness of growth habit, maturity, date, tuber uniformity, tuber skin colour and photoperiodic requirements.

Somaclonal Variations has been used in plant breeding programmes where the genetic variations with desired or improved characters are introduced into the plants and new varieties are created that can exhibit disease resistance, improved quality and yield in plants like cereals, legumes, oil seeds tuber crops etc. Somaclonal variation is applicable for seed

APPLICATIONS OF SOMACLONAL VARIATIONS

- Methodology of introducing somaclonal variations is simpler and easier as compared to recombinant DNA technology.
- Development and production of plants with disease resistance *e.g.* rice, wheat, apple, tomato etc.

- Develop biochemical mutants with abiotic stress resistance *e.g.* aluminium tolerance in carrot, salt tolerance in tobacco and maize.
- Development of somaclonal variants with herbicide resistance *e.g.* tobacco resistant to sulfonylurea
- Development of seeds with improved quality *e.g.* a new variety of Lathyrus sativa seeds (Lathyrus Bio L 212) with low content of neurotoxin.
- Bio-13 – A somaclonal variant of Citronella java (with 37% more oil and 39% more citronellon), a medicinal plant has been released as Bio-13 for commercial cultivation by Central Institute for Medicinal and Aromatic Plants.
- Supertomatoes- Heinz Co. and DNA plant Technology Laboratories (USA) developed Supertomatoes with high solid component by screening somaclones which helped in reducing the shipping and processing costs.

Production of Virus Free Plants

The viral diseases in plants transfer easily and lower the quality and yield of the plants. It is very difficult to treat and cure the virus infected plants therefore te plant breeders are always interested in developing and growing virus free plants.

In some crops like ornamental plants, it has become possible to produce virus free plants through tissue culture at the commercial level.

This is done by regenerating plants from cultured tissues derived from:

- Virus free plants,
- Meristems which are generally free of infection - In the elimination of the virus, the size of the meristem used in cultures play a very critical role because most of the viruses exist by establishing a gradient in plant tissues. The regeneration of virus-free plants through cultures is inversely proportional to the size of the meristem used.,
- Meristems treated with heat shock (34-36°C) to inactivate the virus,
- Callus, which is usually virus free like meristems.
- Chemical treatment of the media- attempts have been made to eradicate the viruses from infected plants by treating the culture medium with chemicals *e.g.* addition of cytokinins suppressed the multiplication of certain viruses.

Among the culture techniques, meristem-tip culture is the most reliable method for virus and other pathogen elimination. Viruses have been eliminated from a number of economically important plant species which has resulted in a significant increase in the yield and production *e.g.* potato virus X from potato, mosaic virus from cassava etc. These virus free plants are not disease resistant so there is a need to maintain stock plants to multiply virus free plants whenever required.

Production of Synthetic Seeds

In synthetic seeds, the somatic embryos are encapsulated in a suitable matrix (*e.g.* sodium alginate), along with substances like mycorrhizae, insecticides, fungicides and herbicides. These artificial seeds can be utilized for the rapid and mass propagation of desired plant species as well as hybrid varieties.

The major benefits of synthetic seeds are:

- They can be stored up to a year with out loss of viability
- Easy to handle and useful as units of delivery
- Can be directly sown in the soil like natural seeds and do not need acclimatization in green house.

Mutant Selection

An important use of cell cultures is in mutant selection in relation to crop improvement. The frequency of mutations can be increased several fold through mutagenic treatments and millions of cells can be screened. A large number of reports are available where mutants have been selected at cellular level. The cells are often selected directly by adding the toxic substance against which resistance is sought in the mutant cells. Using this method, cell lines resistant to amino acid analogues, antibiotics, herbicides, fungal toxins etc. have actually been isolated.

Production of Secondary Metabolites

The most important chemicals produced using cell culture are secondary metabolites, which are defined as' those cell constituents which are not essential for survival'. These secondary metabolites include alkaloids, glycosides (steroids and phenolics), terpenoids, latex, tannins etc. It has been observed that as the cells undergo morphological differentiation and maturation during plant growth, some of the cells specialize to produce secondary metabolites. The in vitro production of secondary metabolites is much higher from differentiated tissues when compared to non-differentiated tissues. The cell cultures contribute in several ways to the production of natural products.

These are:

- A new route of synthesis to establish products *e.g.* codeine, quinine, pyrethroids,
- A route of synthesis to a novel product from plants difficult to grow or establish *e.g.* thebain from Papaver bracteatum,
- A source of novel chemicals in their own right *e.g.* rutacultin from culture of Ruta,
- As biotransformation systems either on their own or as part of a larger chemical process *e.g.* digoxin synthesis.

The Advantages of In Vitro Production of Secondary Metabolites

- The cell cultures and cell growth are easily controlled in order to facilitate improved product formation.
- The recovery of the product is easy.
- As the cell culture systems are independent of environmental factors, seasonal variations, pest and microbial diseases, geographical location constraints, it is easy to increase the production of the required metabolite.
- Mutant cell lines can be developed for the production of novel and commercially useful compounds.
- Compounds are produced under controlled conditions as per the market demands.
- The production time is less and cost effective due to minimal labour involved.

APPLICATIONS OF SECONDARY METABOLITES

Many of these secondary products especially various alkaloids are of immense use in medicine. The yield of these chemicals in cell culture, is though generally lower than in whole plants, it is substantially increased by manipulating physiological and biochemical conditions.

Shikonine is a dye produced by the cells Lithospermum erythrorhizon on a commercial scale. Besides this there are a number of secondary metabolite products that are being widely used for various purposes. Vincristine is used as anticancer agent, digoxin controls cardiovascular disorders, pyrithrins is an insecticide etc. The production of specialty chemicals by plants has become a multibillion industry.

Table. Showing Plant Species and Secondary Metabolites Obtained from them Using Tissue Culture Techniques

Product	Plant Source	Uses
Artemisin	Artemisia spp.	Antimalarial
Azadirachtin	Azadirachta indica	Insecticidal
Berberine	Coptis japonica	Antibacterial, anti inflammatory
Capsaicin	Capsicum annum	Cures Rheumatic pain
Codeine	Papaver spp.	Analgesic
Camptothecin	Campatotheca accuminata	Anticancer
Cephalotaxine	Cephalotaxus harringtonia	Antitumour
Digoxin	Digitalis lanata	Cardiac tonic
Pyrethrin	Chrysanthemum cinerariaefolium	Insecticide (for grain storage)
Morphine	Papaver somniferum	Analgesic, sedative

Quinine	Cinchona officinalis	Antimalarial
Taxol	Taxus spp.	Anticarcinogenic
Vincristine	Cathranthus roseus	Anticarcinogenic
Scopolamine	Datura stramonium	Antihypertensive

Production of Somatic Hybrids and Cybrids

The Somatic cell hybridization/parasexual hybridization or Protoplast fusion offers an alternative method for obtaining distant hybrids with desirable traits significantly between species or genera, which can not be made to cross by conventional method of sexual hybridization.

SOMATIC HYBRIDIZATION

Somatic hybridization broadly involves in vitro fusion of isolated protoplasts to form a hybrid cell and its subsequent development to form a hybrid plant.

The process involves:

- Fusion of protoplasts,
- Selection of hybrid cells,
- Identification of hybrid plants.

During the last two decades, a variety of treatments have been used to bring about the fusion of plant protoplasts. Protoplast fusion can be achieved by spontaneous, mechanical, or induced fusion methods.. These treatments include the use of fusogens like $NaNO_3$, high pH with high Ca^{2++} ion concentration, use of polyethylene glycol (PEG), and electrofusion. These inducing agents used in protoplast fusion are called 'fusogen'.

PEG treatment is the most widely used method for protoplast fusion as it has certain advantages over others.

These are:

- It results in a reproducible high-frequency of heterokaryon formation.,
- The PEG fusion is non specific and therefore can be used for a wide range of plants.,
- It has low toxicity to the cell and
- The formation of binucleate heterokaryons is low.

MECHANISM OF FUSION

The fusion of protoplasts takes place in three phases- agglutination, plasma membrane fusion and formation of heterokaryons. When the two protoplasts come in close contact with each other, they adhere to each other. This agglutination can be induced by PEG, high pH and high Ca^{2+}. The protoplast membranes get fused at localized sites at the point of adhesion. This leads to the formation of cytoplasmic bridges between protoplasts. High pH and high Ca^{2+} ions neutralize the surface charges on the protoplasts which allows closer contact and membrane fusion between agglutinated protoplasts. The fused protoplasts become round as a result of cytoplasmic bridges which leads to the formation of spherical homokaryon or heterokaryon.

APPLICATION OF DNA TECHNOLOGY IN PLANTS

Genetic engineering opens a totally new dimension for bioprospecting. The search for new genes and their application is the primary objective of the biotech industry. Gene technology now enable humans to integrate revolutionary new properties in to cultivated plants through inter-specific or inter-generic gene transfer which was not possible through classical approach of crop improvement.

Genetic engineering is the direct introduction into a plant of an isolated or modified single gene using transformation techniques. Now it is possible to transform almost all of the plants cultivated by man. Where this has not happened, it is more to do with the insignificance of the plant in agriculture or forestry rather than a reflection of a stubborn resistance to all attempts to transform it. A couple of years ago, cereals were regarded as being recalcitrant species. This has new changed and transformation has been reported as being successful for nearly all cereal species.

TRANSGENIC PLANTS

Transgenic is used to describe plants which have had DNA introduced into them by means other than by the transfer of DNA from a sperm cell to an egg. Development achieved through modern biotechnology in the form of transgenic have given new options of controlling insects, diseases and weed pests, while improving the overall integrity and consequences of agricultural practices.

Biotechnology combined with modern crop management techniques and the responsible use of pesticides gives the best tools available to ensure healthy, high yielding crops. However, the benefit expected from the release of genetically modified organisms into environment are quite more which are seems to pay major role in bioremediation, environmental improvement, agriculture, food industry and health care. A number of transgenic plants in various crop species covering wide range of altered characteristics have been approved for commercial cultivation. .

Resistant to Biotic Stresses

Damage to crops by biotic factors like insects-pests and weeds is a major limiting factor in agriculture economic in tropical and temperate regions of the world. Despite large scale investment in the chemical control of pests, they cause great economic loss by damaging or destroying the crops. With a tremendous development in techniques to engineer crops genetically, we now have the ability to make broader use of natural insecticides. The choice of a Bt endotoxin, as the first insecticidal protein for introduction into plants, was based on the extensive knowledge gathered about this class of crystal protein since 1902 and such a strategy has been successfully achieved. The application of glyphosate kills crop plants just as effectively as it kills weeds. The usefulness of glyphosate as a weed control agent in agriculture could be enhanced if resistance to herbicide could be selectively engineered into crop plants. Transgenic have been developed by introducing the corresponding resistance genes from weeds and microbes into crop plants such as tobaccos. In addition genes conferring resistance against herbicides, *viz.*, Sulphobylurea, Imidazolinones, Phosphinothricin, Asulam, Bromoxynil and 2,4-D have been engineered and transgenic plants developed.

Abiotic Resistance

A large area of our country is under stress conditions *i.e.* saling, alkaline water, cold and heat stress etc. transgenic produced from introducing mtl 1D gene from E. coli in tobacco and arabidopsis showed tolerance to high leave of NaCl. Thermoc tolerance was found in Arabidopsis plants engineered with hsf gene. It has been reported that over-experession of sac B gene from Bacillus subtillis leads to high level of fructans in tobacco cells, and this is associated with increased drought tolerance whereas chemically synthesized antifreeze protein genelala 3 leads to improved freezing tolerance of tobacco plants.

Genetic engineering for improved tolerance to abiotic stresses is, therefore, the need of the hour as the existing cultivars in most cases are capable of giving much higher biomass that what we harvest today. While it is true that the response of plants to these stresses is mutagenic, the recent success achieved in genetic engineering against excess salt, high and low temperature as well as less or excess water by altering individual gene is noteworthy.

Product Quality

A group of scientists have reported the ethylene synthesis in transgenic tomato plants, using anti-sense to a gene that has been indentified as ACC oxides. The enzyme convert ACC to ethylene. Ethylene production was inhibited by 97% with a signification reduction in over-ripening and shriveling. Flavour Savr and Endless Summer in tomato, Freedom II in squash, high Laura acid rapeseed (Canola) and Round Up Ready soybean are some examples of the crops that already being commercially grown in developed countries.

Genetic engineering of metabolic pathways of fruits and vegetables has therefore given control over post-harvest process which not only leads to improvement of quality characteristics but also enhances process ability, transportability and prolongation of shelf like.

In addition, engineered plants with change fatty acid composition of edible oil, reduced antinutritional components in many food crops, production of palm oil or animal derived industrial oil into plants like Brassica are some of the good examples of novel products that can be expected from biotechnology. Significant progress has been achieved for production of biodegradable plastics derived from biomass of genetically altered plants having gene from bacteria.

With the advancement in DNA technology and understanding the structure and function of gene and its products more and more transgenics are coming out from various research programme all over the world. However, boon of biotechnology has been confined to few developed countries and developing countries are still in vogue to harvest the benefits of modern technology. More that 48 transgenic has picked up remarkably in last couple of years. During 1986-87, 25000 field trials were conducted with transgenic of more that 60 crops in 45 countries. Sixty per cent of these trials were in first 10 years and 40% in the last two years. Area under transgenic cultivation has also increased tremendously. During 1996, the transgenic crops covered an area of 2.8 million has which increased about 4 times during 1997 and 10 times during 1998.

Table. Area of Transgenic Crops Planted (m ha)

Country	1997	1998
USA	8.1	20.5
Argentina	1.4	4.3
Canada	1.3	2.8
Australia	0.1	0.1
Mexico	<0.1	0.1
Spain	0.0	<0.1
France	0.0	<0.1
South Africa	0.0	<0.1
Total	11.0	27.8

A major challenge for agriculture in the 21st century in developing nations would be to speed up production process economically to fulfill the aspirations of huge popuulace, to achieve diversification and adding value to the primary produce so as to make agriculture enterprise farmers as well as environmental friendly. Biotechnology based advance technologies are expected to materialize many of our expectations in the coming millennium.

Viruses

To minimize the losses due to viral disease and to reduce the use of chemicals, the cross protection phenomenon of viral coat protein and movement proteins have been employed. This method has been used to develop transgenic plants which conferred significant levels of resistance in tobacco plants to TMV or to alfa mosaic virus (AMV) using the AMV coat protein gene. In some viruses it has been found that certain sequences termed satellite RNAs act to reduce disease symptoms in infected plants and as such it is being utilized for developing transgenic with virus resistance. Antisense RNA which binds to sense RNA and artificial ribosome molecules which cleave specific gene transcripts were found very powerful tools for the inhibition of viral disease development. Genetic engineering of antisense or ribosome genes complementary to viral genes whose products are essential may prove one of the best way of viral disease control.

Insects

Transgenic crop plants that produce pesticide proteins such as Bt toxin can offer the growers advantage of increased yields, numerous environmental benefits due to decreased use of conventional insecticidal. A transformed tobacco plants with the gene encoding the cowpea trypsin inhibito, in feeding experiments demonstrated that expression of the gene conferred resistance to a range of insect genera, including Heliothis, Spodoptera, Diabrotica and Tribolium the protein is not however toxic to humans.

Herbicides

The use of herbicide has become an important feature of modern agriculture as a means of controlling weed species, however, it is economically feasible to produce specific herbicides for use with every individual crop species. A more generic approach would be to transform a range of crop plants with a gene which confers tolerance to a single non-selective herbicide.

INVESTMENTS IN PLANT BIOTECHNOLOGY

The implicit rationale for IPR protection is that they promote investments in plant breeding and bio-engineering. This hypothesis is contested by section of analysts. Nevertheless it is a fact that the structure of IPRs do influence investments. IPRs induce their own pattern of innovations. The time taken for R&D to fructify as innovations is a crucial aspect guiding investments.

Often IPR regimes play a key role in influencing the time pattern of innovations. It is a well-known fact that conventional plant breeding methods are dilatory and time consuming. It requires 7 to 8 generations of repeated breeding to reduce heterozygosity of new genotypes. As Vossen observes Arabica Coffee subject to three to four cycles of breeding still display heterozygosity and it is not until 20 to 25 years (spanning 4 to 5 generations) that one arrives at true breeding varieties. This time lag is even more pronounced for the Arabusta coffee variety (inter-specific hybrid of Arabica and Robusta Coffee) where segregation of unfavourable characteristics could still occur even after several generations of backcrossing and selfing.

Another interesting example is the one cited by Evenson about the long gestation period in regard to rice varieties in India. According to Evenson, India released 306 rice varieties for planting during the period 1965-1986 after performing 20,000 crosses over a period of over 15 years since 1950.

It must be understood that this long process of backcrossing was preceded by considerable time spent on location and identification of individuals plants with requisite traits. The speed or pace at which plant varieties qualify for being released for planting have economic implications on investment behaviour in seed/plant breeding industry. Asexual modes of plant breeding which include plant genetic 'engineering' and transgenic techniques could obviate the long gestation period associated with pre-release preparations encountered conventional breeding.

It is a different matter than biosafety concerns may delay releases of these varieties. The time lag in release of plant varieties fulfilling the criteria of novelty, distinctness, uniformity and stability (or alternatively the requirements of novelty, inventiveness and commercial application) raise fundamental issues from the view-point of capital investments. The time lag in varietal release is potentially least in the case of transgenic and non-sexually produced varieties while for traditional or classical methods the time lag can be greater. The UPOV 1991 based plant IPR regimes could induce longer time lags as compared to the situation when an IPR regime is based on the 1978 version of the UPOV. This is due to the fact that the minimum genetic distance concept implicit in UPOV 1991 precludes 'close cousin' varieties from acquiring protection rights, unless the techniques of genetic distance determination by themselves are inadequate.

UPOV 1978 would facilitate capital investments to some extent as the release of plant varieties is periodically more frequent. Even under the plant patent regimes, investments could bring about short-term returns, though the volume of investments will have to be larger on account of capital intensity of advanced plant genetic engineering technologies.

By comparison, UPOV 1991 regime is least conducive to capital investments as plant variety releases can be painfully slow under this regime. Therefore sustenance of conventional plant breeding under UPOV 1991 regimes, can only be ensured by infusion of low discount capital. Paradoxically it is the reduced availability of low discount public funds for plant breeding that has induced countries like India to go in for Plant Varieties Protection Legislations modelled on UPOV lines. The implicit assumption in undertaking this step is that private funds for R&D in plant breeding will 'crowd in', if protection plant varieties rights are accorded. But the reality is that in liberalised deregulated developing economies, private investments have 'crowded in', in short return promising avenues.

In other words, private investment funds for R&D are not likely to be channellised into medium to long return yielding areas. Then the relative rates of discount will be $r1 > rb < rc$ and relative present worth will be $pa < pb > pc$.

Consequently classical plant breeding activities will suffer for want of funds in the UPOV 1991. In other words, UPOV 1991 modelled national legislations will not necessarily hold attraction for private sector investors in R&D projects involving classical plant breeding as compared to say, UPOV 1978 or the Plant Patent regimes. Rather private investments would be catalysed better by *sui-generis* plant protection legislations, which partake of plant patent elements and focus in the main on hybrids and genetically engineered varieties.

Given the linkages between plant breeding and the seed industry these perverse investment behaviour trends can produce major economic consequences. The Seed Laws in various countries, both advanced and developing, are getting to be progressively oriented towards restricting market circulation to varieties that are conferred with plant breeder or patent rights. The UPOV guidelines on DUS are used world wide not only as the basis for establishing varietal distinctiveness and descriptions but also for seed certification purposes. The DUS guidelines are employed for recognising and registering not only the 'basic' seeds used for multiplication of 'pedigree' seeds but also for certification of the pedigree seeds themselves.

Evenson explains how absence of IPR rights will reduce Marginal Variable Costs (MVC) of seed multiplication. Tougher controls by certification of seeds could ensure that seed supply is kept in a 'constricted mode' and MVC of seed multiplication are raised. This in turn, will produce its own supply-demand dynamics. The implications of IPR regimes in terms of the supply curves for seeds.

The situation of 'liberal' plant varieties protection laws (which do not provide for genetic distance) and equally liberal 'seed laws' which permit non-registered varieties (including landraces) to be sold in open markets, the macro-supply curve of seeds will be elastic (S1). When these regimes shift towards 'genetic dependency' PBR regimes and tighter seed laws that prohibit marketing of non-registered varieties, the supply curve will tend to be relatively inelastic (S2).

Analogously, demand curves could shift from D1 (wherein demand for protected varieties will be elastic on account of availability of landraces and non-protected varieties in the markets) to D2 (where the demand for the protected variety will be inelastic on account of ban on sale of landraces and other non-protected varieties). In the D1-S1 situation, supply constraints and demand flexibility could induce low price regimes while in D2-S2 the reverse situation would occur. D2-S2 situation is not desirable for countries like India for other reasons as well.

The experience of transitional economies of East Europe, which had re-modelled their Seed Laws in conformity with European Community Seed Laws, has been sad. Farmers of the East European bloc cultivating and conserving landraces have been adversely affected by new seed laws, which by prohibiting trade in non-registered varieties not fulfilling the DUS criteria have contributed to the erosion of landraces in these countries. It is thus clear that agro-biodiversity can be adversely affected by the uniformity criteria of DUS, which brings about resultant changes in demand and supply position of seeds.

IPR REGIMES FOR PLANT BIOTECHNOLOGY

Article 27(3) of the TRIPs Agreement lays down that members shall provide for the protection of plant varieties either by patents or by an effective *sui-generis* system or by any combination thereof. Article 27(3) in particular is critical to India as this requires us to provide protection of plant varieties by conferring plant breeders rights either by way of patents or through *sui-generis* plant variety protection legislations. Countries with strong R&D base in plant genetic engineering such as USA have robust Plant Utility Patent Legislations.

India is certainly not inclined to adopt patent protection regimes for its plant varieties. India is rather inclined to adopt a *sui-generis* legislation which is non-'patent' based. There are many reasons for this proclivity. India is one of the ten mega-diversity countries of the world and a rich storehouse of landraces of principal agricultural crops. India has a strong R&D base in conventional plant breeding methods. Its strength in plant genetic engineering is impressive but not an overwhelming factor by comparison. The first two strengths explain India's disinclination towards plant patent regimes or towards a *sui-generis* legislation which is 'patent'-driven.

While Plant Utility Patents Act provide for broad patents over plant varieties, traits and genes and even the physical parts of the plants, plant breeders rights provide for IPR only over varieties. As is well known, since 1990s the UPOV (Union for the Protection of Varieties) has been largely viewed by developing countries as offering the best regime for positioning their national legislations.

The central feature of the UPOV is the protection it affords to plant breeders who produce plant varieties that fulfil the criteria of distinctiveness uniformity and stability (DUS). The current version of the UPOV, *viz.* UPOV 1991 had added additional criteria of 'new' to DUS thus rendering DUS as NDUS. Contrary to the popular notion, the NDUS criteria of UPOV 1991 is not substantively different from the principles of 'novelty', 'inventiveness' and 'industrial application' (NII) which applies for patents. The criteria of 'novelty' and 'inventiveness' in Patent Laws are covered by the criteria of 'new' and 'distinct' in UPOV 1991. Thus by distinctness, the UPOV means a variety of plant which is 'clearly distinguishable from other varieties whose existence is a matter of common knowledge'. It is apparent that this term captures the attributes of 'novelty' and 'inventiveness' implicit in Patent Laws. Even in respect of 'uniformity' and 'stability' criteria the UPOV does not offer different recipes.

True, by 'stability', the UPOV conveys that 'relevant characteristics of protected plant variety remain unchanged either for a specified period or after repeated propagations or cycles of propagations'. It is also true that 'stability' is a difficult criteria for a plant breeder to fulfil. Attainment of 'stability' criterion is problematic for cross-pollinated plants and non-single homozygous lines of autogamous plant varieties. This, in turn, reduces the commercialisation potential of the plant variety since inconsistency of genetic quality jeopardises commercial application of the plant variety. The same holds true of the criterion of "uniformity". Therefore, the NDUS criteria of

UPOV 1991 are homologous to the NII criteria implicit in Plant Utility Patent Laws.

The UPOV has undergone two major amendments since its inception in 1961, *i.e.* in 1978 and 1991. The main concern with regard to currently operational UPOV 1991 has been its stringent provisions, which include the additional attribute of 'new' to the DUS criteria. The scope of UPOV 1991 in terms of coverage is larger. While plant varieties of all taxa are covered by UPOV 1991 only plant varieties of nationally designated species were covered by UPOV 1961 and 1978. Further by defining clearly that 'variety' represents a 'group of plants', or 'single' or 'several plants' or 'one' or 'several parts' of a plant(s) the UPOV 1991 has considerably enlarged the scope of its coverage. Going by this definition even sexual and vegetatively propagated varieties are covered by UPOV 1991.

The US PVPA legislation has been quick to adopt the UPOV 1991 formula. The amendments of 1994 to the US, PVPA broadens the term 'plant varieties' to include all materials harvested from the protected varieties. The other drastic feature of UPOV 1991 has been abridgement of the 'plant breeders exemptions' and 'farmers privileges' conferred by the earlier versions. Earlier versions of UPOV allowed farmers to re-use seeds generated from their farms for self-use, and allowed breeders to freely use protected varieties for further improvements. These concessions gave UPOV a 'less stringent' face in comparison to 'patent' laws. UPOV 1991 gives an option to National Governments to disallow the farmers privileges of retaining or re-using seeds for self cultivation. UPOV 1991 also restricts the breeders rights by enlarging the right of the breeder to close variations of his cultivar or 'essentially derived varieties'. In other words, the breeder of a cosmetically bred variety would have to buy genetic dependency rights from the derived variety prior to commercialising the derived version. Further the duration of protection have also been extended under UPOV 1991 to 20 years for crops and to 25 years for trees and vines. Thus in terms of these rigourous features, UPOV 1991 eliminates the arbitrage between plant varieties protection and patent regimes.

POST R&D REGULATIONS AND INVESTMENTS

Post-R&D regulations play a critical role in deciding on the pace of investments in the biotechnology sector. Even if a start-up enterprise enjoys a favourable IPR regime, that rewards its inventions, it is unlikely that it will invest in product development, in case post-R&D regulations are rigid and dilatory in nature. This then forms the backdrop for strategic alliances between start-up ventures and downstream industrial complexes. Multi-layered regulations in the post-R&D phase involve complexities which are expressed in the delays in commercial application/utilization of invented products or technologies.

In the case of third generation plant biotechnologies, such as transgenics, the problem is compounded by the fact that regulations by themselves are complex and less understood by the regulators themselves. A transgenic plant would have to undergo multi-stage regulatory checks and clearances before finding commercial application.

India is uniquely situated as far as biotechnology investments and infrastructure are concerned. In the area of plant biotechnology, advances have been made in transgenics of rice and wheat carrying stress tolerant genes such as CodA, COR47 and HVA1. The larger challenge is to take these products through the regulatory processes. More formidable is the issue of market development and appropriate pricing of transgenic seeds.

There are many imponderables here. Market shares for seeds in India have varied from crop to crop and from region to region. The 'demand' for transgenic plants or seeds may be even more uncertain, given the 'safety' dimensions and risk aversion propensity of Indian farmers.

In the dry agricultural pockets of Central and South India, farmers tend to be attached to traditional drought-hardy varieties of plants, which in turn contributes to their initial 'inertia' to shift to transgenics. Thus the task of marketing and pricing of novel biotechnology plants gets to be complex in such environment of 'demand incoherence'. A high margin pricing strategy for seeds may choke demand for transgenics.

Thus the degree of risks in commercializing transgenic plants could be of a high order. Given these realities, the idea of strategic alliances between start-ups and downstream entities looks to be realistic. The fact that premier R&D institutions such as the Indian Institute of Science, has set up its own start-up biotechnology venture, renders a strategy of tie-ups, a logical option. Much will depend upon the complexities inherent in product development and the willingness of competent downstream entities to signup agreements. To a large extent, this will depend upon enabling policies at the upper end of the value chain.

To sum up, India's agricultural biotechnology industry has the necessary policy and regulatory support to promote active R&D. However, in the critical area of product development and commercialization much more attention needs to be paid.

The multi-tiered regulatory framework that exists could seriously affect the performance of start-up companies that would like to run through the entire life-cycle of biotechnology product development. Consequently, partnerships and strategic alliances should be encouraged between start-up ventures and established companies that have high product development capabilities. The experience of other countries in the Asia Pacific region could be a key input in framing development strategies for biotechnology products in India.

GENOME RELATIONSHIPS AND WIDE-HYBRIDIZATION

Domestication of crops is likely to have captured only part of the allele diversity present in the wild progenitor gene pools, leaving many desirable traits absent from the crop. In some cases, such as barley 3 *Hordeum spontaneum* or maize 3 teosinte, crops can be crossed easily with wild relatives and there are few barriers to the incorporation of novel alleles since the genomes are essentially homologous.

In contrast, recently evolved polyploids such as hexaploid wheat will have had few opportunities to enlarge the gene pool. This is reflected by RFLP studies of wheat, barley and maize which show that wheat is less polymorphic.

The frequency with which allopolyploid species occur in nature shows that the processes of pollen germination, correct pollen tube growth, fertilization of the egg cell, and embryo development are often conserved in related species. This can be exploited to synthesize new combinations of genomes. Such interspecific or intergeneric hybrids are often referred to as wide-hybrids or wide-crosses. Wide-hybrids can be used to enlarge crop gene pools for any character but are most valuable where suitable genetic variation does not exist within a species, or where screening for a desired character would be difficult, time consuming or expensive. Disease resistance or abiotic stress tolerances are the characters most frequently targeted for transfer.

Barriers to Wide-hybridization

The ease of producing wide-hybrids, and the ease with which genetic material may be transferred by backcrossing to a crop parent, varies greatly from cross to cross. Successful gene transfer may be restricted by prefertilization, postfertilization or chromosome instability barriers. These are not necessarily directly related to the degree of genetic divergence between two species and the genetical control of hybridization barriers may be simple.

Prefertilization barriers usually involve failure of pollen germination or inhibition of pollen tube growth, which may be at the surface of the stigma, in the stigma, or in the transmitting tissue of the ovary wall. The best studied incompatibility reactions affecting cereal wide-hybrids are due to the *Kr* genes of wheat where dominant alleles inhibit the growth of pollen tubes of other species at the base of the style.

These genes are best studied in crosses with rye and *Hordeum bulbosum*. No method is known for overcoming the effect of these genes, so wide-hybrids involving bread or durum wheat usually use genotypes with recessive, noninhibiting alleles. In grasses there are, of course, well-characterized genes for self-incompatibility (the *S* and *Z* loci) but it is not known whether there is homology between these and the *Kr* loci or whether a similar mechanism of inhibition operates.

Postfertilization barriers leading to seed abortion are common in widehybrids. In cereals, this is frequently due to defective development and subsequent collapse of the endosperm. Embryos, however, are often viable, and hybrid plants can be recovered if the embryos are excised from developing seeds and cultured on suitable nutrient agar. In the *Triticeae*, using these techniques, a huge catalogue of wide-hybrids have been produced.

For example, hexaploid, tetraploid and diploid wheats and their wild relatives can be intercrossed but embryo rescue is usually unavoidable, especially if different ploidy levels are crossed. Similarly, ovule or embryo culture has been used to recover hybrid plants from *Brassica* crosses. In some crosses, for example between bread and durum wheats, hybrid necrosis kills the developing seedling. This is due to the epistatic interaction of independent genes in the respective species. Crosses within the separate gene pools are unaffected, probably because selection has eliminated necrotic combinations of alleles. Incompatibilities between nuclear genomes and the cytoplasm donated by the female parent may also contribute to lethality or sterility in certain crosses.

In vitro fertilization using isolated gametes, and the subsequent recovery of plants, is a means by which prefertilization and some postfertilization barriers might be overcome. Currently, *in vitro* fertilization has only been used successfully in maize, but such systems have potential for wide-crosses.

A further barrier to the production of wide-hybrids is the phenomenon of chromosome elimination. In such cases fertilization results in the formation of a hybrid zygote, but the complete chromosome complement of one of the parents is lost during subsequent mitotic divisions, leading to the production of a haploid embryo. The best characterized system is the cross between barley and its wild relative *Hordeum bulbosum*. However, the phenomenon occurs in a wide range of other cross combinations in the *Triticeae* including wheat 3 *H. bulbosum*, wheat 3 barley and wheat 3 maize, sorghum or pearl millet.

The mechanism of chromosome elimination is not understood but involves the inability of the chromosomes of one genome to successfully attach to the mitotic spindle. In some crosses, the occurrence of chromosome elimination is affected by the parental genomes since some barley 3 *H. bulbosum* crosses produce a high frequency of barley haploids while others give a high frequency of relatively stable hybrids. The genetical control of this regulation is probably fairly simple. Although chromosome elimination is detrimental to gene transfer, efficient haploid production methods benefit breeding and genetic analysis programmes and are extensively used by workers in these areas.

Utilization of Hybrid Plants

Once hybrid plants have been produced there are usually problems of sterility due to the inability of the hybrid to form viable gametes, particularly male gametes. Usually, this is due to a lack of chromosome pairing and chiasma formation at meiosis between the respective genomes and the subsequent formation of unbalanced gametes from random combinations of univalents. This can occur even in cases where close homoeology between the genomes can be shown by comparative chromosome and genetic analysis. Clearly the mechanisms controlling the pairing of chromosomes can be so precise as to prevent homoeologous pairing between different genomes. Such mechanisms may be genetically simple, as in wheat where pairing between the A, B and D genomes of bread wheat is restricted by a few pairing control genes of which the most powerful is *Ph1* on chromosome 5B.

Fertility can be partially or fully restored in many hybrids by doubling the chromosome number with colchicine or an equivalent spindle inhibitor as this provides identical pairing partners for each chromosome. In such cases new allopolyploid 'species' can be established.

The major motivation for creating wide-hybrids is the introgression of useful genes into cultivated species. Generally, there are three levels at which the transfer of genetic material can be considered. Firstly, a whole genome can be introduced into an existing crop species to create a new allopolyploid species.

For example, the introduction of a diploid rye genome into durum or bread wheat to create hexaploid or octaploid triticales, respectively, has been highly successful. New genome combinations such as *Tritordeum* (the amphiploid between wheat 3 *Hordeum chilense*) are also being investigated as crops for specific environments.

Secondly, whole chromosomes can be introduced to create substitution or addition lines. Thirdly, translocations can be induced between homoeologous or nonhomoeologous chromosomes so that only individual arms or small segments are introduced. The starting point for introgression is to backcross a wide-hybrid to the cultivated species.

In many cases this is possible using the hybrid as the female parent, even if the hybrid is highly sterile, since a proportion of egg cells contain a nonreduced or otherwise viable chromosome complement.

If there is homoeologous pairing between genomes, translocation products can be identified using *in situ* hybridization or molecular markers and stabilized by further backcrossing. Backcrosses can also be tested for the presence of the desired trait or traits in the 'alien' chromosome segment.

Combinations of cytological and marker techniques can then be used to characterize suitable lines.

In some hybrids pairing between homoeologous genomes does not occur, or occurs too infrequently to be useful. In wheat, this can be overcome by using the *ph1* mutant or plants lacking chromosome 5B (carrying the *Ph1* pairing control gene). This allows higher levels of homoeologous chromosome pairing.

Alternatively radiation treatments can be used to induce translocations. In such circumstances it is usually more effective to develop single chromosome addition or substitution lines prior to selecting recombinants, particularly as such stocks can confirm that the desired trait is carried by a single chromosome.

Plant breeders are generally interested in maintaining the characteristics of the cultivated species while adding one or more useful genes from the alien source. Generally the aim is to reduce the size of the alien chromosome segment as much as possible since donor species may have many alleles that are undesirable in the crop. This is greatly assisted by the use of molecular markers which define the alien segment. *In situ* hybridization using total genomic DNA or species-specific repeats is also valuable because it reveals the number and size of the alien segments. *In situ* methods are also increasingly able to locate low-copy sequences on plant chromosomes, thereby linking the molecular and physical mapping.

Comparative mapping information is important for the manipulation of alien chromosome segments because of the importance of producing plants with genetically balanced genomes.

For example, substitution of rye chromosome 1 for a wheat group 1 chromosome is successful because these chromosomes show conserved synteny throughout their length.

In contrast, rye chromosome 7 is translocated with respect to the wheat chromosomes of groups 2, 4 and 5 and substitution of rye chromosome 7 for a wheat group 7 chromosome produces a plant with reduced doses of group 7 segments and extra doses of group 2, group 4 and group 5 segments.

Comparative mapping therefore provides a framework for the assembly of new chromosome segments in genetically balanced combinations. This principle also applies to other crop groups such as *Brassica*.

FRUITS OF TRANSGENIC AGRICULTURE

RICE WITH VITAMIN A

Rice does not contain very much vitamin A. In the poorer parts of Asia, where rice is almost the only food of the rural population, a vitamin A deficiency is common, leading to early blindness. Now Drs. Ingo Potrykus and Peter Beyer, two genetic engineers, have transferred the genes for vitamin A from other species into rice, creating a strain of rice which is rich in vitamin A—the amount of rice in a typical third world diet could provide about fifteen per cent of the recommended daily allowance of vitamin A, sufficient to prevent blindness.

Now that a few plants with this trait have been created, they are being cross bred with other varieties of rice using conventional breeding techniques, as has been done for centuries. Such cross breeding could further increase the vitamin A content.

The development of rice with vitamin A was carried out at the Swiss Federal Institute of Technology, making free use of patented technology and of the earlier research which had established the basic facts about how plants synthesize vitamins.

The corporations holding the various patents all agreed to cost-free use of their patents as long as the rice was to be provided free to poor third-world farmers. The new rice strain was then turned over to the International Rice Research Institute, a non-profit organization based in the Philippines, where it will be evaluated for its adaptability to various growing conditions, food safety, and environmental impacts, etc. The IRRI preserves thousands of varieties of rice with different genetic characteristics, so the new strain can be cross bred to produce varieties suitable for almost any locality. The result is that rural Asians can soon expect to retain normal eyesight. Genetic engineers also intend to produce a rice variety rich in iron, because iron-deficiency anemia is a common problem in the same rural populations. But this is a more difficult problem than increasing rice's vitamin A content. Rice contains a substance called phytate. Phytate prevents the body from absorbing iron, so it does little good to breed for increased iron content, and the rice plant cannot reproduce without adequate phytate in the grains. Dr. Potrykus hopes to be able to find a gene coding for a protein that will break down phytate when the rice is cooked.

No-Till Agriculture

The world's biggest environmental problem is loss of topsoil to wind and drainage. The US experienced its dust bowl during the early part of the 20th century as a result of ploughing up the prairie. The problem is much worse in tropical soils, which may have a thin, inches thin, layer of topsoil above a type of soil which, when ploughed, turns into a non-porous concrete-like substance. One field, one crop, once. The problem, both in the prairie and in the tropics, is deep ploughing, which kills weeds which would otherwise crowd out the desired crop.

The solution is called low-till agriculture. The soil is broken up but not deeply ploughed. Weeds are killed instead by herbicides. A herbicide of choice should be cheap, quickly biodegradable and non-toxic. An excellent choice is a chemical called glyphosate, except that glyphosate kills the crops as well as the weeds. So genetic engineers found a gene which lets plants tolerate glyphosate, and transferred it into soybeans. Today, 63% of the soybeans grown in the US are glyphosate tolerant, allowing soil saving no-till agriculture on half the US soybean acreage.

There is another advantage to no-till agriculture. There are lots of plant residues beneath the ground, both root systems and humus transported by earthworms. Ploughing brings this material to the surface, where it can oxidize. Carbon dioxide is created, a greenhouse gas. So transgenic soybeans are a positive factor in postponing global warming. Approximately four tons of carbon dioxide are retained in the soil per acre per year. This saving is applicable to the accounting of CO_2 reductions in the Kyoto Treaty on Climate.

In fact, any technology which reduces the need to plough, spray, or till crops will reduce carbon dioxide emission. Consider a tractor pulling a ten foot wide harrow over a square mile of agricultural land. Simple arithmetic shows that the tractor will travel 528 miles, all the while burning gasoline.

Perhaps you are thinking that even if ploughing has disadvantages, herbicides don't sound very good either. The very word means "plant killer".

But that is not the choice. Traditional soybeans are also grown using herbicides, most of which are far more toxic than glyphosate. On the average, the transgenic soybeans actually use 30% less total herbicide than conventional soybeans. So environmentally, this is a no-brainer.

WITCHWEED CONTROL

Farmers in east Africa are plagued by a devastating parasitic weed called Striga, or witchweed.

Farmers are used to dealing with weeds that grow in the soil alongside the crop and compete for nutrients. From time immemorial, they have dealt with those weeds by pulling them up by hand. Less neighbour intensive methods like spraying and ploughing are now common. But none of these methods work for the witchweed. Striga attacks plants directly, underground, even before the weed has emerged above the soil surface. It sucks nutrients from the seeds and the roots of the crop. In some parts of Africa, the striga parasite destroys as much as 80% of the crop yield.

But now that a herbicide resistance trait can be transferred to a crop, scientists in Israel and Kenya, working together, have demonstrated a new strategy for striga control. Before planting the crop, they soak its seeds in a herbicide.

The seeds are unharmed, but they become poisonous to the striga parasite. The seed germinates and sprouts without interference. By the time the crop is harvested, the herbicide has decomposed and disappeared.

Their demonstration used herbicide resistant transgenic corn. The same strategy would probably work with Africa's other important grains, sorghum and millet.

Soaking seeds would use far less herbicide than spraying it on the ground, and the complex spraying apparatus would not be needed. This is a significant consideration in Africa, where so many farmers are too poor to own expensive equipment.

CHEESE CHYMOSIN FROM YEASTS

Hard cheeses are made from whole milk by adding an enzyme called chymosin (rennet), which was formerly extracted from the stomachs of calves, a byproduct of veal. The gene for making chymosin was transferred from cows to yeast. Yeast can be grown in vats, as any brewer knows. Although many people consider it wrong to slaughter calves, yeasts have few defenders. Besides, chymosin from yeast is cheaper and purer than chymosin from calves. So today, almost all hard cheese (over 90%) is made from chymosin produced by genetic engineered yeast.

The poet Omar Khayyam wished for *a loaf of bread, a jug of wine, and thou beside me singing in the wilderness.* He owed two of his three pleasures to the working of yeasts. Today he would also be indebted to yeasts for a piece of cheese.

COTTON WITHOUT INSECTICIDES

Cotton farmers are plagued by various insect pests, such as the boll budworm, the tobacco budworm, and the pink bollworm. In the US south, where most of our cotton is raised, these insects were controlled using chemical insecticides. But there is a natural insecticide which has been used for almost a century by organic farmers, a bacterium called Bacillus thuringiensis, Bt for short. The bacterium produces a toxin which is deadly to caterpillars like the three mentioned above, but harmless to almost everything else (except insects of the order lepidoptera, butterflies and moths—even the legendary boll weevil (Anthromonus grandis) is not harmed by the Bt toxin). So genetic engineers transferred the gene for Bt toxin from Bacillus thuringiensis to cotton.

Then the cotton plants, which could make Bt toxin, were cross bred with other varieties in the old fashioned way. Today, much of the US cotton crop is genetic engineered for the Bt toxin trait. The use of chemical insecticides in the cotton belt has declined dramatically, by over a million liters per year. Since the Bt toxin is inside the plant instead of sprayed onto the plant, the only insects which it can harm are those which eat the plant.

The benefit of reduced spraying of cotton is overwhelming. The cotton pesticides replaced are extremely damaging to the environment. They not only kill all insects in a cotton field, harmless or not, but also nearly anything else in the field, thus depriving insectivorous birds of their food. There is no way to keep these pesticides from getting into streams and rivers, where they are a serious hazard to aquatic life. We may think of cotton as a natural material, therefore environmentally friendly, but before there was Bt protected cotton, that was just wrong.

The gene for Bt toxin has been transferred into several other crops, including potatoes and corn. Approximately 30% of US corn is now transgenic, and the most popular transgenic varieties contain the Bt gene. Although we call it a toxin, to humans and other mammals and birds it's just a nutrient.

The principal potato pests are not caterpillars, but beetles, and the Bt toxin that protects cotton and corn doesn't harm beetles. But there is another Bt toxin found in another variety of Bacillus thuringiensis which is deadly to beetles. The gene for that toxin was used in potatoes.

Biotechnology also has overzealous advocates who exploit consumers' fears about pesticides. Except in the case of serious accidents, there's little to worry about. Our bodies deal with many toxic substances in many foods, and as long as the amounts are small enough they give us no problems. The toxic load of pesticide residues on food is completely negligible in comparison with the toxins naturally present.

But as they are applied in the field, these agricultural pesticides are seriously hazardous. Each year many farm workers are poisoned by exposure to pesticides and farmers have nightmares about their children or pets being injured by playing near pesticides. Pesticides can also harm wild birds and small animals, and when they get into waterways they can kill fish and other aquatic life. The Bt transgenic plants reduce or eliminate this danger.

Slow Ripening Fruits

There are many fruits which ripen after picking. After they reach optimum ripeness, they begin to deteriorate. This is necessary for the life cycle of the plant, which relies on the sweet and pulpy parts to nourish the seeds. A ripe fruit literally digests itself.

When this process is rapid, it effectively means that the fruit cannot be enjoyed out of season, or far from its growing area. For example, there is a popular Malaysian papaya variety which is unavailable outside Southeast Asia because it ripens so rapidly that it cannot be shipped very far. But it is quite easy to genetically engineer a fruit so that it does not ripen so rapidly. It doesn't even require a gene from another organism. Instead, a gene involved in the ripening process is copied with the message in reverse order. So now that plant has two genes with mirror image structure.

The way an organism uses the information in a gene to make a protein involves copying the gene (DNA) onto a messenger molecule, known as messenger RNA. The modified plant copies both the original gene and the mirror image gene to produce both types of messenger RNA. But since these messenger RNAs are exact complements of one another, they can wrap about one another just like the two strands of DNA, effectively blocking both messages. This means that the plant makes very little of the enzyme that causes ripening. This genetic engineering trick is called "antisense technology".

The Malaysian papaya was transformed in this way and therefore a slow ripening variety will soon be available.

The very first genetic engineered plant to be commercially developed as a whole food was a slow ripening tomato, called FlavR Savr. It was developed by Calgene, Inc. Because it could remain on store shelves for a long time, it could be left on the tomato plant until optimally ripe, and therefore the FlavR Savr tomatoes sold for a premium compared to other tomatoes. Although consumers initially liked Calgene's tomatoes, they didn't ship well and the variety was eventually dropped.

Controlled Ripening

A coffee bush ripens a few coffee beans each day for many months. The best quality beans must be picked just after ripening, so picking coffee beans is very neighbour intensive. It would obviously be preferable if the beans would all get ripe at the same time.

Genetic engineering will make this possible. There is a coffee gene which turns on to initiate the last stage of ripening. Scientists modified a control gene so that the ripening gene does not turn on until the plant is sprayed with a triggering substance (patented and sold by the company that developed the coffee variety). Therefore all the beans on a bush reach the same not quite ripe stage and stop to wait for the triggering signal. The farmer decides when to spray the bush so it can be picked completely clean a few days later.

This can substantially improve the life of the small farmer. He can take a short vacation without losing part of his livelihood. He can work fewer hours per day, or he can pick all his crop in a few days and increase his income by working at another job. A large scale farmer would need fewer workers to pick the same quantity of coffee beans, and could afford to pay them a higher wage.

The control of when a crop is harvested would be valuable for other crops besides coffee. For example, the quality of grapes declines rapidly after they reach their optimum sugar content. Grape farmers now have to mobilize every available hand to harvest all their crop in a very short time. Their lives would be simpler if they could spread the harvest effort over a few weeks instead of a few days.

Large scale crops are harvested with special equipment. A farmer would not need to own a combine if he could rent it for the few days it was needed. But that wouldn't work if his neighbour needed to rent it for those same few days. If neighboring farmers could control when their crops become ready for harvest, they could share scarce and expensive equipment.

Saving the Banana

Wild bananas have seeds. They reproduce sexually, like beans and oak trees. But they aren't easy to eat. Bananas grown on plantations have no seeds. They are reproduced by taking cuttings from older banana plants.

Cultivated bananas are seedless because they have three of each type of chromosome instead of the normal two of each type. Such plants are called "triploid". They are always sterile. Genetic triploid freaks arise from time to time in nature, but modern breeders can also use chemicals or electric shocks to create triploid mutant cells.

Bananas have been cultivated for many thousands of years and there are about three hundred different banana varieties. Each variety was developed by crossbreeding wild bananas. Whenever a promising variety was been produced, the breeder caused it to be triploid, hence seedless. That plant was cloned by propagating cuttings, and it became the parent of its variety. All bananas plants of the same variety are genetically identical, like identical twins.

Of the three hundred varieties, only one single variety completely dominates international trade. It is called Cavendish. It is possible that you have never seen a banana other than a Cavendish banana.

Certain kinds of fungus can infect and kill banana plants. In many parts of the world, Cavendish banana plants are being attacked by a fungus called black Sigatoka. Since wild bananas can reproduce sexually, they are not all identical and some wild bananas can resist black Sigatoka. If the black Sigatoka fungus is present in a region, the resistant banana types become prevalent. But the Cavendish banana plants have no resistance. They are all identical so they all die.

To grow bananas commercially, growers must spray their plants with fungicides. Year after year, the black Sigatoka fungus has been evolving resistance to these fungicides, so growers have to spray more and more fungicide each year. Approximately one third of the cost of raising a banana is the cost of spraying it with fungicides, and it gets more and more costly each year.

A form of black Sigatoka banana disease, now spreading around the world, can tolerate all known fungicides. Soon it will attack bananas in Central America and the Carribean islands, the heartland of banana culture. The Cavendish banana will become virtually extinct. Agronomists estimate that this will happen within ten years.

This is not a fairy tale. It has happened before. Forty years ago, the most popular variety of banana was one called Gros Michel. But Gros Michel was susceptible to a fungus called "race 1 Panama disease". Now it is gone. Cavendish bananas, which are not susceptible to race 1 Panama disease, replaced them. (A related fungus, race 4 Panama disease, can infect Cavendish banana plants. At present it is only found in Malaysia and some nearby countries.) In turn, some other variety of banana could replace the Cavendish. It would look different and taste different but it would still be a banana.

There is only one practical way to save the Cavendish banana. It must be given some combination of genes from wild bananas which are not susceptible to the fungus. But this can't be done by any natural technique. Cross-breeding can create other banana varieties because wild bananas reproduce sexually. But Cavendish bananas reproduce only by cloning. Conventional breeding would have to rely on rare mutated Cavendish banana plants which can produce seeds and which can therefore be crossbred, in theory. Even the mutated plants produce only a tiny number of viable seeds, as few as two or three seeds in a hundred pounds of bananas. No banana breeding experts think that they can breed fungus resistance into a Cavendish banana variety in only ten years. Biotechnology can rescue the Cavendish banana. Scientists in Belgium have used genetic engineering to transfer some fungal resistance genes from wild bananas. The transformed Cavendish plants are not susceptible to the fungus and they can then be reproduced into nursery stock by the usual method of taking cuttings.

BIOTECHNOLOGY: PUTTING CLONES IN CONTEXT

Cloning of mammals using somatic cells is big news in biotechnology. A nucleus from a diploid, mammalian cell culture derived from an organ other than the gonads has been substituted for the nucleus of a mammalian egg, the egg implanted into a surrogate mother and a viable offspring produced.

This article will examine cloning in the context of how it might relate to products regulated by CVM.

News media attention to cloning reached new heights in February 1997 with the announcement from the Roslin Institute in Scotland that they had successfully created Dolly, a clone from a somatic cell derived from cultured mammary gland cells of an adult sheep. Instead of evaluating the impact of this advance upon animal biotechnology, the media and the public leaped ahead to consider the bioethical and social implications of cloned humans. The President's National Bioethics Advisory Commission last June proposed a 5-year moratorium on cloning directed at creating a human child, but not the cloning of human or other DNA in cell lines or the cloning of animals.

In an interview with The Washington Post published January 20, 1998, the Acting Commissioner of FDA, Dr. Michael Friedman, stated "Through the Food, Drug, and Cosmetic Act we do have the authority to regulate human cloning, and we are prepared to assert that authority." He added that human cloning is a form of cellular or genetic therapy that requires prior FDA approval. The same article goes on to further cite Dr. Friedman as saying that anyone wanting to attempt human cloning legally must file a formal application with FDA which would undertake a lengthy review. The FDA would initiate legal action against anyone failing to file that application.

Cloning of humans has long been recognized as an ethical and social challenge. Westerners have been conditioned by books like Brave New World and movies like Boys from Brazil to reject human cloning. Our society will probably still be debating the permissibility of cloning humans long after cloning becomes technically and economically feasible.

On the other hand, clones of plants have long been prominent members of agriculture and horticulture. Genetic uniformity is an important agricultural plant trait that, among other things, facilitates efficient machine harvesting of crops and helps standardize product quality. For example, each apple tree of a single variety is genetically identical to every other individual of the same variety apples and many other fruits are cloned through vegetative propagation, i.e., stem cuttings, because they do not "breed true". In other words, the reshuffling of genes that occurs during sexual reproduction of apples results in apple offspring that are too dissimilar from the parents to be reliable for propagation of apples for commercial production. However, there is no movement afoot to ban the cloning of apples or to consider the bioethical dilemmas posed by cloned apples on the fabric of human society. Cloned plants are accepted as a normal part of human society.

Aside from Dolly the sheep, clones of domesticated animals are rare compared to clones of plants in our everyday environment. One might well argue that this is only because clones of animals are more difficult to establish than plants. The goals of many selective breeding programs, many of which feature repetitive back-crossing and inbreeding, include establishing a population of animals, e.g., dogs, horses, cattle, fish, etc., that possesses desirable traits that "breed true." Selective breeding reduces genetic diversity in "breeds of domesticated animals. The extreme case of this selective breeding can be found in the many strains of laboratory rats and mice used for product testing. These strains are so genetically homogeneous that there is no tissue rejection when tissue from one mouse is transplanted to another of the same strain. They are practically genetically identical. This uniformity is important for animal testing because statistically powerful testing may be conducted with small numbers of test individuals. There is less statistical "noise" caused by individual variation. Since the results are similar, cloning of animals could be looked upon as an alternative to selective breeding, especially useful where selective breeding is not practical.

So, is Dolly the logical extension of sheep selective breeding, possessing all the desirable traits sought by sheep breeders of the past? Well, no. Dolly is proof that the process of cloning apparently can work in sheep. Why were sheep selected to clone? Economics. Sheep have shorter generation times than cattle and they still produce milk in reasonable quantities. Why the interest in milk? The mammary gland is a marvelous protein synthesis and secreting organ whose products cannot always be duplicated by bacterial fermentation or chemical synthesis. Domesticated strains of transgenic sheep or cattle might be used to secrete bioactive proteins in milk that can be used as drugs, biologics and animal drugs.

Transgenic animals are genetically modified to contain a grouping of inserted genes (transgenes) from practically any species or combination of species. One application of transgenic animals is for the expression of proteins or enzymes that are not normally present or for the expression of higher concentrations of proteins that are normally present. Expression may be throughout the animal or directed to a specific organ, for example, the mammary gland. Transgenic animals have already been created through various genetic modification techniques that secrete proteins of great value in their milk, e.g., drugs that can be extracted and purified. The only trouble has been that, like the apple, most transgenic animals do not breed true. A major application for the technique of cloning will be in reproducing, in one generation, copies of transgenic animals that yield a high value product.

January 20, 1998, also brought a report from the University of Massachusetts that transgenic Holstein calves had been cloned and born in Texas as part of a project sponsored by Advanced Cell Technology, Inc., and Genzyme Transgenics Corporation. The stated goal of the project is producing recombinant human albumin in cows milk. (This can be purified for use as a human blood product.) Like Dolly, these calves are clones and test models, rather than production animals. Unlike Dolly, the calves are genetically modified. They contain only a marker gene, however, instead of a human albumin gene. Then, too, they are a male clone. Dairy cattle are also capable of more efficient milk production than sheep, so the experiment is an advance toward commercial production in this respect, too.

Cloning of somatic cells in animals is more a tool to reproduce valuable animals, like transgenic animals that produce drugs or biologics, or endangered species, or the best results of selective breeding programs, like the dairy herd improvement plan. The societal benefit to cost ratio for cloning of animals is decidedly different from the ratio for cloning humans, and the use of cloning in animals deserves to be considered separately.

Will the successful somatic cell cloning of mammals open the product floodgates from animal biotechnology companies? Like almost everything in the field of biotechnology, this is hard to predict. For the time being, the answer is "probably not."

Cloning will probably always be an expensive way to reproduce mammals compared to sexual reproduction and so will be limited to extremely valuable animals producing high value products.

Does CVM plan to regulate cloning of animals under the Food, Drug, and Cosmetic Act? The answer is probably yes and no. CVM regulates products. Cloning is a process. In the sense that cloning is a means to "manufacture" more transgenic animals and if those transgenic animals are producing an animal drug, then CVM could be involved in examining cloning carried out as part of the production process for the animal drug. CVM also regulates the drugs used on animals as part of the cloning process, for example, drugs for super-ovulation and estrus regulation. Other FDA Centers could similarly be indirectly regulating the cloning process as it is applied to transgenic animals producing a human drug or biologic.

Still under consideration is whether CVM should play a role in determining the safety of food derived from cloned animals that are not genetically modified, for example, clones of a high-producing dairy cow. Does the process of cloning have the potential to introduce changes in the quality of milk or meat derived from clones of normal dairy cows? While clones of food-producing animals are now starting to be developed, there are not yet enough of them to answer this question with hard data. Strictly on a theoretical basis, clones of normal animals should be indistinguishable from their parent in nearly all respects and, therefore, present no food safety concern.

CLONAL PROPAGATION OR MICROPROPAGATION

It has been demonstrated that a variety of plant species can be conveniently propagated through the techniques of cell, tissue or organ culture. This is popularly described as clonal propagation ormicropropagation.

The major benefits of this method include the following;

- Rapid multiplication of superior clones and maintenance of uniformity;
- Multiplication of disease free plants and
- Multiplication of sexually derived sterile hybrids.

In most cases, clonal propagation is achieved by placing sterilized shoot tips or auxillary buds onto a culture medium that is sufficient to induce formation of multiple buds. This method has already been used to propagate a large number of marketable ornamentals.

Following stages are involved in the method of clonal propagation:

(i) Stage I involves establishment of tissue in vitro.

(ii) Stage II involves multiplication of shoots (often media are not changed between stage I and stage II). (iii)

(iii) Stage III concerns root formation and conditioning of propagules prior to transfer to the green house; this stage requires alteration of media for promotion of root formation.

(iv) Stage IV involves growth in pots followed by field trials. The number of steps in micropropagation may sometimes be reduced to three or two as shown in diagram.

Three Different Methods of Micropropagation of Plant Material

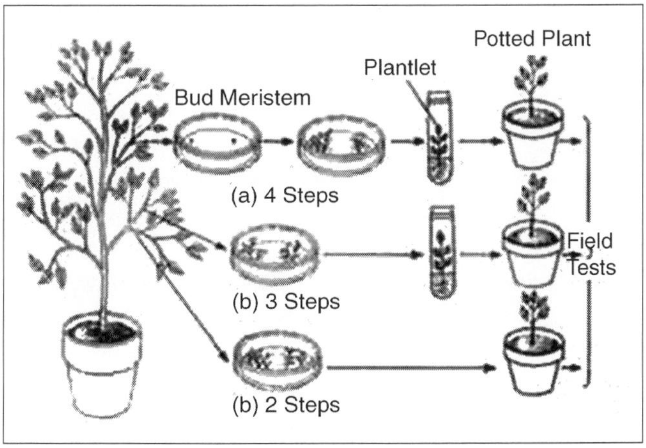

A wide range of plants have now been regenerated through technique of tissue culture. This has been found particularly useful for propagation of tree species, so that a large number of plant species have been successfully grown by tissue culture.

LARGE-SCALE PROPAGATION

To bridge the gap between research and application, the Department of Biotechnology, Government of India sponsored the setting-up of two pilot-scale facilities for large-scale propagation of elite planting material of forest trees through tissue culture. One of these facilities has been established at TERI's 36-hectare-campus in Gual Pahari, Haryana with an annual capacity of a million plantlets. Research at these facilities focuses exclusively on developing new protocols for mass cloning of elite planting material, mainly of trees. Till date, over 4 million plants have been dispatched for field plantation from these facilities. The tissue culture raised plants are presently being evaluated under field conditions. This is being done in tandem with the forest departments of Haryana, Uttar Pradesh, Madhya Pradesh, Bihar, Jammu and Kashmir, and Orissa. For initial screening for phenotypically superior trees only a few hundred plantlets of the same are raised and tested under various agroclimatic zones. The best clones are then mass multiplied and monitored regularly for their performance. Field data suggest a survival percentage of more than 90% even in the harsh conditions of Aravalis without the life-saving irrigation. At half the rotation age some of the selected clones of Eucalyptus are showing a significant increase in productivity as compared to the conventional seed raised progenies.

PRODUCING CLONES: PLANT LIFE

Nature- has been cloning organisms for billions of years. For example, when a strawberry plant sends out a runner (a form of modified stem), a new plant grows where the runner takes root. That new plant is a clone. Similar cloning occurs in grass, potatoes and onions.

People have been cloning plants in one way or another for thousands of years. For example, when you take a leaf cutting from a plant and grow it into a new plant (vegetative propagation), you are cloning the original plant because the new plant has the same genetic makeup as the donor plant. Vegetative propagation works because the end of the cutting forms a mass of non-specialized cells called a callus. With luck, the callus will grow, divide and form various specialized cells (roots, stems), eventually forming a new plant.

More recently, scientists have been able to clone plants by taking pieces of specialized roots, breaking them up into root cells and growing the root cells in a nutrient-rich culture. In culture, the specialized cells become unspecialized (dedifferentiated) into calluses. The calluses can then be stimulated with the appropriate plant hormones to grow into new plants that are identical to the original plant from which the root pieces were taken.

This procedure, called tissue culture propagation, has been widely used by horticulturists to grow prized orchids and other rare flowers.

PRODUCING CLONES: ANIMAL KINGDOM

Plants are not the only organisms that can be cloned naturally. The unfer-tilized eggs of some animals (small invertebrates, worms, some species of fish, lizards and frogs) can develop into full-grown adults under certain environmental conditions -- usually a chemical stimulus of some kind. This process is called parthenogenesis, and the offspring are clones of the females that laid the eggs.

Another example of natural cloning is identical twins. Although they are genetically different from their parents, identical twins are naturally occurring clones of each other.

Scientists have experimented with animal cloning, but have never been able to stimulate a specialized (differentiated) cell to produce a new organism directly. Instead, they rely on transplanting the genetic information from a specialized cell into an unfertilized egg cell whose genetic information has been destroyed or physically removed. In the 1970s, a scientist named John Gurdon successfully cloned tadpoles. He transplanted the nucleus from a specialized cell of one frog (B) into an unfertilized egg of another frog (A) in which the nucleus had been destroyed by ultraviolet light. The egg with the transplanted nucleus developed into a tadpole that was genetically identical to frog B.

While Gurdon's tadpoles did not survive to grow into adult frogs, his experiment showed that the process of specialization in animal cells was reversible, and his technique of nuclear transfer paved the way for later cloning successes.

Dolly

In 1996, cloning was revolutionized when Ian Wilmut and his colleagues at the Roslin- Institute in Edinburgh, Scotland, successfully cloned a sheep named Dolly. Dolly was the first cloned mammal.

Wilmut and his colleagues transplanted a nucleus from a mammary gland cell of a Finn Dorsett sheep into the enucleated egg of a Scottish blackface ewe. The nucleus-egg combination was stimulated with electricity to fuse the two and to stimulate cell division. The new cell divided and was placed in the uterus of a blackface ewe to develop. Dolly was born months later.

Dolly was shown to be genetically identical to the Finn Dorsett mammary cells and not to the blackface ewe, which clearly demonstrated that she was a successful clone (it took 276 attempts before the experiment was successful). Dolly has since grown and reproduced several offspring of her own through normal sexual means. Therefore, Dolly is a viable, healthy clone.

Since Dolly, several university laboratories and companies have used various modifications of the nuclear transfer technique to produce cloned mammals, including cows, pigs, monkeys, mice and Noah.

THE CRITICS OF AGRICULTURE

The recent history of agriculture in industrialized countries is a history of technological change. Machines, chemical inputs, and genetic improvements were developed at an advancing pace throughout the twentieth century, and many of these technologies were widely adopted by farmers in industrial countries. During the 1940s, a group of scientific, business and political leaders conceptualized the Green Revolution, a massive effort of technology development that was to reproduce what they took to be the success of industrialized agriculture in developing countries. The Green Revolution strategy of aggressive applied scientific research, followed by equally aggressive efforts of technology transfer, was intended to improve agriculture by improving the tools and materials of farming. It was a distilled version of the philosophy behind the laws that established the agricultural experiment stations in the United States in 1887, and the state and national agricultural extension services in 1914. The two laws placed land grant universities in each state in harness with research and extension efforts at the US Department of Agriculture and created what has come to be known as the USDA/land grant system. The rationale for these laws was the general betterment of rural communities, but they have gradually been implemented in ways that focus ever more narrowly on technology.

Whether one speaks of the USDA/land grant philosophy that gave rise to agricultural technologies in the United States or the Green Revolution philosophy that spread a somewhat narrower view of technology transfer across the globe, agricultural science is linked to technology adoption by farmers. In many cases, the technology eventually adopted by farmers must be supplied by firms that manufacture the tractors, cultivators, chemicals, or seed varieties that may have been originally researched in universities or public agencies such as the Agricultural Research Service.

Therefore, the recent history of agriculture is also the history of emerging agribusiness firms that supply farmers with the technologies needed for food and fibre production. The transformation that has taken place in agriculture is dramatic. When industrialized production systems are measured in terms of productivity, the success of this philosophy is startling. Farming technology has increased the productivity of agriculture, and the result is that people in industrial countries expend far less of their income on food and fibre than do those in countries where farming continues to be the primary occupation of most people.

Most participants in the USDA/land grant, Green Revolution complex (including scientists, public officials, agribusiness firms, and adopting producers) would regard these changes as a successful application of technology, but a chorus of voices has arisen in criticism. The primary purpose of this chapter is to examine the environmental critics, but it is helpful to realise that environmental criticisms of agriculture are but one voice in a four part harmony of critique. Perhaps the first themes were sounded by critics of the Green Revolution itself. According to DeWalt, Carl Sauer wrote as early as 1941 that Mexican agriculture cannot be pointed towards "standardization of a few commercial types without upsetting native economy and culture hopelessly." Published critiques of the Green Revolution began to appear in the late 1960s and early 1970s, with Keith Griffin's book, *The Political Economy of Agrarian Change: An Essay on the Green Revolution,* being the first in a series of studies that documented the social turmoil associated with agricultural technology transfer in developing countries.

It distorts the criticism of the Green Revolution to generalize, but it is fair to say that critics have consistently applied two themes. First, critics reject the assumption that changes in agricultural production technology can be evaluated in terms of aggregated indicators such as increased food production or total rural income. They insist that inequitable distribution of benefits and harms overturns the judgment that agricultural technologies have produced success in the developing world. Second, the presumption that scientific research can produce beneficial changes in any cultural and political environment is replaced with the view that Western science is deeply dependent on the social institutions of developed economies. By altering these two presumptions of the Green Revolution, critics conclude that the loss of local autonomy and indigenous knowledge far outweighs any benefit from increased agricultural production. Frequently, critics apply revolutionary political rhetoric in placing the ultimate blame for Green Revolution failures on capitalist or free market ideology.

A second group of critics have noted the social implications of agricultural technology within developed economies. In short, critics think industrial technology is inimical to small farms and rural quality of life. The origin of this theme may be a study by anthropologist Walter Goldschmidt. Originally published in 1947, Goldschmidt's *As You Sow* prefigured many criticisms that were to mount in the following four decades, as family farmers in rural communities became increasingly aware of their plight. The attack was leveled directly at USDA/land grant organizations in an influential 1973 book, *Hard Tomatoes, Hard Times* by Jim Hightower, who was later to become Texas Commissioner of Agriculture. Hightower's book and a series of papers on the development of the mechanical tomato harvester in California laid the responsibility for the lost employment and farms that ensued at the doorstep of the University of California. The result was a lawsuit, filed on behalf of displaced field workers, that was settled in favour of the University in 1991.

The general theme of domestic critics is a populist one. Government, including the USDA/land grant system, should protect the "little guy" from the forces of impersonal industrialization, in general, and from big business in particular. The trend in agriculture has manifestly been towards larger, more specialized farms and farm input businesses, and this trend is antithetical to the populist ethic of family farming. These critics often mix environmental critiques with their populist themes, so Marty Strange's book, *Family Farming: A New Economic Vision* incorporates the critique of center pivot irrigation alluded to earlier. The environmental dimension of populist critique will surface in later, but one must assess these critics as being *primarily* concerned about the loss of small farms, and generally of the opinion that, if small farms were preserved, the environmental problems of agriculture would take care of themselves.

The final voice of criticism is the most obscure, and is, like the lowest of bass notes, indiscernible to the casual listener. It is, however, arguably the most effective in influencing the direction of change within agricultural institutions. In 1972, the National Research Council (NRC) of the National Academy of Science issued what has come to be known as the "Pound Report." The Pound Report took agricultural universities and the USDA to task for not being scientific enough. The substance of the report attacked needless replications and duplication of studies from state to state, and noted that many agricultural scientists appeared to be working on subjects of little scientific merit or interest.

Other NRC reports have been critical of agricultural universities, including the 1989 report, *Alternative Agriculture,* which criticized USDA/land grant administrators for failing to investigate alternatives to the mechanically and chemically intensive technologies of conventional agriculture. While there has been grousing about the accuracy of NRC studies, they have arguably done more to promote change in agricultural institutions than have all the other critics combined.

The philosophical importance of NRC criticisms is quite different from those of the Green Revolution, which stress equity and autonomy, or of the populists, which stress family farms. The NRC reports question the academic or scientific integrity of the agricultural research system, arguing that the methods for identifying research priorities and funding agricultural science continue to be too much influenced by parochial and non-scientific interests. Primary among these interests are agribusiness firms. Chemical companies, for example, fund many graduate assistantships to perform blind tests on the efficacy of new pesticides. Such tests are of no scientific interest.

They may be limited to blind data collection which renders them useless even as learning experiences. However, the family farmers, Green Revolution critics, and environmentalists who also seek to influence agriculture and agricultural research are no less parochial and non-scientific than agribusiness. The National Research Council would no more like to see the research agenda in agriculture controlled by the Sierra Club than by the Monsanto Company. As such, though NRC criticisms are effective in changing research directions, their effectiveness is only accidentally related to environmental issues.

The final point before considering some environmental critics in more detail is to note that the choir of critics makes the evaluation of agriculture and its impacts exceedingly complex. Environmental criticisms can run badly afoul of equity concerns in agriculture. Each of these themes is sometimes reinforced, sometimes undercut by critics who want agriculture to be more scientific. Agricultural ethics is always an exercise in balancing these multiple themes against one another, and there are seldom only two sides to an issue.

The Critics of Pesticide

The implicit ethical basis for agricultural research in the USDA/land grant system is *utilitarian* in that it defines the value of research in terms of its capacity to improve the balance of costs and benefits associated with agricultural production. It is also *anthropocentric* insofar as the balancing act is limited to costs and benefits to human beings.

According to this implicit ethics, the most fully justified research project is the one that promises to achieve the greatest good for the greatest number of people. When distributional issues as well as costs to non-human animals and the environment are ignored, the utilitarian view of social ethics makes it easy to think of ethical evaluation as a form of calculation in which all the benefits and costs of various options are weighed, because the ethically justified course of action is that which best satisfies the rule of maximizing aggregate good (or in situations where all options are unattractive, of minimizing evil).

Even when distributional issues and costs to non-human animals and the environment are bracketed, however, there is one general problem with optimizing decision procedures such as the utilitarian ethic described above: it is impossible in practice to obtain complete and reliable information on all the relevant consequences of a policy decision. Some factors are inevitably left out, and when these factors affect human health and safety, the economic well-being of minority groups, or the quality of the human environment, the entire moral calculation of relative benefits can be drastically skewed. Costs or harms that are simply left out of a utilitarian calculation can be called *externalities*.

Costs are sometimes left out because the decision maker does not have a reliable way to measure or compare them, and costs are also left out when persons or groups deciding on behalf of their own interests do not have to bear them. From the point of view of the self-interested decision maker, the costs are truly external. Decision makers entrusted with the public good must make stringent efforts to reflect all such costs in any estimate of total social benefit or harm, or they cannot truly be said to have optimized outcomes.

The majority of environmentally based criticisms of agriculture and agricultural research clearly take the form of noting the factors and impacts that have simply been left out of the assessment of costs and benefits. The most celebrated of all works in the critical literature, Rachel Carson's *Silent Spring,* follows the pattern of citing externalities by providing an extensive list of unintended consequences associated with the use of insecticides. Carson was a gifted writer with a flair for evoking the beauty and dignity of wildlife and natural habitat and for expressing outrage at practices that place nature at risk. Nevertheless, her basic philosophical strategy in the book was simply to identify unintended and negative consequences. *Silent Spring* is a laundry list of unwanted consequences and risks that had not been accounted for in making assessments of pesticide use in agricultural production.

Silent Spring is just the first in a long line of critiques identifying environmental impacts of chemicals used in agricultural production. Carson's arguments were reiterated and extended in Frank Graham's 1970 book *Since Silent Spring* and in Robert van den Bosch's *The Pesticide Conspiracy,* to cite two of the most prominent titles. While Carson noted toxic effects upon non-target species, Van den Bosch identified unintended consequences that typify the class of ecosystem outcomes of intense interest to environmentalists. According to Van den Bosch, chemical pest control ignores ecological forces that control insects. The number of insect species classified as pests doubled from 1962 to 1978, despite increasingly efficient chemical control, skyrocketing insect control costs, and worsening environmental impacts, a phenomenon that Van den Bosch labels an "insecticide treadmill." Insecticides or biocides kill natural enemies of insect pests, and eradicate the natural predators and parasites. A biotic vacuum is then created where the surviving pests thrive without predators or parasites. Continued spraying becomes a necessity.

Van den Bosch makes an explicit ethical argument in his book when he indicts chemical company salespersons and advertisers for their effect upon a farmer s decision to use more pesticides and contribute to the treadmill. Scientific societies and administrators of land grant universities, where the nations pest control research is conducted, are also implicated when pressures and political reprisals from chemical companies affect USDA/land grant research programmes. Van den Bosch's condemnation of public sector agricultural research is based upon a judgment that administrators and scientists have been "captured" by the commercial pesticide industry, and have thus failed in an ethical obligation to conduct research in the public interest.

His criticism is still consistent with Carson's original way of identifying uncounted costs, however, for Van den Bosch clearly understands the public interest in terms of maximizing benefits for human beings. The problem he has with chemical insecticides is that their costs outweigh their benefits. It is profitable for chemical companies to sell pesticides and for farmers to use them only so long as important long-term costs are not included in the overall assessment (or, to say the same thing, so long as the costs are "externalized"). In this case costs are externalized either in the sense that they are borne by individuals whose interests are not included in the tabulation of consequences or in the sense that costs occur beyond the time frame for which consequences have been assessed. Van den Bosch's criticism of insecticide came forward within a complicated political context. On the one hand, concern about the human health effects of pesticide had become commonplace among urban consumers. Publications from the Rodale Press sounded the themes that pesticide-free foods were better for consumers, and that organic production was a realistic possibility. Opposition to pesticide use modeled on *Silent Spring* was expressed routinely in popular articles on environmental themes.

Pesticide had become emblematic of what was wrong with the culture of consumption. Van den Bosch's criticisms, on the other hand, spread a different anti-pesticide message that began to be internalized by entomologists. Working with Texas cotton producers, Perry Lee Adkisson and Ray Frisbee documented the phenomenon of acquired pesticide resistance in the field. Texas growers using heavy sprays to control the boll weevil unintentionally created a new pest when the pink boll worm, previously not a serious problem in Texas, became resistant to the chemicals being used on cotton. They were among a group of agricultural scientists who began to develop a strategy called Integrated Pest Management (IPM). Opposition to pesticides oscillates between the two poles of critique represented by Carson and IPM.

Many recent critics of pesticide have further modified the claims of Carson, Graham, and Van den Bosch, but they continue to list a broad range of unwanted consequences from chemical agriculture. David Pimentel has produced a series of papers written with a variety of coauthors documenting unwanted effects of pesticides. Pimentel s early work was based on an energy audit of pesticide use, noting that the energy used in the manufacture, transport, and application of chemicals severely compromised the energy efficiency of farming.

He has documented the growth in consumption of pesticides, collected citations of scientific studies indicating the risks to human health, and mounted an argument for drastic reductions in pesticide use. Another stream of criticism notes the use of pesticides in developing countries. Pesticides long banned in the United States were used extensively in developing countries for many years, and continue to be used in some applications. Critics have argued that the export of these banned chemicals causes significant human health risk to agricultural workers in other countries, and eventually to developed country consumers, who consume fruits, coffee, and other products that may contain residue of long banned chemicals. When continued use of long banned chemicals is factored into the argument, the list of unwanted impacts from pesticide use cannot fail to impress one with the continuing seriousness of pesticides' unintended consequences.

Advocates of IPM stress a much narrower range of unintended consequences, so much so that they represent an alternative to the Carson-Pimentel line of argument. The IPM story deserves attention from anyone contemplating an ethic of the environment. Insect life in farm fields is a model in miniature of wildlife ecology. Some insects feed on plant matter; some are predators that feed on other insects. Insects become agricultural pests only when they do economic damage to crops, and this happens only when their numbers are not sufficiently controlled by predation.

Insects that feed on pests are called *beneficial insects,* so the farm field is an ecosystem where pest and beneficial insects stand in some form of balance with respect to one another. As pest populations grow, food for beneficial insects is easier to find. Eventually, the population of beneficial insects will grow in response to the easy pickings, and the number of pests will decline. The balance between pest and beneficial insects is far from perfect from the farmer's perspective, but there are, in any given field, likely to be a few species that feed on the crop, but which do not become serious pests in virtue of the fact that their populations are controlled effectively by beneficial insects.

Chemical pesticides are not selective; they kill pest and beneficial insects. After a pesticide application has lost its effectiveness (usually within a few days or weeks), insect populations begin to rebound. For plant eaters, there is plenty of food, so population builds quickly. Beneficial insects cannot begin to rebound until there is an adequate amount of prey. This creates a window of opportunity for insects normally controlled by beneficials. With their natural enemies in disarray, their populations can grow rapidly. Although the balance between pests and beneficials will eventually be restored, farmers can expect a surge of pest insects that will take place before beneficial populations have an opportunity to rebound. Given the pesticide practices in use prior to 1980, farmers would notice the surge and make a new application of pesticide. Beneficial insects never had an opportunity to rebound.

Pesticides are costly to buy and to apply. Entomologists began to discover that sometimes farmers could do better financially by accepting crop damage from pests than by getting on the pesticide treadmill. The treadmill phenomenon becomes even more serious when an insect acquires resistance to chemicals being applied. Pesticide resistance is a textbook example of natural selection. When pesticides are applied, a few insects in the population may be resistant to their toxic effects.

These insects will constitute a much higher percentage of the total population after spraying than before. The reason is obvious enough: most of the non-resistant insects are dead. If the farmer sprays again, the percentage of resistant insects will be greater still. As their percentage grows, these insects begin to pass resistance on to subsequent generations. Under the selection pressure of toxic chemical sprays, insect populations can acquire widespread resistance to the toxic effects with surprising rapidity, rendering the pesticide useless.

Now, IPM entomologists do not reject the use of pesticides. The IPM philosophy holds that on some occasions chemical use is economically viable, and even necessary Pesticides will only be effective at these needed times when insect populations have not become resistant to them. Minimizing pesticide use limits the number of resistant insects in the population, and can significantly extend the amount of time that it takes for a population to become resistant to a given chemical compound. This is an extremely significant fact for ethics, for it converts IPM from being a simple norm of financial prudence to being a general social norm for farmers of a given crop. The reasoning for this conclusion deserves careful attention even by those who reject the premise that pesticides will continue to be useful and necessary.

In the first exposition of the pesticide treadmill, a farmer is wise to limit spraying when the cost of chemicals exceeds the value of the crop protected, particularly when successive sprays will be needed. On this level, IPM is just good business sense. A farmer who wants to waste money with multiple sprays is foolish, but there has never been a moral injunction against foolishness, at least as long as it is one s own money that is being wasted. The decision to use IPM has social implications because even the farmer who practices IPM will want the pesticide to remain effective against pests so that it can be used when necessary. If the IPM farmer has neighbours who spray wastefully, the insect pests in the region are likely to develop resistance, anyway. Insects that become resistant in the fields of the wasteful will find their way to the practitioners of IPM. Avoiding pesticide resistance requires participation in IPM by all (or most) farmers. It therefore becomes plausible to say that farmers should participate in IPM out of a moral duty that they have to their neighbours.

While it has become difficult to find an entomologist who will not privately confide that farmers collectively use too many pesticides, many continue to reject the stronger claims of Rachel Carson and her heirs. It is the IPM view that dominates among agricultural scientists. Supported by a series of scientific studies, agricultural scientists generally question the seriousness of human and environmental health risks associated with chemical pesticides. The recent work of Bruce Ames is frequently noted. Ames has discovered that food crops naturally contain a complex mixture of mutagens as part of their natural defence mechanism. He has argued that any cancer risk associated with pesticide residues is overshadowed by risks from these naturally occurring substances. This is not the place to undertake a discussion of whether pesticides do or do not cause harm to humans, wildlife, or other ecosystem impacts. Whether they do or don't is an empirical question, not a philosophical one. However, the pesticide controversy does have philosophical implications.

In framing the pesticide issue as they have, Rachel Carson's heirs make an argument that depends entirely on the factual accuracy of the allegation that pesticide does or may cause harm. This claim admits of three possible responses. One is to deny the accuracy of the claim. A second is to accept the accuracy of the claim, and to look for new technology that mitigates the risk. The third is to accept the accuracy of the allegation, but to argue that both risks and actual harms are outweighed by the benefits of pesticide use.

The first two responses are scientific and technological; they do not raise any ethical issues at all. The third response points towards difficult questions of acceptable risk, but even these questions are entirely consistent with the utilitarian framework discussed above. Risk issues, furthermore, are often dominated by empirical efforts to ascertain accurate measurements of the probability of harm. None of the responses to critics involve the defenders of pesticide in serious reflection on the values and goals appropriate to agriculture. The pesticide controversy is preoccupied with empirical questions, and has failed to generate much discussion that is fruitful for environmental ethics. If critics of pesticide have hoped to draw upon farmers' sense of moral responsibility for the environmental implications of agriculture, they have failed spectacularly.

CRITICS OF AGRICULTURAL BIOTECHNOLOGY

Chemical pesticides represent an ideal case study for environmental criticisms of agriculture. The case against pesticide is largely an environmental one. Pesticides have not been prominently implicated as technologies that contribute to social dislocation, so populists have given pesticide only incidental attention. Although entomologists doing pesticide work are vulnerable to some of the NRC criticisms, it is also true that several entomologists, including Perry Adkisson, have attained a high level of recognition among scientists for their work on acquired resistance and IPM. Critics of the Green Revolution philosophy certainly object to the use of pesticide in settings formerly committed to peasant subsistence farming, but irrigation, fertilizer, new seeds, and mechanization have been more prominent than pesticide in technology transfer. The case against pesticides has been made largely by environmental critics, and one would expect that it should provide the clearest signal for identifying environmental criticisms of agriculture.

Biotechnology, by contrast, presents a tangled jumble of criticisms. For one thing, although recombinant techniques for moving genetic material are becoming commonplace in agricultural research settings, few technologies developed through the use of recombinant DNA are currently used in farming or the food industry. One success has been the commercial use of a genetically modified organism (GMO) that produces a very pure form of rennet, a substance used in cheese making. Since rennet has historically been harvested from the entrails of slaughtered calves, the modified bacterium that produces pure chymosin, the active enzyme in rennet, has produced few opponents.

Other agricultural biotechnologies have produced a firestorm of criticism, however. Jack Doyle produced an extended environmentalist critique of biotechnology in his 1985 book, *Altered Harvest.* Doyle described how plant breeders had produced varieties of maize that shared a genetic trait called Texas T cytoplasm. The varieties were planted extensively across the United States, and in 1972 a virus emerged that attacked plants sharing this trait. The result was a disaster for the US corn crop. Doyle used the case to illustrate why it is important that agricultural crops maintain a diversity of genetic traits, and suggested that one risk of agricultural biotechnology would be to increase the chance of a repeat performance.

Defenders of crop biotechnology take the thrust of the criticism seriously, but argue that recombinant techniques give them greater ability to minimize the probability that an entire crop would be susceptible to a given disease. Texas T cytoplasm was produced through conventional crosses that transferred a package of traits, some beneficial, some not, into many varieties of maize. Plant scientists argue that recombinant techniques would have allowed them to home in on beneficial traits with more precision, thus producing more genetic diversity rather than less.

This pattern of claim and counter claim, already evident in the debate over pesticides, also characterizes debates over GMOs. By far the most debate over biotechnology has revolved around recombinant bovine somatotropin (BST), a growth hormone used to increase dairy production. The debate over BST is complex, and serious students of agricultural biotechnology will want to examine it in some detail. BST, however, has not aroused the ire of environmental critics. The main point of contention has been the effects of the new technology on small dairy farms, a theme associated with populist, rather than environmental critique.

Environmental themes have figured in discussion of two technologies, ice-nucleating bacteria and herbicide tolerance, however, and it is worth reviewing each in more detail. By 1981 the commercial potential of gene transfer for agriculture had begun to be recognized. By 1983, one of the first products of these new techniques, the "ice minus" bacteria, was ready for field testing. Intended to inhibit the growth of ice-nucleating bacteria by crowding their ecological niche with genetically altered competitors, the "ice minus" bacteria were expected to extend the growing season for a variety of crops, including potatoes and strawberries, by reducing the likelihood of a crop loss due to freezing temperatures. At the time that field tests for "ice minus" were proposed, regulatory authority for release of genetically engineered organisms was thought to reside in the National Institutes of Health's Recombinant DNA Advisory Committee (RAC).

The RAC was the outgrowth of a decade of concern over potential health effects of genetically engineered organisms, but the "ice minus" case was novel in important respects. NIH involvement in regulation of recombinant DNA experiments was a legacy of the moratorium on gene transfer of the early 1970s, the Asilomar conference of 1974, and the stringent guidelines for labouratory research that had been established in its wake. By 1983 experience with recombinant DNA had allayed many fears, and NIH guidelines had been successively weakened. NIH had become comfortable with the vast majority of ongoing basic and biomedical research.

Indeed, many of the RAC s most difficult cases dealt not with *safety* of gene transfer but with ethical questions such as the permissibility of altering the human genome. The "ice minus" experiment, however, deviated from the basic and biomedical types of research over which the RAC was understood to have clear regulatory authority. It was also, rightly or wrongly, among the first recombinant DNA experiments thought to have potential for unwanted environmental consequences. By late 1983, then, the regulatory authority of NIH had been questioned, both in the courtroom and by the Environmental Protection Agency.

The saga of "ice minus" grew increasingly complex. The original experiment, proposed under the auspices of the University of California, was blocked by Judge John Sirica on May 16 1984, who in the same ruling held that a private company proposing the same experiment would not be bound by laws requiring environmental impact analysis. Within two weeks of Sirica's ruling, the RAC recommended approval for UC scientist David Lindow's experiment, this time submitted under the auspices of Advanced Genetic Resources, escaping the force of Sirica's ruling. This action merely precipitated a lawsuit, delaying the experiment again.

Both suits were eventually resolved in a manner that permitted the "ice minus" experiment, but NIH also came under sharp criticism in one decision for failing to "sufficiently analyse the potential for the bacteria to be used in the California experiment to disperse or survive in the environment". The "ice minus" experiment had by now received enough publicity to generate public opposition at the strawberry test site near Tulelake, California. County boards in both communities voted to prohibit the experiment, and the California Superior Court issued a restraining order on August 6 1986, delaying the experiment until spring 1987. Delayed by four years, the "ice minus" field test commenced on April 23 1987 at a third site near Brentwood in Contra Costa County.

Regulatory policy for agricultural biotechnology made little progress in the meantime. The overlap between NIH and EPA continued throughout 1984, as environmental scientists stressed the need to assess ecological risks before permitting release of engineered organisms. By 1985, EPA had conducted its own favourable review of the "ice minus" experiment, but regulatory confusion had only deepened as the US Department of Agriculture and the Food and Drug Administration were proposed as addition-al partners with NIH and EPA in a "Biotechnology Science Coordinating Committee."

By 1986, the biotechnology industry itself had begun to call for government involvement in the regulatory process, partially as a way to stifle opposition arising from unwarranted fears and speculations, and the Reagan White House announced a plan that supported NIH guidelines but transferred regulatory authority to EPA and USDA. The new guidelines established the principle that risks of genetically manipulated products should be evaluated on the basis of product characteristics, not manufacturing processes. No sooner had these guidelines been proposed than they were the target of yet another lawsuit. In October 1986, *Science* reported significant inconsistencies between EPA and USDA, with ecologists expressing concern about the government s intention categorically to exempt certain types of genetically created organisms from environmental impact assessment. In the following month there were similar conflicts between USDA and NIH.

Since the furor over "ice minus," government oversight of agricultural biotechnology has increased, but confusion over requirements and regulatory authority still occurs. At present, EPA and USDA continue to claim authority to regulate recombinant organisms that may affect the environment. Working relationships among agencies have been facilitated through an Agricultural Biotechnology Regulatory Advisory Committee (ABRAC). The ABRAC itself makes no regulatory decisions, but advises several agencies on policy, and by doing so functions as a focal point for coordination. Even within USDA, however, regulatory authority is not always clear.

USDA has an Office of Biotechnology, whose primary purpose has been to foster development of recombinant techniques for agriculture, and a National Biotechnology Impact Assessment Programme, designed to develop procedures for risk analysis. Neither has regulatory authority, however. The

Food and Drug Administration (FDA), the Animal Plant Health Inspection Service (APHIS), and the Food Safety Inspection Service (FSIS) all have authority, but would not normally be concerned with research, such as was involved in the "ice minus" case. Research sponsored by USDA through the Cooperative State Research Service (CSRS) is regulated as part of the proposal review process, and universities that conduct publicly funded research are required to have an Institutional Biosafety Committee (IBC). The IBC will normally review all research involving recombinant DNA, yet it remains the case that privately conducted research is not subject to regulatory review.

To sum up, regulatory policy for agricultural biotechnology lies buried deep in the forest of government acronyms. A venture into the woods may or may not turn up a clear answer as to what is permitted, or when risks are too great. When regulatory authority is unclear, cases ultimately wind up in the courts. For activists such as Jeremy Rifkin, the goal of a lawsuit may be simply to slow down experiments so that affected parties have adequate opportunity to ensure that their interests are adequately protected. The success of litigation initiated by Rifkin's Foundation on Economic Trends (FET) in opposing the "ice minus" experiment complicated that case with jurisdictional issues, but the underlying ethical issue was acceptable risk.

With respect to "ice minus" bacteria, the question had little to do with whether the experiment poses a serious hazard; every review of the proposed research had concluded that it did not. The FET lawsuits, however, exposed a general confusion over what, in fact, is meant by "acceptable risk" and whether one agency's judgment of acceptability is binding on another. The series of lawsuits represents a classic use of uncertainty arguments to raise a succession of doubts, first about the safety of a practice itself, then about the reliability of methods for assessing risk, and finally about the integrity and reliability of experts conducting the analysis.

At the time of the "ice minus" case, regulation of agricultural biotechnology had not found a way to manage uncertainty arguments. In the minds of citizens and affected parties, uncertainty escalates the risk of new products and procedures. The problem is typically described as one of managing public perception of risk, but this is misleading. It is the public perception of the scientific community (and of their methods for analyzing technology's unwanted consequences) that was the basis for the judgment that risks were unacceptable in the "ice minus" experiment. While the scientific community focuses intently on the characteristics of the organism itself, the public, with little basis for making a judgment on the probable consequences of deliberate release, focuses instead on the characteristics of the scientists.

The potential for unwanted impact due to field testing and commercial use of modified plants has been taken relatively seriously by agricultural scientists, though not seriously enough in the eyes of critics. As the debate has moved on to the technical provisions of regulations and protocols for field testing, and commercial release, the issues have become too arcane for general public consumption. Like pesticide questions, these issues depend heavily on the measurement and weighing of risks, and in balancing risks against expected benefits.

The difference of opinion between advocates of biotechnology and opponents boils down to differing estimates of the probability and degree of harm, and value of compensating benefits. This difference of opinion depends largely on issues that can only be settled by empirical inquiry. It is the expense and difficulty of gathering the data to settle the issues that keeps them alive.

The case of herbicide tolerance is quite different. Like insecticides, herbicides typically kill both beneficial plants (*e.g.* crops) and pests (*e.g.* weeds). Obviously, this limits the applicability of herbicides rather dramatically, generally to early season use, before crops come up, or to highly targeted use, away from crop roots and foliage. The idea behind genetic engineering of herbicide resistance is that if crops acquire resistance, one can use herbicides with impunity. Plant scientists have had some success in identifying the genes that code for herbicide tolerance, so recombinant techniques can be used to move these genes to the beneficial crop plants.

This is anathema to environmental critics of agriculture, for it seems that biotechnology is being used in a way that will exacerbate the problems of chemical agriculture. Defenders of herbicide tolerant crops argue that they will allow farmers to apply principles of IPM to weeds, as well to as insect pests. Now, they note, farmers must use herbicides early, before they know whether they even have a weed problem. With herbicide tolerant crops, they could wait until weed infestation threatens to cause economic damage exceeding the cost of spraying before using herbicides at all.

Farmers could quit using the wide variety of herbicides now used that target specific weeds but avoid damage to crops, and could switch to a broad spectrum herbicide that kills all the weeds, but not a genetically modified resistant crop. In addition, of course, they argue that farmers will get better weed control and increase yields. The argument, then, is that contrary to the claims of environmentalists, herbicide tolerant crops may reduce the amount of herbicide used, and will, in any case, produce benefits that more than compensate for chemical risks.

Philosophical dimensions of the case for and against genetically engineered herbicide tolerance have been reviewed thoughtfully by Gary Comstock and a paper by Comstock and molecular biologist Jack Dekker presents the reasoning that led Dekker to discontinue his research in the area. In the present context, a 1990 document, *Biotechnology's Bitter Harvest,* prepared for the Biotechnology Working Group by Rebecca Goldberg, Jane Rissler, Hope Shand, and Chuck Hassebrook is particularly instructive for the way it combines all four voices of criticism noted above.

The report echoes Green Revolution critics by noting Third World impacts of agricultural biotechnology, and cites the NRC report *Alternative Agriculture* in support of its conclusions. The specific complaints against herbicide tolerant crops are noted in a chapter entitled, "The Human Health, Environmental, Social, and Economic Impacts of Herbicides and Herbicide-Tolerant Crops." The chapter text follows the laundry list model implied in its title by listing and documenting a series of unwanted consequences that could follow the introduction of herbicide tolerance into crops and trees.

Among outcomes noted are the suspected carcinogenic properties of specific herbicides, food safety concerns associated with herbicide residues and with the consumption of the modified crops, contaminated drinking water, and interbreeding with other weeds. The emphasis in this list is clearly on health and environmental impact, though questions about a farmer's increasing dependence on private sector technology link herbicide tolerant crops to family farm issues.

The report provides equally extensive listing and documentation of how herbicide tolerance research is organized and funded. Research is underway both within the private sector and at agricultural universities and the USDA. University and government research is funded by a mixture of public and private funding. Funding receives extensive discussion in the report because it is crucial to three central points of criticism. First, the amount of money spent on genetically engineered herbicide tolerance vastly exceeds the amounts spent on sustainable alternatives that would be preferred by the reports authors. The report states, "Perhaps the greatest problem with herbicide tolerance, however, is that it diverts us from paths that really could lead to reduced chemical dependency in agriculture".

Second, public funds are being spent in a manner that effectively subsidizes research costs for chemical companies, or that benefits directly corporations by increasing the market for their herbicides. The implicit premise is that money spent to benefit small farmers would be in the public interest, while money that benefits the input industry does not. Third, the authors argue that chemical corporations are supporting research on herbicide tolerance (both directly and by lobbying public officials) because it helps them gain control of the research agenda in agricultural biotechnology.

While these are all important and interesting arguments in their own right, in this context they are remarkable for the way in which they integrate specific health and environmental criticisms with other forms of concern. Populist themes, in particular, emerge in the implied criticism of links between government and big business. None of the three objections notes environmental impacts. They disparage herbicide tolerance research, but not in virtue of unintended consequences.

Yet, the arguments would have little force *without the* environmental consequences noted above. The authors' review of funding priorities leaves readers with the impression that there is something fundamentally skewed about agriculture and agricultural research. The report is explicit in noting that a turn towards sustainable agriculture would help put things right, and hints that the difference between industrial and sustainable agriculture has something to do with a persons "mindset". However, there is little more than the list of unwanted outcomes to differentiate sustainable from non-sustainable agriculture. As such, it is far from clear that a truly philosophical shift is what the authors mean by "mindset."

The debate between critics and defenders of herbicide tolerance reprises the debate over pesticides. Despite attempts to bring in "mindset," the facts are what is at issue, and there is little of ethical significance to debate. Again, the form of the criticisms invites defenders to reply in ways that fail to generate philosophical reflection or thought. Furthermore, although the controversy is nominally about agricultural biotechnology, it is really a debate about chemicals. Agricultural biotechnology is being criticized for failing to move agriculture away from dependence on chemical inputs, and for, in fact, offering nothing more than symptomatic relief of the problems engendered by industrial agriculture.

However, if critics express their dissatisfaction in terms of risk to human health, and wild plant or animal species, they should expect a response that focuses on minimizing these risks, or that compensates for harm done. If the problem with chemically intensive agriculture is something *other* than the risks associated with chemical use, critics should not expend so much effort predicting harmful consequences from chemical use. If the problem is that modern industrial agriculture is founded on philosophical premises that are fundamentally flawed, it is those premises that should be exposed and criticized. In short, one should not grumble about responses that address only symptoms when one s original complaint has itself remained at the level of symptoms. Both *Altered Harvest* and *Biotechnology's Bitter Harvest* merge environmental criticism with some of the populist themes that have been prominent in the attack on BST. Critics create the distinct impression that something is fundamentally amiss in industrial agriculture, and blame the unwanted impacts they cite upon agriculture's dominance by commercial interests, seeking profit from the sale of commodities and from input technologies. Agriculture, however, has been controlled by commercial interests at least since the decline of the feudal system in Europe.

The critics want a return to a more humane agriculture, such as what may have existed in the late nineteenth and early twentieth centuries, but that agriculture was thoroughly commercial. Furthermore, collectivization experiments in the Soviet Union and other socialist countries have produced both human and environmental problems. As such, the commercial orientation of industrialized agriculture cannot be either a necessary nor sufficient condition for the unwanted impacts of modern agricultural technology, including biotechnology.

Environmental Ethics

The environmental critics of insecticides and agricultural biotechnology provide ample documentation of agriculture's importance for environmental policy. The criticisms they mount, however, do not constitute philosophical problems for agriculture, nor do they represent points of philosophical interest for those constructing an environmental ethic. This is not to say that the environmental criticisms of agriculture have no philosophical implications at all. Indeed, three points of significance illustrate why agricultures environmental impact has been of little interest to environmental philosophers. These points also indicate a line of inquiry for a more philosophical review of agriculture.

First, the critics of pesticide and of agricultural biotechnology recite laundry lists of unwanted impacts, but provide little insight into how or why impacts on nature differ from harm and risk to human beings. The food safety risks associated with chemical residue are as prominent in the environmental critique of pesticides as the impact on wildlife or biological diversity. Setting empirical questions aside, this pattern of criticism neglects a distinction of keen interest to environmental ethicists. One does not need an environmental ethic to explain why harming people are wrong.

An ethic of minimizing suffering or respecting human rights is perfectly capable of accounting for the wrongness of human health risks. It is not clear, however, that traditional ethical theories explain why we should be concerned about impact upon wildlife or biological diversity. Environmental ethicists have dedicated themselves to the task in a manner that is reviewed at several junctures elsewhere in this volume. Far from providing any unique or unifying environmental theme for philosophers to consider, the laundry list style of criticism appears to lack sophistication in themes that have already been well covered by philosophers.

Second, by stressing unwanted outcomes, the critics are working within, rather than against, the existing utilitarian philosophical framework of industrial agriculture. Industrial agriculture is committed to an ethic of optimizing the trade-off between costs and benefits. It has no intrinsic commitment to chemicals or to molecular technology. Business and scientific practices made it easy to overlook some of the costs of pesticide technology, but if they are true to their utilitarian principles, researchers and planners must be cognizant of all consequences. To the extent that the critics help decision makers attain cognizance, they assist utilitarian evaluation of agricultural practices. They make no philosophical objections to the optimization philosophy of utilitarian agriculture, in any case.

Third, the critics implicitly invite farmers, businesses, and researchers to solve environmental problems by developing alternatives that avoid or compensate for unwanted impacts. If the problem with pesticide or agricultural biotechnology consists in unintended consequences, why not keep doing what we're doing, but get rid of the consequences no one intends?

As already noted, the criticism provokes a technical response rather than philosophical reflection on the part of producers or scientists. If environmental critics truly want agriculture to rethink its philosophical bent towards production, as they seem to, they will have to mount an attack that goes beyond a list of unwanted outcomes. Such lists not only serve the existing ethic of industrial agriculture by requesting that producers correct their cost accounting, but they present little of interest to environmental philosophers, who might help envision an alternative agriculture.

Leopold and Schumacher

Two of the formative intellectual figures in the environmental movement, Aldo Leopold and E.F. Schumacher, did include some discussion of agricultural philosophy in their writings. Leopold is best known among philosophers for his essay "The Land Ethic" from *A Sand County Almanac.* The essay begins with a passage in which Leopold describes the rejection of human slavery as one of the key instances of moral progress in history. The key to this event, he thinks, was in ceasing to understand human beings as property, in extending the scope of the moral community to include all human beings. Leopold's message is that we must now find a way to think of our relation to land, understood again to mean the general biosphere, as something other than mere property.

Leopold finds any attempt to reflect conservationist concerns within the kind of optimizing calculations that underlie a traditional approach to agricultural decision making hopelessly lacking. In Leopold's view there is ample basis for care and concern about ecological values, but the problem is that the importance people place upon nature cannot be reflected in monetary terms. He writes:

When one of these non-economic categories is threatened, and if we happen to love it, we invent subterfuges to give it economic importance. At the beginnings of the century song-birds were supposed to be disappearing. Ornithologists jumped to the rescue with some shaky evidence to the effect that insects would eat us up if birds failed to control them. The evidence had to be economic in order to be valid.

Here Leopold would also seem to be rejecting the notion that unwanted outcomes of agricultural production decisions can be accommodated by a broader framework of benefits and costs, and including some constraints. Indeed, it is property rights, Leopold's target, that serve as the model for constraints. Instead, we must rethink our lives and our values so as to attain a fuller appreciation of the interdependence between human and natural communities.

One of the chief sources for understanding these links is agriculture. Early on in his book Leopold writes, "There are two spiritual dangers in not owning a farm. One is the danger of supposing that breakfast comes from the grocery, and the other that heat comes from a furnace". One who lives on a farm cannot, in Leopold's view, long forget the dependence of human action upon the underlying natural ecology. Written in the 1940s, *A Sand County Almanac* does not reflect more recent critics' concern that agriculture is on the verge of destroying its ecological base, but Leopold does express cynicism about the optimizing strategies of experiment station research:

The State College tells farmers that Chinese elms do not clog screens, and are hence preferable to cottonwoods. It also pontificates on cherry preserves, Bang's disease, hybrid corn, and beautifying the farm home. The only thing it does not know about farms is where they came from. Its job is to make Illinois safe for soybeans.

Leopold's land ethic, thus, rejects the optimizing strategy that takes increasing income, increasing production, and increasing benefits to consumers as its core. Instead, Leopold urges us to:

Examine each question in terms of what is ethically and aesthetically right, as well as what is economically expedient. A thing is right when it tends to preserve the integrity, stability, and beauty of the biotic community. It is wrong when it tends otherwise.

E.F. Schumacher's 1972 book *Small is Beautiful* followed *Silent Spring by* a decade, and combined Carson's concern for agricultural technology with Leopold's s distaste for making moral evaluations by calculating costs and benefits. The central theme of the book was widely taken to be an attack upon technologies that consumed relatively large quantities of fossil fuels and required large investments of fixed capital. However, in the chapter entitled "The Proper Use of Land," Schumacher takes up a central question in agricultural ethics. The argument of the chapter is first a criticism of what Schumacher calls "the philosophy of the townsman" and, second, a description of an alternative programme. The "townsmen" see agricultures economic woes as evidence that farming or ranching is a declining enterprise, and see the central problem of agriculture as one of improving farm income. Schumacher finds this view deficient. He writes:

We know too much about ecology today to have any excuse for the many abuses that are currently going on in the management of the land, in the management of animals, in food storage, food processing, and in heedless urbanization. If we permit them, this is not due to poverty, as if we could not afford to stop them; it is due to the fact that, as a society, we have no firm basis of belief in any meta-economic values, and when there is no such belief the economic calculus takes over.

In Schumacher's view, the problem arises when agriculture is understood as essentially defined by its capacity to produce and market saleable commodities. In making a statement of the wider goals of agriculture he writes:

A wider view sees agriculture as having to fulfill at least three tasks:

- To keep man in touch with living nature, of which he is and remains a highly vulnerable part;
- To humanize and ennoble man's wider habitat; and
- To bring forth the foodstuffs and other materials which are needed for a becoming life.

I do not believe that a civilization which recognizes only the third of these tasks, and which pursues it with such ruthlessness and violence that the other two tasks are not merely neglected but systematically counteracted, has any chance of long-term survival.

These remarks on agriculture must be understood in the light of Schumacher's overall attack upon "economic values" and his campaign to substitute a norm of "Buddhist economics" in its place. In criticizing economic values Schumacher means to attack the utilitarian emphasis upon increasing incomes; by interposing "Buddhist economics" in place of this emphasis, he means to suggest that there is an alternative way of conceptualizing economic activity, one that would trace production, distribution, and exchange according to the long-term impact of these activities upon the natural systems needed to support all. Economic policies that encourage consumption in order to promote economic growth are, on the view of Buddhist economics, incompatible with the goal of a permanent and stable society.

Although Schumacher's choice of words has the ring of late sixties hippie jargon, his point should be understood as a shift in philosophical perspective. Political theorist Paul Diesing has argued that Schumacher's critique is a complete rejection of the traditional utilitarian perspective on agricultural production. On this traditional view, Diesing writes: Nature appears in three forms: natural resources, cultivated land..., and externalities of production. Natural resources are free goods, *res nullius,* nothings, having no value until they are "produced" and made available for exchange.

When the central goal of agriculture is understood in terms of production, agricultural land is a form of fixed capital, and this, in turn, suggests that this land should be devoted to its most productive use. Although unwanted outcomes can be factored into the optimizing equation either as costs or as constraints, the result looks a bit like pre-Copernican models of the solar system, where epicycles and reversing rotations were added on to the charts for planetary motion in order to preserve a theory that falsely placed the earth at the center of the universe. In Diesing's view, Schumacher rejects this strategy when he insists that agriculture is not a form of industry and should not be viewed as fixed capital or even as a factor of production at all. Instead, land is the basis for life itself, a precondition for productive economic life, and not merely one among many factors available for productive appropriation. In Diesing's view, the agrarian component of Schumacher's thought is its essential philosophical theme. The more celebrated work on appropriate technology flows from Schumacher's view of agriculture, rather than the reverse.

The citations from Leopold and Schumacher indicate how each had a view of agriculture that was inconsistent both with the utilitarian orientation typical of producers or agricultural researchers, and with the presumptions of academic environmental ethicists. They make comments hinting at a philosophy of agriculture that includes an environmental ethic drawn from the very practices of farming itself. The development of such an ethic, and an evaluation of its significance to broader questions in agriculture, environmental policy, and environmental philosophy is the task that now awaits.

MODERN PLANT BIOTECHNOLOGY

Plants are made of millions of cells all working together. Every cell of a plant has a complete "instruction manual" or genome that is inherited from the parents of the plant as a combination of their genomes.

Genes are found within the genome and serve as the "words" of the instruction manual. When a cell reads a word, or in scientific terms "expresses a gene," a specific protein is produced. Proteins give an individual cell, and therefore the plant, its form and function. Genes (words) are written using the four-letter alphabet A, C, G, T. The letters are abbreviations for four chemicals called bases, which together make up DNA. DNA is universal in nature, meaning that the four chemical bases of DNA are the same in all living organisms. Consequently, a gene from one organism can function in any other organism.

The ability to move genes into plants from other organisms, thereby producing new proteins in the plant, has resulted in significant achievements in plant biotechnology that were not possible using traditional breeding practices.

Clearly, the ultimate solution to each of these problems is reducing population growth, a difficult challenge that is further complicated by social, political, economic and religious considerations. In the hope that national and international efforts will help to stabilize world population in the next few decades, our challenge is to use the power of plant biotechnology towards the solution of the numerous problems caused by population growth by increasing productivity, by reducing crop losses, and by protecting and conserving the environment. Plant biotechnology is not a magic bullet that will solve all of these problems, yet it is becoming abundantly clear that it is the best tool that we have and it can, if used wisely and in a timely fashion, make significant contributions.

It is rather ironic that at a time when international agriculture is under increasing pressure to meet the food needs of the ever increasing population, and when plant biotechnology is beginning to make significant contributions to food productivity and environmental safety, it has become the target of well coordinated and sustained attacks by many environmental and self-appointed watchdog groups, particularly in Western Europe and in some of the developing countries. In the United States, we had faced similar attacks during the 1970's and 1980's.

However, after extensive public debate, protest demonstrations, court challenges and congressional hearings, a federal regulatory framework was developed that has served the public and the private interest well. It has allowed the plant biotechnology industry to grow and introduce its products into the market place. The American consumers and farmers have accepted and benefited from transgenic products. Transgenic crops are being grown this year on nearly 100 million acres of American farmland, accounting for 74 per cent of our soybean, 71 per cent of our cotton and 32 per cent of our corn acreages.

It is fortunate and encouraging that China and India, the two most populous countries in the world, with increasing demand for food and worsening environmental problems, have recognized the importance of plant biotechnology in agriculture and have established active and successful research and development programmes in plant biotechnology, targeting many regional vegetable and fruit crops, in addition to such staples as wheat, rice, maize, soybean, various pulses, canola, and cotton. It is not too far-fetched to expect that within the next few years these two countries will plant the largest acreages of transgenic crops in the world.

Argentina and South Africa are two other developing countries that are increasing their planting of transgenic crops. Indeed, at the present time, nearly 10 per cent of the global acreage of transgenic crops is planted in the developing countries. Thus the argument that plant biotechnology is a tool of the industrialized countries for the exploitation of the developing world is no longer sustainable. It is our hope that the success of plant biotechnology in these countries will encourage similar efforts in other parts of the developing world.

It is true that the first generation of transgenic crops, which contain genes for resistance to herbicides and insects, did not provide any direct benefits to the consumer. Nevertheless, there are numerous indirect benefits, such as the reduced use of pesticides and herbicides, reduced tillage leading to soil conservation, reduced use of natural resources such as petrochemical products and water for the manufacture, transport and application of agro-chemicals, and reduced labour costs. By producing more food on the same amount of land transgenic crops promote conservation and biodiversity by saving wildlife habitats and precious forests from being converted into farmland. Reduced use of pesticides has already shown a marked decrease in illness and death caused by pesticide poisonings in China and South Africa (nearly 500 cotton farmers in China die each year of acute pesticide poisoning).

The vital role of agriculture and food production in human health and nutrition, in poverty alleviation, and in social and political stability, is well known. This was recognized as far back as 1970, when Norman Borlaug was awarded the Nobel Peace Prize for his work that led to the Green Revolution and that helped to save hundreds of millions of lives in the developing countries.

Plant biotechnology too can contribute to international peace and security by increasing food production, producing safer and healthier foods, protecting our rather finite natural resources and the environment, and improving human health. It is, therefore, morally and socially irresponsible and indefensible to prevent or delay the applications of plant biotechnology to problems of hunger, health and protection of the environment. The opponents of plant biotechnology would have us believe that it is an unnatural and unsafe process that produces harmful products, and that it is totally different from plant breeding and selection that account for almost all of our modern crops. The indisputable fact, however, is that humans have engineered crops for nearly 10,000 years. Almost all of our major crops-such as maize, wheat, potato, tomato and others-are man made. Indeed, none of our modern crops are capable of surviving in the wild without human care.

The molecular and genetic principles of plant biotechnology and plant breeding and selection are the same. Plant biotechnology is no different from breeding and selection or for that matter from radiation and chemically induced mutation breeding, except that it is extraordinarily precise and predictable, and is not restricted by taxonomic boundaries. No compelling evidence has ever been presented to show that transgenic crops are innately different the non-transgenic products of breeding and selection.

In retrospect, however, we must share some of the blame for the perception that plant biotechnology is different from plant breeding and selection. During the 1970's and 1980's, when there was a great deal of euphoria over the production of somatic hybrids, doubled-haploid breeding lines and transgenic plants, the plant biotechnology community made a serious error in strategy and judgement when it distanced itself from breeding and selection and established a separate identity for itself. It was us who placed the spotlight on the process and not the product. This has come to haunt us now as it has attracted undue and undeserved attention and opposition. It was also an error not to engage early in the debate on transgenic plants, and to permit the opponents of plant biotechnology to dictate the agenda.

As responsible members of the world community, and as scientists, we cannot, and should not, be silent observers of this debate. We must play an active role in the debate on plant biotechnology and make an informed and professional contribution to the public dialogue, emphasizing the many benefits of transgenic crops to human health and the environment. The opposition to transgenic foods in Europe and elsewhere is based exclusively on political and ideological differences rather than on any credible scientific evidence. On two rare occasions an attempt was made to present scientific arguments against the use of transgenic crops. These involved the allegedly harmful effect of pollen from Bt maize plants on the larvae of the Monarch butterfly, and the alleged transgene contamination of maize in Mexico.

More detailed investigations by several research groups have since refuted these claims and showed them to be of dubious scientific value. Indeed, in the Mexican maize story, the journal Nature, in an unprecedented action in its more than 100 year history, was forced to disown the paper published in its own pages. The consumer, the farmer and the biotechnology industry have been ill served by the sustained campaign of misinformation and unsubstantiated claims of dangers to public health and the environment by transgenic crops and their products. After more than ten years and thousands of field trials in many countries, after nearly a decade of commercial plantings on hundreds of millions of acres, and after transgenic food products having being used by hundreds of millions of humans and farm animals, there is not a single documented instance of illness reported in any human or animal, or of ecological or environmental damage. What then is the basis and rationale for the many restrictions still placed on the field planting and human use of transgenic foods?

The enviable and unblemished record of transgenic crops and their products is the strongest evidence for their safety and wholesomeness. The opponents of plant biotechnology should compare this record with that of the many drugs approved for human use in the United States. In an exhaustive study published recently in the Journal of the American Medical Association, it was reported that 20 per cent of the 548 drugs approved for human use during the past 25 years were later found to have serious or life-threatening side effects. Seven of the drugs possibly contributed to 1002 deaths, and 16 were forced to be withdrawn from the market. In comparison, not a single transgenic food product has ever been shown to have any harmful effects, and none has been withdrawn because of adverse reactions in humans or animals.

The plant biotechnology community has already done more for the environment and the developing countries than the self-proclaimed environmental groups and the so-called friends of the poor. Indeed, the opponents of plant biotechnology have done much harm to their professed cause by slowing down and/or preventing the planting and utilization of transgenic crops around the world. The contributions of the plant biotechnology community, on the other hand, are socially and morally responsible and of considerable humanitarian value. We have every reason to be proud of these contributions.

The rules and regulations adopted for transgenic crops in the 1980's were both prudent and necessary. At that time there were many unknowns about transgenic crops and about their possible effect on humans and the environment. There was a need to establish a database to satisfy the concerns of the general public as well as the scientific community. Three federal agencies, the United States Department of Agriculture, the Environmental Protection Agency and the Food and Drug Administration, were given oversight responsibilities for transgenic crops. The resulting open and transparent system established in the United States has worked well and has served its purpose.

It has done much to gain the confidence and support of the American public for plant biotechnology and its products. In light of the demonstrated safety of transgenic crops to humans, animals and the environment, the question must be asked whether it is any more necessary, or even advisable, to continue the expensive, time consuming and burdensome requirements for the public release of transgenic crops and their products (field trials of transgenic crops are 10-20 times more expensive than of similar plants developed by conventional means).

We propose that based on our considerable experience and on the vast amount of information gathered about the safety of transgenic crops over the past decade, it is time for our regulatory agencies to consider whether some or all of the current regulatory requirements can be gradually relaxed and ultimately suspended, except in those rare instances where there is the clear likelihood of risk to human health and the environment. Genuine concerns about gene flow and development of resistance to antibiotics, and pests or pathogens, can be met adequately with currently available and emerging technologies.

The process of deregulation of transgenic crops, controversial and difficult as it may be, needs to begin now because the continuation of the present rules and regulations is entirely unnecessary, unjustified and counterproductive. In order to be effective and acceptable, the process should be open to all points of view. The decisions, however, must be based on science and facts and not on political or ideological considerations. Nearly two decades ago, the United States played a leading and useful role in establishing the rules and regulations for the field planting, evaluation and human use of transgenic crops. It should now play a similar role in having these restrictions relaxed and removed.

It is clear that the challenges we face in the 21st century are greater than those we faced in the last century. Of all the available technologies, plant biotechnology offers the best hope for producing more and better food, fibre and pharmaceuticals, and for protecting, preserving and improving the environment for the benefit of humankind.

My own confidence in plant biotechnology comes from knowing that the science behind it is sound, that it is well tested and proven, that it benefits the consumer, the farmer and the industry, and that it protects and conserves the environment. It is for these reasons that I am convinced that plant biotechnology will within the next two decades become an integral part of the international agricultural system. With the United States, China, and lately India, three of the most populous countries in the world serving as examples, we have taken the first steps towards achieving that objective.

2

Photosynthesis and Productivity of Agriculture

PHOTOSYNTHESIS

Photosynthesis is a process that converts carbon dioxide into organic compounds, especially sugars, using the energy from sunlight. Photosynthesis occurs in plants, algae, and many species of bacteria, but not in archaea. Photosynthetic organisms are called photoautotrophs, since they can create their own food. In plants, algae, and cyanobacteria, photosynthesis uses carbon dioxide and water, releasing oxygen as a waste product. Photosynthesis is vital for all aerobic life on Earth.

As well as maintaining the normal level of oxygen in the atmosphere, nearly all life either depends on it directly as a source of energy, or indirectly as the ultimate source of the energy in their food (the exceptions are chemoautotrophs that live in rocks or around deep sea hydrothermal vents). The rate of energy capture by photosynthesis is immense, approximately 100 terawatts, which is about six times larger than the power consumption of human civilization. As well as energy, photosynthesis is also the source of the carbon in all the organic compounds within organisms' bodies. In all, photosynthetic organisms convert around 100–115 teragrams of carbon into biomass per year.

Although photosynthesis can happen in different ways in different species, some features are always the same. For example, the process always begins when energy from light is absorbed by proteins called photosynthetic reaction centers that contain chlorophylls. In plants, these proteins are held inside organelles called chloroplasts, while in bacteria they are embedded in the plasma membrane. Some of the light energy gathered by chlorophylls is stored in the form of adenosine triphosphate (ATP). The rest of the energy is used to remove electrons from a substance such as water. These electrons are then used in the reactions that turn carbon dioxide into organic compounds. In plants, algae and cyanobacteria, this is done by a sequence of reactions called the Calvin cycle, but different sets of reactions are found in some bacteria, such as the reverse Krebs cycle in Chlorobium. Many photosynthetic organisms have adaptations that concentrate or store carbon dioxide. This helps reduce a wasteful process called photorespiration that can consume part of the sugar produced during photo-synthesis.

The first photosynthetic organisms probably evolved about 3,500 million years ago, early in the evolutionary history of life, when all forms of life on Earth were microorganisms and the atmosphere had much more carbon dioxide. They most likely used hydrogen or hydrogen sulfide as sources of electrons, rather than water. Cyanobacteria appeared later, around 3,000 million years ago, and drastically changed the Earth when they began to oxygenate the atmosphere, beginning about 2,400 million years ago.

This new atmosphere allowed the evolution of complex life such as protists. Eventually, no later than a billion years ago, one of these protists formed a symbiotic relationship with a cyanobacterium, producing the ancestor of many plants and algae. The chloroplasts in modern plants are the descendants of these ancient symbiotic cyanobacteria.

Photosynthetic organisms are photoautotrophs, which means that they are repositories of energy, they are able to synthesize food directly from carbon dioxide using energy from light.

They accrue it as part of their potential energy. However, not all organisms that use light as a source of energy carry out photosynthesis, since photoheterotrophs use organic compounds, rather than carbon dioxide, as a source of carbon. In plants, algae and cyanobacteria, photosynthesis releases oxygen.

This is called oxygenic photosynthesis. Although there are some differences between oxygenic photosynthesis in plants, algae and cyanobacteria, the overall process is quite similar in these organisms. However, there are some types of bacteria that carry out anoxygenic photosynthesis, which consumes carbon dioxide but does not release oxygen.

Carbon dioxide is converted into sugars in a process called carbon fixation. Carbon fixation is a redox reaction, so photosynthesis needs to supply both a source of energy to drive this process, and the electrons needed to convert carbon dioxide into a carbohydrate, which is a reduction reaction. In general outline, photosynthesis is the opposite of cellular respiration, where glucose and other compounds are oxidized to produce carbon dioxide, water, and release chemical energy. However, the two processes take place through a different sequence of chemical reactions and in different cellular compartments.

The general equation for photosynthesis is therefore:

$$2n\ CO_2 + 2n\ H_2O + \text{photons} \rightarrow 2(CH_2O)_n + n\ O_2 + 2n\ A$$

Carbon dioxide + electron donor + light energy → carbohydrate + oxygen + oxidized electron donor

Since water is used as the electron donor in oxygenic photosynthesis, the equation for this process is:

$$2n\ CO_2 + 2n\ H_2O + \text{photons} \rightarrow 2(CH_2O)_n + 2n\ O_2$$

carbon dioxide + water + light energy → carbohydrate + oxygen

Other processes substitute other compounds (such as arsenite) for water in the electron–supply role; the microbes use sunlight to oxidize arsenite to arsenate: The equation for this reaction is:

$$(AsO_3^{3-}) + CO_2 + \text{photons} \rightarrow CO + (AsO_4^{3-})$$

arsenite + carbon dioxide + light energy → carbon monoxide + arsenate

Photosynthesis occurs in two stages. In the first stage, light–dependent reactions or light reactions capture the energy of light and use it to make the energy–storage molecules ATP and NADPH. During the second stage, the light–independent reactions use these products to capture and reduce carbon dioxide. Most organisms that utilize photosynthesis to produce oxygen use visible light to do so, although at least three use infrared radiation.

PLANT GROWTH AND LEAF ANATOMY

We have discussed how plant growth and leaf anatomy are influenced by light. Now we turn our attention to how CO_2 concentration affects photosynthesis. CO_2 diffuses from the atmosphere into leaves—first through stomata, then through the intercellular air spaces, and ultimately into cells and chloroplasts. In the presence of adequate amounts of light, higher CO_2 concentrations support higher photosynthetic rates. The reverse is also true; that is, low CO_2 concentration can limit the amount of photosynthesis. The concentration of atmospheric CO_2 in recent history, and its availability for carbon-fixing processes. Then we'll consider the limitations that CO_2 places on photosynthesis and the impact of the CO_2-concentrating mechanisms of C_4 plants.

ATMOSPHERIC CO_2 CONCENTRATION KEEPS RISING

Carbon dioxide is a trace gas in the atmosphere, presently accounting for about 0.037%, or 370 parts per million (ppm), of air. The partial pressure of ambient CO_2 (Ca) varies with atmospheric pressure and is approximately 36 pascals (Pa) at sea level. Water vapor usually accounts for up to 2% of the atmosphere and O_2 for about 20%. The bulk of the atmosphere, nearly 80%, is nitrogen. The current atmospheric concentration of CO_2 is almost twice the concentration that has prevailed during most of the last 160,00 years, as measured from air bubbles trapped in glacial ice in Antarctica. Except for the last 200 years, CO_2 concentrations during the recent geological past have been low, fluctuating between 180 and 260 ppm.

These low concentrations were typical of times extending back to the Cretaceous, when Earth was much warmer and the CO_2 concentration may have been as high as 1200 to 2800 ppm. The current CO_2 concentration of the atmosphere is increasing by about 1 ppm each year, primarily because of the burning of fossil fuels. Since 1958, when systematic measurements of CO_2 began at Mauna Loa, Hawaii, atmospheric CO_2 concentrations have increased by more than 17%, and by 2020 the atmospheric CO_2 concentration could reach 600 ppm.

The Greenhouse Effect

The consequences of this increase in atmospheric CO_2 are under intense scrutiny by scientists and government agencies, particularly because of predictions that the greenhouse effect is altering the world's climate. The term *greenhouse effect* refers to the resulting warming of Earth's climate, which is caused by the trapping of long-wavelength radiation by the atmosphere. A greenhouse roof transmits visible light, which is absorbed by plants and other surfaces inside the greenhouse. The absorbed light energy is converted to heat, and part of it is re-emitted as long-wavelength radiation.

Because glass transmits long-wavelength radiation very poorly, this radiation cannot leave the greenhouse through the glass roof, and the greenhouse heats up. Certain gases in the atmosphere, particularly CO_2 and methane, play the same role as the glass roof in a greenhouse. The increased CO_2 concentration and temperature associated with the greenhouse effect can influence photosynthesis. At current atmospheric CO_2 concentrations, photosynthesis in C_3 plants is CO_2 limited, but this situation could change as atmospheric CO_2 concentrations continue to rise. Under laboratory conditions, most C_3 plants grow 30 to 60% faster when CO_2 concentration is doubled (to 600–700 ppm), and the growth rate changes depend on nutrient status. In some plants the enhanced growth is only temporary. For many crops, such as tomatoes, lettuce, cucumbers, and roses growing in greenhouses under optimal nutrition, carbon dioxide enrichment in the greenhouse environment results in increased productivity. The photosynthetic performance of C_3 plants under elevated CO_2 is enhanced because photorespiration decreases.

Diffusion of CO_2 to the Chloroplastis Essential to Photosynthesis

For photosynthesis to occur, carbon dioxide must diffuse from the atmosphere into the leaf and into the carboxylation site of rubisco. Because diffusion rates depend on concentration gradients, appropriate gradients are needed to ensure adequate diffusion of CO_2 from the leaf surface to the chloroplast. The cuticle that covers the leaf is nearly impermeable to CO_2, so the main port of entry of CO_2 into the leaf is the stomatal pore. CO_2 diffuses through the pore into the substomatal cavity and into the intercellular air spaces between the mesophyll cells.

This portion of the diffusion path of CO_2 into the chloroplast is a gaseous phase. The remainder of the diffusion path to the chloroplast is a liquid phase, which begins at the water layer that wets the walls of the mesophyll cells and continues through the plasma membrane, the cytosol, and the chloroplast. Each portion of this diffusion pathway imposes a resistance to CO_2 diffusion, so the supply of CO_2 for photosynthesis meets a series of different points of resistance. An evaluation of the magnitude of each point of resistance is helpful for understanding CO_2 limitations to photosynthesis. Carbon dioxide enters the intercellular air spaces of the leaf through the stomatal pores. From the air spaces it dissolves in the water of wet cell walls and diffuses into the cell and chloroplast.

The same path is traveled in the reverse direction by H_2O. The sharing of this pathway by CO_2 and water presents the plant with a functional dilemma. In air of high relative humidity, the diffusion gradient that drives water loss is about 50 times larger than the gradient that drives CO_2 uptake. In drier air, this gradient can be even larger. Therefore, a decrease in stomatal resistance through the opening of stomata facilitates higher CO_2 uptake but is unavoidably accompanied by substantial water loss. The gas phase of CO_2 diffusion into the leaf can be divided into three components—the boundary layer, the stomata, and the intercellular spaces of the leaf—each of which imposes a resistance to CO_2 diffusion.

The boundary layer consists of relatively unstirred air at the leaf surface, and its resistance to diffusion is called the boundary layer resistance. The magnitude of the boundary layer resistance decreases with leaf size and wind speed. The boundary layer resistance to water and CO_2 diffusion is physically related to the boundary layer resistance to sensible heat loss discussed earlier. Smaller leaves have a lower boundary layer resistance to CO_2 and water diffusion, and to sensible heat loss. Leaves of desert plants are usually small, facilitating sensible heat loss. The large leaves often found in the humid Tropics can have large boundary layer resistances, but these leaves can dissipate the radiation heat load by evaporative cooling because of the high transpiration rates made possible by the abundant water supply in these habitats.

After diffusing through the boundary layer, CO_2 enters the leaf through the stomatal pores, which impose the next type of resistance in the diffusion pathway, the stomatal resistance. Under most conditions in nature, in which the air around a leaf is seldom completely still, the boundary layer resistance is much smaller than the stomatal resistance, and the main limitation to CO_2 diffusion is imposed by the stomatal resistance.

There is also a resistance to CO_2 diffusion in the air spaces that separate the substomatal cavity from the walls of the mesophyll cells, called the intercellular air space resistance. This resistance is also usually small—causing a drop of 0.5 Pa or less in partial pressure of CO_2, compared with the 36 Pa outside the leaf. The resistance to CO_2 diffusion of the liquid phase—the liquid phase resistance, also called mesophyll resistance—encompasses diffusion from the intercellular leaf spaces to the carboxylation sites in the chloroplast. This point of resistance to CO_2 diffusion has been calculated as approximately one-tenth of the combined boundary layer resistance and stomatal resistance when the stomata are fully open. This low resistance value can be attributed in part to the large surface area of mesophyll cells exposed to the intercellular air spaces, which can be as much as 10 to 30 times the projected leaf area.

In addition, the localization of chloroplasts near the cell periphery minimizes the distance that CO_2 diffuses to carboxylation sites within the chloroplast. The positioning of chloroplasts and the relatively large percentage of intercellular air space (about 20–40%) are special anatomic features that facilitate the internal diffusion and uptake of CO_2 by leaves.

Because the stomatal pores usually impose the largest resistance to CO_2 uptake and water loss in the diffusion pathway, this regulation provides the plant with an effective way to control gas exchange between the leaf and the atmosphere. In experimental measurements of gas exchange from leaves, the boundary layer resistance and the intercellular air space resistance are usually ignored, and the stomatal resistance is used as the single parameter describing the gas phase resistance to CO_2.

Patterns of Light Absorption GenerateGradients of CO_2 Fixation within the Leaf

We have discussed how leaf anatomy is specialized for capturing light and how it also facilitates the internal diffusion of CO_2, but where in the leaf do maximum rates of photosynthesis occur? In most leaves, light is preferentially absorbed at the upper surface, whereas CO_2 enters through the lower surface. Given that light and CO_2 enter from opposing sides of the leaf, does photosynthesis occur uniformly within the leaf tissues, or is there a gradient in photosynthesis across the leaf?

The photosynthetic properties of a leaf are determined by the following:

- Profiles of light absorption across the mesophyll
- Photosynthetic capacity of those tissues
- Internal CO_2 supply

For most leaves, internal CO_2 diffusion is rapid, so limitations on photosynthetic performance within the leaf are imposed by factors other than CO_2 supply. When white light enters the upper surface of a leaf, blue and red photons are preferentially absorbed by chloroplasts near the irradiated surface, owing to the strong absorption bands of chlorophyll in the blue and red regions of the spectrum. Green light, on the other hand, penetrates deeper into the leaf. Compared to blue and red, chlorophyll absorbs poorly in the green, yet green light is very effective in supplying energy for photosynthesis in the tissues within the leaf depleted from blue and red photons.

The capacity of the leaf tissue for photosynthetic CO_2 assimilation depends to a large extent on its rubisco content. In spinach and the faba bean (*Vicia faba*), rubisco content starts out low at the top of the leaf, increases toward the middle, and then decreases again toward the bottom. As a result, the distribution of carbon fixation within the leaf is bell shaped. The spongy mesophyll fixes about 40% of the total carbon in spinach. The functional significance of the rubisco distribution and the profiles of carbon assimilation within leaves is not yet known, although it is likely that photosynthesis profiles vary in leaves with different anatomy and in leaves adapted to different environments.

CO_2 Imposes Limitations on Photosynthesis

Expressing photosynthetic rate as a function of the partial pressure of CO_2 in the intercellular air space (Ci) within the leaf makes it possible to evaluate limitations to photosynthesis imposed by CO_2 supply. At very low intercellular CO_2 concentrations, photosynthesis is strongly limited by the low CO_2, while respiratory rates are unaffected. As a result, there is a negative balance between CO_2 fixed by photosynthesis and CO_2 produced by respiration, and a net efflux of CO_2 from the plant. Increasing intercellular CO_2 to the concentration at which these two processes balance each other defines the CO_2 compensation point, at which the net efflux of CO_2 from the plant is zero. This concept is analogous to that of the light compensation point discussed earlier in the chapter: *The CO_2 compensation point reflects the balance between photosynthesis and respiration as a function of CO_2 concentration, and the light compensation point reflects that balance as a function of photon flux.*

In C_3 plants, increasing CO_2 above the compensation point stimulates photosynthesis over a wide concentration range. At low to intermediate CO_2 concentrations, photosynthesis is limited by the carboxylation capacity of rubisco. At high CO_2 concentrations, photosynthesis is limited by the capacity of Calvin cycle to regenerate the acceptor molecule ribulose-1,5-bisphosphate, which depends on electron transport rates. By regulating stomatal conductance, most leaves appear to regulate their Ci (internal partial pressure for CO_2) such that it is intermediate between limitations imposed by carboxylation capacity and the capacity to regenerate ribulose-1,5-bisphosphate. A plot of CO_2 assimilation as a function intercellular partial pressures of CO_2 tells us how photosynthesis is regulated by CO_2, independent of the functioning of stomata.

Inspection of such a plot for C_3 and C_4 plants reveals interesting differences between the two types of carbon metabolism:

1. In *C_4 plants,* photosynthetic rates saturate at Ci values of about 15 Pa, reflecting the effective CO_2-concentrating mechanisms operating in these plants.
2. In *C_3 plants,* increasing Ci levels continue to stimulate photosynthesis over a much broader range.

These results indicate that C_3 plants may benefit more from ongoing increases in atmospheric CO_2 concentrations. In contrast, photosynthesis in C_4 plants is CO_2 saturated at low concentrations, and as a result C_4 plants do not benefit from increases in atmospheric CO_2 concentrations. Plants with C_4 metabolism have a CO_2 compensation point of zero or nearly zero, reflecting their very low levels of photorespiration. This difference between C_3 and C_4

plants is not seen when the experiments are conducted at low oxygen concentrations because oxygenation is also suppressed in C_3 plants.

CO_2-Concentrating MechanismsAffect Photosynthetic Responses of Leaves

Because of the operating CO_2-concentrating mechanisms in C_4 plants, CO_2 concentration at the carboxylation sites within C_4 chloroplasts is often saturating for rubisco activity. As a result, plants with C4 metabolism need less rubisco than C3 plants need to achieve a given rate of photosynthesis, and require less nitrogen to grow. In addition, the CO_2-concentrating mechanism allows the leaf to maintain high photosynthetic rates at lower Ci values, which require lower rates of stomatal conductance for a given rate of photosynthesis.

Thus, C_4 plants can use water and nitrogen more efficiently than C_3 plants can. On the other hand, the additional energy cost of the concentrating mechanism makes C_4 plants less efficient in their utilization of light. This is probably one of the reasons that most shade-adapted plants are C_3 plants. Many cacti and other succulent plants with CAM metabolism open their stomata at night and close them during the day. The CO_2 taken up during the night is fixed into malate. Because air temperatures are much lower at night than during the day, water loss is low and a significant amount of water is saved relative to the amount of CO_2 fixed.

The main constraint on CAM metabolism is that the capacity to store malic acid is limited, and this limitation restricts the amount of CO_2 uptake. However, many CAM plants can fix CO_2 via the Calvin cycle at the end of the day, when temperature gradients are less extreme. Cladodes (flattened stems) of cacti can survive after detachment from the plant for several months without water. Their stomata are closed all the time, and the CO_2 released by respiration is refixed into malate. This process, which has been called *CAM idling,* allows the plant to survive for prolonged periods of time while losing remarkably little water.

ROLE OF PHOTOSYNTHESIS

Photosynthesis is the physico-chemical process by which plants, algae and photosynthetic bacteria use light energy to drive the synthesis of organic compounds. In plants, algae and certain types of bacteria, the photosynthetic process results in the release of molecular oxygen and the removal of carbon dioxide from the atmosphere that is used to synthesize carbohydrates (oxygenic photosynthesis).

Other types of bacteria use light energy to create organic compounds but do not produce oxygen (anoxygenic photosynthesis). Photosynthesis provides the energy and reduced carbon required for the survival of virtually all life on our planet, as well as the molecular oxygen necessary for the survival of oxygen consuming organisms1.In addition, the fossil fuels currently being burned to provide energy for human activity were produced by ancient photosynthetic organisms. Although photosynthesis occurs in cells or organelles that are typically only a few microns across, the process has a profound impact on the earth's atmosphere and climate.

Each year more than 10% of the total atmospheric carbon dioxide is reduced to carbohydrate by photosynthetic organisms. Most, if not all, of the reduced carbon is returned to the atmosphere as carbon dioxide by microbial, plant and animal metabolism, and by biomass combustion. In turn, the performance of photosynthetic organisms depends on the earth's atmosphere and climate.

Over the next century, the large increase in the amount of atmospheric carbon dioxide created by human activity is certain to have a profound impact on the performance and competition of photosynthetic organisms. Knowledge of the physico-chemical process of photosynthesis is essential for understanding the relationship between living organisms and the atmosphere and the balance of life on earth. Several books on photosynthesis are available for the uninitiated or advanced student. Taiz and Zeiger place the photosynthetic process in the context of over all plant physiology, and Cramer and Knaff describe the bioenergetic foundation of photosynthesis.

The overall equation for photosynthesis is deceptively simple. In fact, a complex set of physical and chemical reactions must occur in a coordinated manner for the synthesis of carbohydrates. To produce a sugar molecule such as sucrose, plants require nearly 30 distinct proteins that work within a complicated membrane structure.

Research into the mechanism of photosynthesis centers on understanding the structure of the photosynthetic components and the molecular processes that use radiant energy to drive carbohydrate synthesis. The research involves several disciplines, including physics, biophysics, chemistry, structural biology, biochemistry, molecular biology and physiology, and serves as an outstanding example of the success of multidisciplinary research. As such, photosynthesis presents a special challenge in understanding several interrelated molecular processes.

BRIEF HISTORY

In the 1770s Joseph Priestley, an English chemist and clergyman, performed experiments showing that plants release a type of air that allows combustion. He demonstrated this by burning a candle in a closed vessel until the flame went out. He placed a sprig of mint in the chamber and after several days showed that the candle could burn again. Although Priestley did not know about molecular oxygen, his work showed that plants release oxygen into the atmosphere. It is noteworthy that over 200 years later, investigating the mechanism by which plants produce oxygen is one of the most active areas of photosynthetic research. Building on the work of Priestley, Jan Ingenhousz, a Dutch physician, demonstrated that sunlight was necessary for photosynthesis and that only the green parts of plants could release oxygen. During this period Jean Senebier, a Swiss botanist and naturalist, discovered that CO_2 is required for photosynthetic growth and Nicolas-The odore de Saussure, a Swiss chemist and plant physiologist, showed that water is required. It was not until 1845 that Julius Robert von Mayer, a German physician and physicist, proposed that photosynthetic organisms convert light energy into chemical free energy. An interesting time line of the history of photosynthesis has been presented by Huzisige and Ke. By the middle of the nineteenth century the key features of plant photosynthesis were known, namely, that plants could use light energy to make carbohydrates from CO_2 and water. The empirical equation representing the net reaction of photosynthesis for oxygen evolving organisms is :

$$CO_2 + 2H_2O + \text{Light Energy} \rightarrow [CH_2O] + O_2 + H_2O, \ (1)$$

where [CH2O] represents a carbohydrate (e.g., glucose, a six-carbon sugar). The synthesis of carbohydrate from carbon and water requires a large input of light energy. The standard free energy for the reduction of one mole of CO_2 to the level of glucose is +478 kJ/mol. Because glucose, a six carbon sugar, is often an intermediate product of photosynthesis, the net equation of photosynthesis is frequently written as:

$$6CO_2 + 12H_2O + \text{Light Energy} \rightarrow C_6H_{12}O_6 + 6O_2 + 6H_2O. \ (2)$$

The standard free energy for the synthesis of glucose is +2,870 kJ/mol. Not surprisingly, early scientists studying photosynthesis concluded that the O_2 released by plants came from CO_2, which was thought to be split by light energy. In the 1930s comparison of bacterial and plant photosynthesis lead Cornelis van Niel to propose the general equation of photosynthesis that applies to plants, algae and photosynthetic bacteria. Van Niel was aware that some photosynthetic bacteria could use hydrogen sulfide (H_2S) instead of water for photosynthesis and that these organisms released sulfur instead of oxygen. Van Niel, among others, concluded that photosynthesis depends on electron donation and acceptor reactions and that the O_2 released during photosynthesis comes from the oxidation of water. Van Niel's generalized equation is :

$$CO_2 + 2H_2A + \text{Light Energy} \rightarrow [CH_2O] + 2A + H_2O. \ (3)$$

In oxygenic photosynthesis, 2A is O_2, whereas in anoxygenic photosynthesis, which occurs in some photosynthetic bacteria, the electron donor can be an inorganic hydrogen donor, such as H2S (in which case A is elemental sulfur) or an organic hydrogen donor such as succinate (in which case, A is fumarate). Experimental evidence that molecular oxygen came from water was provided by Hill and Scarisbrick (1940) who demonstrated oxygen evolution in the absence of CO_2 in illuminated chloroplasts and by Ruben et al. (1941) who used 18O enriched water.

The biochemical conversion of CO_2 to carbohydrate is a reduction reaction that involves the rearrangement of covalent bonds between carbon, hydrogen and oxygen. The energy for the reduction of carbon is provided by energy rich molecules that are produced by the light driven electron transfer reactions. Carbon reduction can occur in the dark and involves a series of biochemical reactions that were elucidated by Melvin Calvin, Andrew Benson and James Bassham in the late 1940s and 1950s. Using the radioisotope 14C, most of the intermediate steps that result in the production of carbohydrate were identified. Calvin was awarded the Nobel Prize for Chemistry in 1961 for this work.

In 1954 Daniel Arnon and coworkers discovered that plants, and A. Frenkel discovered that photosynthetic bacteria, use light energy to produce ATP, an organic molecule that serves as an energy source for many biochemical reactions. During the same period L.N.M. Duysens showed that the primary photochemical reaction of photosynthesis is an oxidation/ reduction reaction that occurs in a protein complex (the reaction centre). Over the next few years the work of several groups, including those of Robert Emerson, Bessel Kok, L.N.M. Duysens, Robert Hill and Horst Witt, combined to prove that plants, algae and cyanobacteria require two reaction centers, photosystem II and photosystem I, operating in series.

In 1961 Peter Mitchell suggested that cells can store energy by creating an electric field or a proton gradient across a membrane. Mitchell's proposal that energy is stored as an electrochemical gradient across a vesicular membrane opened the door for understanding energy transformation by membrane systems. He was awarded the Nobel Prize in Chemistry in 1978 for his theory of chemiosmotic energy transduction.

Most of the proteins required for the conversion of light energy and electron transfer reactions of photosynthesis are located in membranes. Despite decades of work, efforts to determine the structure of membrane bound proteins had little success.

This changed in the 1980s when Johann Deisenhofer, Hartmut Michel, Robert Huber and co-workers determined the structure of the reaction centre of the purple bacterium Rhodospeudomonas viridis.. They were awarded the Nobel Prize for Chemistry in 1988 for their work, which has provided insight into the relationship between structure and function in membrane-bound proteins.

A key element in photosynthetic energy conversion is electron transfer within and between protein complexes and simple organic molecules. The electron transfer reactions are rapid (as fast as a few picoseconds) and highly specific. Much of our current understanding of the physical principles that guide electron transfer is based on the pioneering work of Rudolph A. Marcus, who received the Nobel Prize in Chemistry in 1992 for his contributions to the theory of electron transfer reaction in chemical systems.

PHOTOSYNTHETIC MEMBRANES AND ORGANELLES

The proteins that gather light for photosynthesis are embedded within cell membranes. The simplest way these are arranged is in photosynthetic bacteria, where these proteins are held within the plasma membrane. However, this membrane may be tightly folded into cylindrical sheets called thylakoids, or bunched up into round vesicles called intracytoplasmic membranes. These structures can fill most of the interior of a cell, giving the membrane a very large surface area and therefore increasing the amount of light that the bacteria can absorb.

In plants and algae, photosynthesis takes place in organelles called chloroplasts. A typical plant cell contains about 10 to 100 chloroplasts. The chloroplast is enclosed by a membrane. This membrane is composed of a phospholipid inner membrane, a phospholipid outer membrane, and an intermembrane space between them. Within the membrane is an aqueous fluid called the stroma.

The stroma contains stacks (grana) of thylakoids, which are the site of photosynthesis. The thylakoids are flattened disks, bounded by a membrane with a lumen or thylakoid space within it. The site of photosynthesis is the thylakoid membrane, which contains integral and peripheral membrane protein complexes, including the pigments that absorb light energy, which form the photosystems. Plants absorb light primarily using the pigment chlorophyll, which is the reason that most plants have a green colour. Besides chlorophyll, plants also use pigments such as carotenes and xanthophylls.

Algae also use chlorophyll, but various other pigments are present as phycocyanin, carotenes, and xanthophylls in green algae, phycoerythrin in red algae (rhodophytes) and fucoxanthin in brown algae and diatoms resulting in a wide variety of colours.

These pigments are embedded in plants and algae in special antenna–proteins. In such proteins all the pigments are ordered to work well together. Such a protein is also called a light–harvesting complex. Although all cells in the green parts of a plant have chloroplasts, most of the energy is captured in the leaves. The cells in the interior tissues of a leaf, called the mesophyll, can contain between 450,000 and 800,000 chloroplasts for every square millimeter of leaf. The surface of the leaf is uniformly coated with a water–resistant waxy cuticle that protects the leaf from excessive evaporation of water and decreases the absorption of ultraviolet or blue light to reduce heating. The transparent epidermis layer allows light to pass through to the palisade mesophyll cells where most of the photosynthesis takes place.

LIGHT REACTIONS

In the light reactions, one molecule of the pigment chlorophyll absorbs one photon and loses one electron. This electron is passed to a modified form of chlorophyll called pheophytin, which passes the electron to a quinone molecule, allowing the start of a flow of electrons down an electron transport chain that leads to the ultimate reduction of NADP to NADPH. In addition, this creates a proton gradient across the chloroplast membrane; its dissipation is used by ATP synthase for the concomitant synthesis of ATP. The chlorophyll molecule regains the lost electron from a water molecule through a process called photolysis, which releases a dioxygen (O_2) molecule. The overall equation for the light–dependent reactions under the conditions of non–cyclic electron flow in green plants is:

$$2\ H_2O + 2\ NADP^+ + 3\ ADP + 3\ P_i + \text{light} \rightarrow 2\ NADPH + 2\ H^+ + 3\ ATP + O_2$$

Not all wavelengths of light can support photosynthesis. The photosynthetic action spectrum depends on the type of accessory pigments present. For example, in green plants, the action spectrum resembles the absorption spectrum for chlorophylls and carotenoids with peaks for violet–blue and red light. In red algae, the action spectrum overlaps with the absorption spectrum of phycobilins for blue–green light, which allows these algae to grow in deeper waters that filter out the longer wavelengths used by green plants. The non–absorbed part of the light spectrum is what gives photosynthetic organisms their colour (*e.g.*, green plants, red algae, purple bacteria) and is the least effective for photosynthesis in the respective organisms.

Z Scheme

In plants, light–dependent reactions occur in the thylakoid membranes of the chloroplasts and use light energy to synthesize ATP and NADPH. The light–dependent reaction has two forms: cyclic and non–cyclic. In the non–cyclic reaction, the photons are captured in the light–harvesting antenna complexes of photosystem II by chlorophyll and other accessory pigments. When a chlorophyll molecule at the core of the photosystem II reaction centre obtains sufficient excitation energy from the adjacent antenna pigments, an electron is transferred to the primary electron–acceptor molecule, pheophytin, through a process called photoinduced charge separation.

These electrons are shuttled through an electron transport chain, the so called Z–scheme shown in the diagram, that initially functions to generate a chemiosmotic potential across the membrane. An ATP synthase enzyme uses the chemiosmotic potential to make ATP during photophosphorylation, whereas NADPH is aproduct of the terminal redox reaction in the Z–scheme.

The electron enters a chlorophyll molecule in Photosystem I. The electron is excited due to the light absorbed by the photosystem. A second electron carrier accepts the electron, which again is passed down lowering energies of electron acceptors. The energy created by the electron acceptors is used to move hydrogen ions across the thylakoid membrane into the lumen. The electron is used to reduce the co–enzyme NADP, which has functions in the light–independent reaction. The cyclic reaction is similar to that of the non–cyclic, but differs in the form that it generates only ATP, and no reduced NADP (NADPH) is created. The cyclic reaction takes place only at photosystem I. Once the electron is displaced from the photosystem, the electron is passed down the electron acceptor molecules and returns back to photosystem I, from where it was emitted, hence the name cyclic reaction.

WATER PHOTOLYSIS

The NADPH is the main reducing agent in chloroplasts, providing a source of energetic electrons to other reactions. Its production leaves chlorophyll with a deficit of electrons (oxidized), which must be obtained from some other reducing agent. The excited electrons lost from chlorophyll in photosystem I are replaced from the electron transport chain by plastocyanin. However, since photosystem II includes the first steps of the Z–scheme, an external source of electrons is required to reduce its oxidized chlorophyll a molecules. The source of electrons in green–plant and cyanobacterial photosynthesis is water.

Two water molecules are oxidized by four successive charge–separation reactions by photosystem II to yield a molecule of diatomic oxygen and four hydrogen ions; the electron yielded in each step is transferred to a redox–active tyrosine residue that then reduces the photoxidized paired–chlorophyll a species called P680 that serves as the primary (light–driven) electron donor in the photosystem II reaction centre.

The oxidation of water is catalyzed in photosystem II by a redox–active structure that contains four manganese ions and a calcium ion; this oxygen–evolving complex binds two water molecules and stores the four oxidizing equivalents that are required to drive the water–oxidizing reaction. Photosystem II is the only known biological enzyme that carries out this oxidation of water.

The hydrogen ions contribute to the transmembrane chemiosmotic potential that leads to ATP synthesis. Oxygen is a waste product of light–dependent reactions, but the majority of organisms on Earth use oxygen for cellular respiration, including photosynthetic organisms.

LIGHT–INDEPENDENT REACTIONS

The Calvin Cycle

In the Light–independent or dark reactions the enzyme RuBisCO captures CO_2 from the atmosphere and in a process that requires the newly formed NADPH, called the Calvin–Benson Cycle, releases three–carbon sugars, which are later combined to form sucrose and starch. The overall equation for the light–independent reactions in green plants is:

$$3\ CO_2 + 9\ ATP + 6\ NADPH + 6\ H^+ \rightarrow$$
$$C_3H_6O_3\text{–phosphate} + 9\ ADP + 8\ P_i + 6\ NADP^+ + 3\ H_2O$$

To be more specific, carbon fixation produces an intermediate product, which is then converted to the final carbohydrate products. The carbon skeletons produced by photosynthesis are then variously used to form other organic compounds, such as the building material cellulose, as precursors for lipid and amino acid biosynthesis, or as a fuel in cellular respiration. The latter occurs not only in plants but also in animals when the energy from plants gets passed through a food chain.

The fixation or reduction of carbon dioxide is a process in which carbon dioxide combines with a five–carbon sugar, ribulose 1,5–bisphosphate (RuBP), to yield two molecules of a three–carbon compound, glycerate 3–phosphate (GP), also known as 3–phosphoglycerate (PGA).

GP, in the presence of ATP and NADPH from the light–dependent stages, is reduced to glyceraldehyde 3–phosphate (G3P). This product is also referred to as 3–phosphogl-yceraldehyde (PGAL) or even as triose phosphate.

Triose is a 3–carbon sugar. Most (5 out of 6 molecules) of the G3P produced is used to regenerate RuBP so the process can continue. The 1 out of 6 molecules of the triose phosphates not "recycled" often condense to form hexose phosphates, which ultimately yield sucrose, starch and cellulose. The sugars produced during carbon metabolism yield carbon skeletons that can be used for other metabolic reactions like the production of amino acids and lipids.

CLASSIFICATION OF PHOTOSYNTHETIC ORGANISMS

All life can be divided into three domains, Archaea, Bacteria and Eucarya, which originated from a common ancestor. Historically, the term photosynthesis has been applied to organisms that depend on chlorophyll (or bacteriochlorophyll) for the conversion of light energy into chemical free energy. These include organisms in the domains Bacteria (photosynthetic bacteria) and Eucarya (algae and higher plants). The most primitive domain, Archaea, includes organisms known as halobacteria, that convert light energy into chemical free energy.

However, the mechanism by which halobacteria convert light is fundamentally different from that of higher organisms because there is no oxidation/reduction chemistry and halobacteria cannot use CO_2 as their carbon source. Consequently some biologists do not consider halobacteria as photosynthetic. This chapter will follow the historical definition of photosynthesis and omit halobacteria.

Oxygenic Photosynthetic Organisms

The photosynthetic process in all plants and algae as well as in certain types of photosynthetic bacteria involves the reduction of CO_2 to carbohydrate and removal of electrons from H_20, which results in the release of O_2. In this process, known as oxygenic photosynthesis, water is oxidized by the photosystem II reaction centre, a multisubunit protein located in the photosynthetic membrane.

Years of research have shown that the structure and function of photosystem II is similar in plants, algae and certain bacteria, so that knowledge gained in one species can be applied to others. This homology is a common feature of proteins that perform the same reaction in different species.

This homology at the molecular level is important because there are estimated to be 300,000-500,000 species of plants. If different species had evolved diverse mechanisms for oxidizing water, research aimed at a general understanding of photosynthetic water oxidation would be hopeless.

Anoxygenic Photosynthetic Organisms

Some photosynthetic bacteria can use light energy to extract electrons from molecules other than water. These organisms are of ancient origin, presumed to have evolved before oxygenic photosynthetic organisms. Anoxygenic photosynthetic organisms occur in the domain Bacteria and have representatives in four phyla - Purple Bacteria, Green Sulfur Bacteria, Green Gliding Bacteria, and Gram Positive Bacteria.

CARBON CONCENTRATING MECHANISMS

On Land

In hot and dry conditions, plants close their stomata to prevent the loss of water. Under these conditions, CO_2 will decrease, and oxygen gas, produced by the light reactions of photosynthesis, will decrease in the stem, not leaves, causing an increase of photorespiration by the oxygenase activity of ribulose–1,5–bisphosphate carboxylase/oxygenase and decrease in carbon fixation. Some plants have evolved mechanisms to increase the CO_2 concentration in the leaves under these conditions.

C_4 plants chemically fix carbon dioxide in the cells of the mesophyll by adding it to the three–carbon molecule phosphoenolpyruvate (PEP), a reaction catalyzed by an enzyme called PEP carboxylase, creating the four–carbon organic acid oxaloacetic acid.

Oxaloacetic acid or malate synthesized by this process is then translocated to specialized bundle sheath cells where the enzyme RuBisCO and other Calvin cycle enzymes are located, and where CO_2 released by decarboxylation of the four–carbon acids is then fixed by RuBisCO activity to the three–carbon sugar 3–phosphoglyceric acids.

The physical separation of RuBisCO from the oxygen–generating light reactions reduces photorespiration and increases CO_2 fixation and, thus, photosynthetic capacity of the leaf. C_4 plants can produce more sugar than C_3 plants in conditions of high light and temperature.

Many important crop plants are C_4 plants, including maize, sorghum, sugarcane, and millet. Plants that do not use PEP–carboxylase in carbon fixation are called C_3 plants because the primary carboxylation reaction, catalyzed by RuBisCO, produces the three–carbon sugar 3–phosphoglyceric acids directly in the Calvin–Benson cycle. Over 90% of plants use C_3 carbon fixation, compared to 3% that use C_4 carbon fixation. Xerophytes, such as cacti and most succulents, also use PEP carboxylase to capture carbon dioxide in a process called Crassulacean acid metabolism (CAM).

In contrast to C_4 metabolism, which physically separates the CO_2 fixation to PEP from the Calvin cycle, CAM temporally separates these two processes. CAM plants have a different leaf anatomy from C_3 plants, and fix the CO_2 at night, when their stomata are open.

CAM plants store the CO_2 mostly in the form of malic acid via carboxylation of phosphoenolpyruvate to oxaloacetate, which is then reduced to malate.

Decarboxylation of malate during the day releases CO_2 inside the leaves, thus allowing carbon fixation to 3–phosphoglycerate by RuBisCO. Sixteen thousand species of plants use CAM.

In Water

Cyanobacteria possess carboxysomes, which increase the concentration of CO_2 around RuBisCO to increase the rate of photosynthesis. This operates by carbonic anhydrase, producing hydrocarbonate ions (HCO_3^-), which are then pumped into the carboxysome, before being processed by a different carbonic anhydrase to produce CO_2. Pyrenoids in algae and hornworts also act to concentrate CO_2 around rubisco.

ORDER AND KINETICS

Plants usually convert light into chemical energy with a photosynthetic efficiency of 3–6%. Actual plants' photosynthetic efficiency varies with the frequency of the light being converted, light intensity, temperature and proportion of carbon dioxide in the atmosphere, and can vary from 0.1% to 8%. By comparison, solar panels convert light into electric energy at an efficiency of approximately 6–20% for mass–produced panels, and up to 41% in a research laboratory.

Evolution

Early photosynthetic systems, such as those from green and purple sulfur and green and purple nonsulfur bacteria, are thought to have been anoxygenic, using various molecules as electron donors. Green and purple sulfur bacteria are thought to have used hydrogen and sulfur as an electron donor. Green nonsulfur bacteria used various amino and other organic acids. Purple nonsulfur bacteria used a variety of nonspecific organic molecules. The use of these molecules is consistent with the geological evidence that the atmosphere was highly reduced at that time. Fossils of what are thought to be filamentous photosynthetic organisms have been dated at 3.4 billion years old.

The main source of oxygen in the atmosphere is oxygenic photosynthesis, and its first appearance is sometimes referred to as the oxygen catastrophe. Geological evidence suggests that oxygenic photosynthesis, such as that in cyanobacteria, became important during the Paleoproterozoic era around 2 billion years ago. Modern photosynthesis in plants and most photosynthetic prokaryotes is oxygenic. Oxygenic photosynthesis uses water as an electron donor, which is oxidized to molecular oxygen (O_2) in the photosynthetic reaction centre.

SYMBIOSIS AND THE ORIGIN OF CHLOROPLASTS

Several groups of animals have formed symbiotic relationships with photosynthetic algae. These are most common in corals, sponges and sea anemones. It is presumed that this is due to the particularly simple body plans and large surface areas of these animals compared to their volumes. In addition, a few marine mollusks Elysia viridis and Elysia chlorotica also maintain a symbiotic relationship with chloroplasts they capture from the algae in their diet and then store in their bodies. This allows the molluscs to survive solely by photosynthesis for several months at a time. Some of the genes from the plant cell nucleus have even been transferred to the slugs, so that the chloroplasts can be supplied with proteins that they need to survive.

An even closer form of symbiosis may explain the origin of chloroplasts. Chloroplasts have many similarities with photosynthetic bacteria, including a circular chromosome, prokaryotic–type ribosomes, and similar proteins in the photosynthetic reaction centre. The endosymbiotic theory suggests that photosynthetic bacteria were acquired (by endocytosis) by early eukaryotic cells to form the first plant cells. Therefore, chloroplasts may be photosynthetic bacteria that adapted to life inside plant cells. Like mitochondria, chloroplasts still possess their own DNA, separate from the nuclear DNA of their plant host cells and the genes in this chloroplast DNA resemble those in cyanobacteria. DNA in chloroplasts codes for redox proteins such as photosynthetic reaction centers. The CoRR Hypothesis proposes that this Co–location is required for Redox Regulation.

Cyanobacteria and the Evolution of Photosynthesis

The biochemical capacity to use water as the source for electrons in photosynthesis evolved once, in a common ancestor of extant cyanobacteria. The geological record indicates that this transforming event took place early in Earth's history, at least 2450–2320 million years ago (Ma), and, it is speculated, much earlier. Available evidence from geobiological studies of Archean (>2500 Ma) sedimentary rocks indicates that life existed 3500 Ma, but the question of when oxygenic photosynthesis evolved is still unanswered. A clear paleontological window on cyanobacterial evolution opened about 2000 Ma, revealing an already–diverse biota of blue–greens. Cyanobacteria remained principal primary producers throughout the Proterozoic Eon (2500–543 Ma), in part because the redox structure of the oceans favoured photoautotrophs capable of nitrogen fixation.

Green algae joined blue–greens as major primary producers on continental shelves near the end of the Proterozoic, but only with the Mesozoic (251–65 Ma) radiations of dinoflagellates, coccolithophorids, and diatoms did primary production in marine shelf waters take modern form. Cyanobacteria remain critical to marine ecosystems as primary producers in oceanic gyres, as agents of biological nitrogen fixation, and, in modified form, as the plastids of marine algae. A 2010 study by researchers at Tel Aviv University discovered that the Oriental hornet (Vespa orientalis) converts sunlight into electric power using a pigment called xanthopterin. This is the first scientific evidence of a member of the animal kingdom engaging in photosynthesis.

Although some of the steps in photosynthesis are still not completely understood, the overall photosynthetic equation has been known since the 19th century. Jan van Helmont began the research of the process in the mid–17th century when he carefully measured the mass of the soil used by a plant and the mass of the plant as it grew. After noticing that the soil mass changed very little, he hypothesized that the mass of the growing plant must come from the water, the only substance he added to the potted plant. His hypothesis was partially accurate—much of the gained mass also comes from carbon dioxide as well as water. However, this was a signaling point to the idea that the bulk of a plant's biomass comes from the inputs of photosynthesis, not the soil itself.

Joseph Priestley, a chemist and minister, discovered that, when he isolated a volume of air under an inverted jar, and burned a candle in it, the candle would burn out very quickly, much before it ran out of wax. He further discovered that a mouse could similarly "injure" air. He then showed that the air that had been "injured" by the candle and the mouse could be restored by a plant. In 1778, Jan Ingenhousz, court physician to the Austrian Empress, repeated Priestley's experiments. He discovered that it was the influence of sunlight on the plant that could cause it to revive a mouse in a matter of hours.

In 1796, Jean Senebier, a Swiss pastor, botanist, and naturalist, demonstrated that green plants consume carbon dioxide and release oxygen under the influence of light. Soon afterward, Nicolas–Théodore de Saussure showed that the increase in mass of the plant as it grows could not be due only to uptake of CO_2 but also to the incorporation of water. Thus, the basic reaction by which photosynthesis is used to produce food (such as glucose) was outlined. Cornelis Van Niel made key discoveries explaining the chemistry of photosynthesis. By studying purple sulfur bacteria and green bacteria he was the first scientist to demonstrate that photosynthesis is a light–dependent redox reaction, in which hydrogen reduces carbon dioxide.

Robert Emerson discovered two light reactions by testing plant productivity using different wavelengths of light. With the red alone, the light reactions were suppressed. When blue and red were combined, the output was much more substantial. Thus, there were two photosystems, one aborbing up to 600 nm wavelengths, the other up to 700. The former is known as PSII, the latter is PSI. PSI contains only chlorophyll a, PSII contains primarily chlorophyll a with most of the available chlorophyll b, among other pigments.

Further experiments to prove that the oxygen developed during the photosynthesis of green plants came from water, were performed by Robert Hill in 1937 and 1939. He showed that isolated chloroplasts give off oxygen in the presence of unnatural reducing agents like iron oxalate, ferricyanide or benzoquinone after exposure to light.

The Hill reaction is as follows:

$$2\ H_2O + 2\ A + \text{(light, chloroplasts)} \rightarrow 2\ AH_2 + O_2$$

where A is the electron acceptor. Therefore, in light, the electron acceptor is reduced and oxygen is evolved. Cyt b_6, now known as a plastoquinone, is one electron acceptor. Samuel Ruben and Martin Kamen used radioactive isotopes to determine that the oxygen liberated in photosynthesis came from the water.

Melvin Calvin and Andrew Benson, along with James Bassham, elucidated the path of carbon assimilation (the photosynthetic carbon reduction cycle) in plants. The carbon reduction cycle is known as the Calvin cycle, which ignores the contribution of Bassham and Benson.

Many scientists refer to the cycle as the Calvin–Benson Cycle, Benson–Calvin, and some even call it the Calvin–Benson–Bassham (or CBB) Cycle. Nobel Prize–winning scientist Rudolph A. Marcus was able to discover the function and significance of the electron transport chain. Otto Heinrich Warburg and Dean Burk discovered the I–quantum photosynthesis reaction that splits the CO2, activated by the respiration.

Factors

There are three main factors affecting photosynthesis and several corollary factors. The three main are:

- Light irradiance and wavelength
- Carbon dioxide concentration
- Temperature.

Light Intensity (Irradiance),Wavelength and Temperature

In the early 20th century, Frederick Frost Blackman along with Albert Einstein investigated the effects of light intensity (irradiance) and temperature on the rate of carbon assimilation.

- At constant temperature, the rate of carbon assimilation varies with irradiance, initially increasing as the irradiance increases. However, at higher irradiance, this relationship no longer holds and the rate of carbon assimilation reaches a plateau.
- At constant irradiance, the rate of carbon assimilation increases as the temperature is increased over a limited range. This effect is seen only at high irradiance levels. At low irradiance, increasing the temperature has little influence on the rate of carbon assimilation.

These two experiments illustrate vital points: First, from research it is known that, in general, photochemical reactions are not affected by temperature. However, these experiments clearly show that temperature affects the rate of carbon assimilation, so there must be two sets of reactions in the full process of carbon assimilation. These are, of course, the light–dependent 'photochemical' stage and the light–independent, temperature–dependent stage. Second, Blackman's experiments illustrate the concept of limiting factors. Another limiting factor is the wavelength of light. Cyanobacteria, which reside several meters underwater, cannot receive the correct wavelengths required to cause photoinduced charge separation in conventional photosynthetic pigments. To combat this problem, a series of proteins with different pigments surround the reaction centre.This unit is called a phycobilisome.

Carbon Dioxide Levels and Photorespiration

As carbon dioxide concentrations rise, the rate at which sugars are made by the light–independent reactions increases until limited by other factors. RuBisCO, the enzyme that captures carbon dioxide in the light–independent reactions, has a binding affinity for both carbon dioxide and oxygen. When the concentration of carbon dioxide is high, RuBisCO will fix carbon dioxide. However, if the carbon dioxide concentration is low, RuBisCO will bind oxygen instead of carbon dioxide. This process, called photorespiration, uses energy, but does not produce sugars.

RuBisCO oxygenase activity is disadvantageous to plants for several reasons:

- One product of oxygenase activity is phosphoglycolate (2 carbon) instead of 3–phosphoglycerate (3 carbon). Phosphoglycolate cannot be metabolized by the Calvin–Benson cycle and represents carbon lost from the cycle. A high oxygenase activity, therefore, drains the sugars that are required to recycle ribulose 5–bisphosphate and for the continuation of the Calvin–Benson cycle.
- Phosphoglycolate is quickly metabolized to glycolate that is toxic to a plant at a high concentration; it inhibits photosynthesis.
- Salvaging glycolate is an energetically expensive process that uses the glycolate pathway, and only 75% of the carbon is returned to the Calvin–Benson cycle as 3–phosphoglycerate. The reactions also produce ammonia (NH_3), which is able to diffuse out of the plant, leading to a loss of nitrogen.

A highly simplified summary is:

2 glycolate + ATP $\rightarrow$ 3–phosphoglycerate + carbon dioxide + ADP +NH_3

The salvaging pathway for the products of RuBisCO oxygenase activity is more commonly known as photorespiration, since it is characterized by light–dependent oxygen consumption and the release of carbon dioxide.

OVERALL EQUATION OF PHOTOSYNTHESIS

Photosynthesis is the intracellular anabolic process, characteristic of the green cells of plants in which carbohydrates are synthesized from carbon dioxide and water in the presence of light and chlorophyll. In this process, light energy is converted into chemical energy and stored in carbohydrate molecules, while oxygen is liberated.

The overall process of photosynthesis is represented by the following general chemical equation.

$$6CO_2 + 12H_2O \xrightarrow[\text{Chlorophyll-a}]{\text{Light}} C_6H_{12}O_6 + 6H_2O + 6O_2$$

PRIMARY PROCESSES OF PHOTOSYNTHESIS

Nature of Light

As stated earlier, solar radiation is the only natural source of light for all organisms. Green plants can utilize light belonging to the visible spectrum only for photosynthesis. This visible part of the electromagnetic spectrum from the sun is confined between wavelengths of 390 nm (violet) and 760 nm (red).

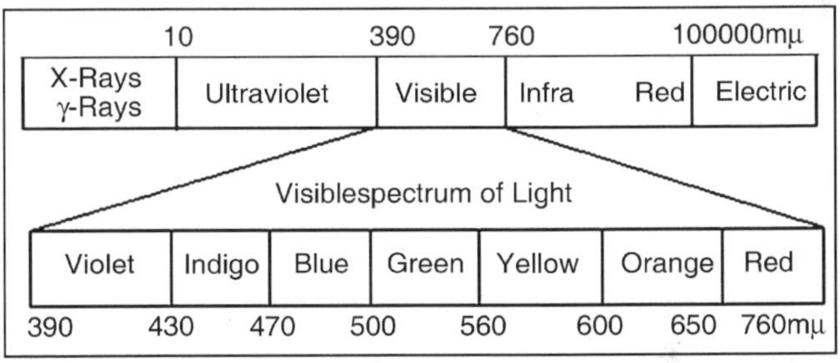

Light is a form of energy. It appears to travel as a stream of discrete particles called photons. Each photon contains one quantum (unit) of light energy. The energy quantity of each type of light depends on its wavelength. Light with a shorter wavelength has greater energy; light with a longer wavelength has lesser energy.

Rate of Photosynthesis

Rate of photosynthesis is measured in terms of the amount of carbon dioxide used (reduced) or oxygen released. It has been estimated that 8 quanta (photons) of light are required to reduce each molecule of carbon dioxide (or to release each molecule of oxygen) during photosynthesis (Emerson and Lewis, 1943). This is called the quantum requirement in photosynthesis.

Light Trapping Systems

During the photochemical phase, light is trapped by the photosynthetic pigments present in the quantasomes of the grana thylakoids. These pigments are organized into two pigment systems called pigment system I (PS I) and pigment system II (PS II).

(I) PS I is composed of the following pigment molecules:

Chl-a 700 (P700)	– one molecule	– Reaction center
Chl-a 683	– 200 molecules	– Antenna chlorophyll
Carotenoids	– 50 molecules	– Accessory pigments

(II) PS II is composed of the following pigment molecules:

Chl- a 680	– one molecule	– Reaction center
Chl- a 670	– 200 molecules	– Antenna chlorophyll
Chl- b	– up to 200 molecules	} Accessory pigments
Carotenoids	– up to 50 molecules	

In both the systems, all pigment molecules help in trapping light energy. However, all other molecules transfer their energy to the reaction center to be used for the photochemical reactions in photosynthesis.

Photo-excitation of Chlorophyll-a

When a molecule of chlorophyll-a (acting as the reaction center) receives light energy in the form of photons (quanta), it becomes energy-rich and activated. This is called the excited state of chlorophyll. In this excited state, the chlorophyll-a expels one electron. With the loss of an electron, the chlorophyll-a develops a positive charge. This is called ionized chlorophyll a.

The expelled electron contains an extra amount of energy received from light. In other words, the energy-rich expelled electron represents light energy. This electron energy is used for the formation of ATP during photosynthesis.

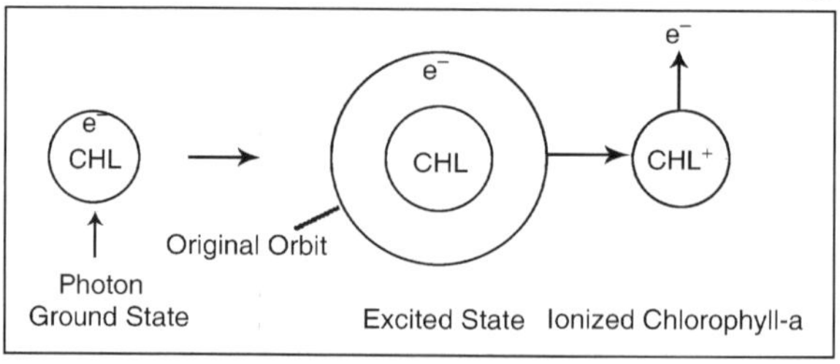

Photolysis of Water

In the photosynthesis of green plants, water is used as a source of hydrogen required for the reduction of carbon dioxide to form carbohydrates.

$$CO_2 + 2H_2O \xrightarrow[\text{Chlorophyll}]{\text{light}} \underset{\text{carbohydrate}}{(CH_2O)} + H_2O + O_2$$

In this process, water is oxidized in presence of light and chlorophyll. Hydrogen is removed from water, and oxygen is released as follows:

$$4H_2O \xrightarrow[\text{Chlorophyll}]{\text{light}} 2H_2O + 4H^+ + 4e^- + O_2$$

$$\text{or } 2H_2O \rightarrow 4H^+ + 4e^- + O_2$$

This is described as photolysis of water, or photochemical oxidation of water. This was first indicated by Van Niel (1931). It was demonstrated with the help of an experiment by R. Hill (1937). Hill suggested that during the photolysis of water, hydrogen combines with some hydrogen acceptor in the plant, while oxygen is released.

$$2H_2O + 2A \xrightarrow[\text{Chlorophyll}]{\text{light}} 2AH_2 + O_2$$

(where A = unknown H-acceptor in plants).

Later on, Arnon (1951) discovered this hydrogen acceptor to be a coenzyme called NADP (nicotinamide adenine dinucleotide phosphate).

Substance A = NADP

$$2H_2O + 2NADP \xrightarrow{\text{light}} 2NADPH_2 + O_2$$
$$\text{Chlorophyll reduced co-enzyme}$$

Similarly, Ruben, Kamen, *et al.* (1941) confirmed that the oxygen evolved during photosynthesis comes from the splitting of water only. They used heavy isotope of oxygen (O_{18}) to demonstrate the fact as follows:

$$\text{(I) } 6CO^{18}{}_2 + 12H_2O \xrightarrow[\text{Chlorophyll}]{\text{Light}} C6H_{12}O^{18}{}_6 + 6H_2O^{18} + 6O_2$$

$$\text{(II) } 6CO_2 + 12H_2O^{18} \xrightarrow[\text{Chlorophyll}]{\text{Light}} C6H_{12}O_6 + 6H_2O + 6O^{18}{}_2$$

HILL REACTION

R. Hill (1937) suspended isolated chloroplasts in water in a beaker. He added some known hydrogen accepting compound (*e.g.* benzoquinone or ferric salt) to the water and exposed the beaker to the light. Immediately the oxidation of water took place and bubbles of oxygen were evolved. When the hydrogen acceptor in the beaker was examined, it was found to be reduced.

$$2H_2O + 2A \xrightarrow[\text{Chlorophyll}]{\text{Light}} 2AH_2 + O_2$$

(A = hydrogen acceptor)

This is photolysis or the photochemical oxidation of water and is commonly called the Hill reaction.

Photophosphorylation

The formation of ATP molecules (chemical energy) from ADP and H_3PO_4 in presence of light and chlorophyll-a during the photochemical phase of photosynthesis is called photophosphorylation or photosynthetic phosphorylation. More commonly, it is described as the conversion of light energy into chemical energy (ATP).

$$ADP + H_2PO_4 \xrightarrow[\text{Chlorophyll-a}]{\text{Light}} ATP$$

(ADP = adenosine diphosphate)

During the light reactions (the primary process or the photochemical phase), ATP formation takes place through two types of phosphorylation reactions.

These are:

The non-cyclic process involves both PS I and PS II. The cyclic process involves only PS I.

- Non-cyclic photophosphorylation

In the light reaction of photosynthesis, the formation of ATP from ADP and H_3PO_4 in presence of light and chlorophyll-a during the non-cyclic transfer of electrons is called non-cyclic photophosphorylation.

Important Features:

- It is the major pathway of the light reaction of photosynthesis.
- It is a photochemical reaction.
- It takes place in the grana of chloroplasts.
- It requires participation of both PS I and PS II.
- The transfer of electrons through the ETS is unidirectional or non-cyclic.
- The non-cyclic process involves:
  - Photophosphorylation (ATP formation)
  - Photolysis of water
  - Formation of assimilatory power (ATP and $NADPH_2$)
  - Liberation of oxygen

The non-cyclic photophosphorylation takes place as follows:

Photoexcitation of PS I

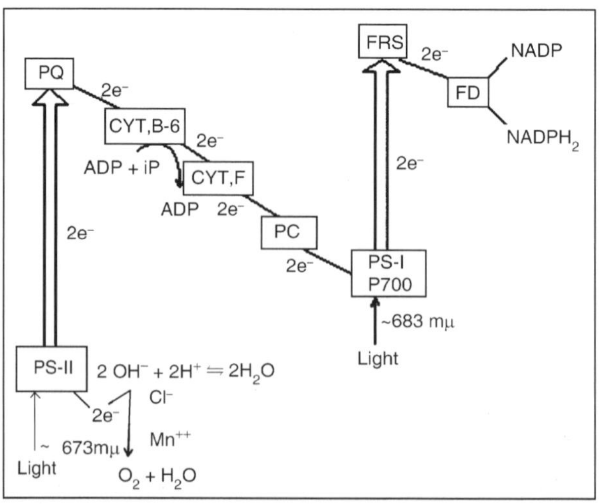

Fig. Diagrammatic Representation of the Non-cyclic Process (Z Scheme)

The chlorophyll-a (P 700) of PS I is activated on receiving photons of light and so it expels electrons. As a result of this loss, it becomes ionized chl-a^+. The electrons from PS I are first accepted by FRS (ferredoxin reducing substance) and are transferred to co-enzyme NADP via Fd (ferredoxin). NADP retains the electrons and is thus reduced.

Photoexcitation of PS II

Similarly, on receiving photons of light, the chl-a (680) of PS II also gets activated and expels electrons. As a result, it becomes ionized chl-a^+. The high energy containing electrons from PS II are first accepted by an electron acceptor PQ (plastoquinone) and are then transferred along stet Electron Transport System consisting of plastoquinone, cytochrome b6, Cytochrome f and plastocyanin. Finally, the electrons are accepted by the ionized chl-a^+ of PS I. In this way, ionized PS I comes back to the ground state and can again participate in the process.

Photolysis of Water

The ionized chlorophyll-a^+ of PS II is brought back to the ground state with the help of the electrons made available through the photolysis of water.

$$2H_2O \xrightarrow[\text{Chlorophyll}]{\text{light}} 4e^- + 4H^+ + O_2 \,(\text{Photolysis of water})$$

Formation of $NADPH_2$ and Liberation of O_2

During photolysis of water, oxygen is released as a by-product. The 2 NADP receive $4e^-$ from PS I and $4H^+$ from the photolysis of water to form reduced coenzyme $NADPH_2$

$$2NADP + 4e^- + 4H^+ \rightarrow 2NADPH_2$$

Non-Cyclic Electron Transfer

In this process, the movement of electrons is considered to be non-cyclic, or unidirectional. This is because, the final electron acceptor is different from the initial electron donor.

Initial electron donor Final electron acceptor:

(i) $\text{Chl} - \text{a of PSI} \xrightarrow{e^-} \text{NADP}$

(ii) $\text{Chl} - \text{a of PSII} \xrightarrow{e^-} \text{Chl.a} + \text{of PSI}$

(iii) $\underset{\text{(Photolysis)}}{\text{Water}} \xrightarrow{e^-} \text{Chl.a} + \text{of PSII}$

In this way, the light-induced electron flow is maintained continuously in the non-cyclic manner from water to NADP as follows

$$\text{Water} \xrightarrow{e^-} \text{PSII} \xrightarrow{e^-} \text{PSI} \xrightarrow{e^-} \text{NADP}$$

Because of the zig-zag path of the electrons, the non cyclic process is also called 'Z' scheme electron transport.

ATP Formation

During the non-cyclic transfer of energy-rich electrons through the Electron Transport System from PS II to PS I, energy from the electrons is released. This is used for the formation of ATP from ADP and H_3PO_4. This is called non-cyclic photophosphorylation.

End Products of Non-cyclic Process and the Assimilatory Power

The end products are ATP, $NADPH_2$ and O_2. Of these, O_2 is liberated from the green plants. ATP and $NADPH_2$ are used in the following dark reactions (phase II) for carbon assimilation(*i.e.,* reduction of CO_2 to form carbohydrates). ATP and $NADPH_2$ of light reaction are called 'assimilatory power.

PHOTOSYNTHETIC RESPONSES TO TEMPERATURE

When photosynthetic rate is plotted as a function of temperature, the curve has a characteristic bell shape. The ascending arm of the curve represents a temperature- dependent stimulation of photosynthesis up to an optimum; the descending arm is associated with deleterious effects, some of which are reversible while others are not.

Temperature affects all biochemical reactions of photosynthesis, so it is not surprising that the responses to temperature are complex. We can gain insight into the underlying mechanisms by comparing photosynthetic rates in air at normal and at high CO_2 concentrations.

At high CO_2, there is an ample supply of CO_2 at the carboxylation sites, and the rate of photosynthesis is limited primarily by biochemical reactions connected with electron transport. In these conditions, temperature changes have large effects on fixation rates.

At ambient CO_2 concentrations, photosynthesis is limited by the activity of rubisco, and the response reflects two conflicting processes: an increase in carboxylation rate with temperature and a decrease in the affinity of rubisco for CO_2 as the temperature rises.

These opposing effects dampen the temperature response of photosynthesis at ambient CO_2 concentrations.

Respiration rates also increase as a function of temperature, and the interaction between photorespiration and photosynthesis becomes apparent in temperature responses. Changes in quantum yield as a function of temperature in a C_3 plant and in a C_4 plant. In the C_4 plant the quantum yield remains constant with temperature, reflecting typical low rates of photorespiration. In the C_3 plant the quantum yield decreases with temperature, reflecting a stimulation of photorespiration by temperature and an ensuing higher energy demand per net CO_2 fixed.

At low temperatures, photosynthesis is often limited by phosphate availability at the chloroplast. When triose phosphates are exported from the chloroplast to the cytosol, an equimolar amount of inorganic phosphate is taken up via translocators in the chloroplast membrane. If the rate of triose phosphate utilization in the cytosol decreases, phosphate uptake into the chloroplast is inhibited and photosynthesis becomes phosphate limited. Starch synthesis and sucrose synthesis decrease rapidly with temperature, reducing the demand for triose phosphates and causing the phosphate limitation observed at low temperatures.

The highest photosynthetic rates seen in temperature responses represent the so-called *optimal temperature response*. When these temperatures are exceeded, photosynthetic rates decrease again.

It has been argued that this optimal temperature is the point at which the capacities of the various steps of photosynthesis are optimally balanced, with some of the steps becoming limiting as the temperature decreases or increases. Optimal temperatures have strong genetic and physiological components.

Plants of different species growing in habitats with different temperatures have different optimal temperatures for photosynthesis, and plants of the same species, grown at different temperatures and then tested for their photosynthetic responses, show temperature optima that correlate with the temperature at which they were grown. Plants growing at low temperatures maintain higher photosynthetic rates at low temperatures than plants grown at high temperatures.

These changes in photosynthetic properties in response to temperature play an important role in plant adaptations to different environments. Plants are remarkably plastic in their adaptations to temperature. In the lower temperature range, plants growing in alpine areas are capable of net CO_2 uptake at temperatures close to 0°C; at the other extreme, plants living in Death Valley, California, have optimal rates of photosynthesis at temperatures approaching 50°C.

Photosynthetic activity in the intact leaf is an integral process that depends on many biochemical reactions. Different environmental factors can limit photosynthetic rates. Leaf anatomy is highly specialized for light absorption, and the properties of palisade and mesophyll cells ensure uniform light absorption throughout the leaf.

In addition to the anatomic features of the leaf, chloroplast movements within cells and solar tracking by the leaf blade help maximize light absorption. Light transmitted through upper leaves is absorbed by leaves growing beneath them. Many properties of the photosynthetic apparatus change as a function of the available light, including the light compensation point, which is higher in sun leaves than in shade leaves.

The linear portion of the lightresponse curve for photosynthesis provides a measure of the quantum yield of photosynthesis in the intact leaf. In temperate areas, quantum yields of C_3 plants are generally higher than those of C_4 plants. Sunlight imposes a substantial heat load on the leaf, which is dissipated back into the air by long-wavelength radiation, by sensible heat loss, or by evaporative heat loss. Increasing CO_2 concentrations in the atmosphere are increasing the heat load on the biosphere.

This process could cause damaging changes in the world's climate, but it could also reduce the CO_2 limitations on photosynthesis. At high photon flux, photosynthesis in most plants is CO_2 limited, but the limitation is substantially lower in C_4 and CAM plants because of their CO_2-concentrating mechanisms. Diffusion of CO_2 into the leaf is constrained by a series of different points of resistance.

The largest resistance is usually that imposed by the stomata, so modulation of stomatal apertures provides the plant with an effective means of controlling water loss and CO_2 uptake. Both stomatal and nonstomatal factors affect CO_2 limitations on photosynthesis.

Temperature responses of photosynthesis reflect the temperature sensitivity of the biochemical reactions of photosynthesis and are most pronounced at high CO_2 concentrations.

Because of the role of photorespiration, the quantum yield is strongly dependent on temperature in C_3 plants but is nearly independent of temperature in C_4 plants. Leaves growing in cold climates can maintain higher photosynthetic rates at low temperatures than leaves growing in warmer climates.

Leaves grown at high temperatures perform better at high temperatures than leaves grown at low temperatures do. Functional changes in the photosynthetic apparatus in response to prevailing temperatures in their environment have an important effect on the capacity of plants to live in diverse habitats.

CONCEPTS AND UNITS IN THEMEASUREMENT OF LIGHT

Three light parameters are especially important in the measurement of light:

1. Spectral quality,
2. Amount, and
3. Direction.

Adiscussion of the amount and direction of light reaching the plant requires consideration of the geometry of the part of the plant that receives the light: Is the plant organ flat or cylindrical? Flat, or planar, light sensors are best suited for flat leaves.

The light reaching the plant can be measured as energy, and the amount of energy that falls on a flat sensor of known area per unit time is quantified as irradiance. Units can be expressed in terms of energy, such as watts per square meter (Wm^{-2}). Time (seconds) is contained within the term watt: 1 W = 1 joule (J) s^{-1}. Light can also be measured as the number of incident quanta (singular *quantum*). In this case, units can be expressed in moles per square meter per second (mol m^{-2} s^{-1}), where *moles* refers to the number of photons (1 mol of light = 6.02×10^{23} photons, Avogadro's number). This measure is called photon irradiance. Quanta and energy units can be interconverted relatively easily, provided that the wavelength of the light, λ, is known.

The energy of a photon is related to its wavelength as follows:

$$E = \frac{hc}{\lambda}$$

where *c* is the speed of light (3×10^8 ms^{-1}), *h* is Planck's constant (6.63×10^{-34} Js), and λ is the wavelength of light, usually expressed in nm (1 nm = 10^{-9} m). From this equation it can be shown that a photon at 400 nm has twice the energy of a photon at 800 nm. Now let's turn our attention to the direction of light. Light can strike a flat surface directly from above or obliquely. When light deviates from perpendicular, irradiance is proportional to the cosine of the angle at which the light rays hit the sensor.

There are many examples in nature in which the lightintercepting object is not flat (*e.g.*, complex shoots, whole plants, chloroplasts). In addition, in some situations light can come from many directions simultaneously (*e.g.*, direct light from the sun plus the light that is reflected upward from sand, soil, or snow). In these situations it makes more sense to measure light with a spherical sensor that takes measurements omnidirectionally (from all directions).

The term for this omnidirectional measurement is fluence rate, and this quantity can be expressed in watts per square meter (W m^{-2}) or moles per square meter per second (mol m^{-2} s^{-1}). The units clearly indicate whether light is being measured as energy (W) or as photons (mol).

In contrast to a flat sensor's sensitivity, the sensitivity to light of a spherical sensor is independent of direction. Depending on whether the light is collimated (rays are parallel) or diffuse (rays travel in random directions), values for fluence rate versus irradiance measured with a flat or a spherical sensor can provide different values.

Photosynthetically Active Radiation

PAR, 400–700 nm may also be expressed in terms of energy (W m^{-2}) or quanta (mol m^{-2} s^{-1}). Note that PAR is an irradiance-type measurement. In research on photosynthesis, when PAR is expressed on a quantum basis, it is given the special term photosynthetic photon flux density (PPFD). However, it has been suggested that the term *density* be discontinued because within the International System of Units (SI units, where *SI* stands for *System International*), *density* can mean area or volume. In summary, when choosing how to quantify light, it is important to match sensor geometry and spectral response with that of the plant. Flat, cosine-corrected sensors are ideally suited to measure the amount of light that strikes the surface of a leaf; spherical sensors are more appropriate in other situations, such as in studies of a chloroplast suspension or a branch from a tree.

How much light is there on a sunny day, and what is the relationship between PAR irradiance and PAR fluence rate? Under direct sunlight, PAR irradiance and fluence rate are both about 2000 $\mu mol\ m^{-2}\ s^{-1}$, though higher values can be measured at high altitudes. The corresponding value in energy units is about 400 $W\ m^{-2}$.

Leaf Anatomy Maximizes Light Absorption

Roughly 1.3 $kW\ m^{-2}$ of radiant energy from the sun reaches Earth, but only about 5% of this energy can be converted into carbohydrates by a photosynthesizing leaf. The reason this percentage is so low is that a major fraction of the incident light is of a wavelength either too short or too long to be absorbed by the photosynthetic pigments. Of the absorbed light energy, a significant fraction is lost as heat, and a smaller amount is lost as fluorescence. Radiant energy from the sun consists of many different wavelengths of light. Only photons of wavelengths from 400 to 700 nm are utilized in photosynthesis, and about 85 to 90% of this PAR is absorbed by the leaf; the remainder is either reflected at the leaf surface or transmitted through the leaf. Because chlorophyll absorbs very strongly in the blue and the red regions of the spectrum, the transmitted and reflected light are vastly enriched in green—hence the green color of vegetation.

The anatomy of the leaf is highly specialized for light absorption. The outermost cell layer, the epidermis, is typically transparent to visible light, and the individual cells are often convex. Convex epidermal cells can act as lenses and can focus light so that the amount reaching some of the chloroplasts can be many times greater than the amount of ambient light. Epidermal focusing is common among herbaceous plants and is especially prominent among tropical plants that grow in the forest understory, where light levels are very low.

Below the epidermis, the top layers of photosynthetic cells are called palisade cells; they are shaped like pillars that stand in parallel columns one to three layers deep. Some leaves have several layers of columnar palisade cells, and we may wonder how efficient it is for a plant to invest energy in the development of multiple cell layers when the high chlorophyll content of the first layer would appear to allow little transmission of the incident light to the leaf interior. In fact, more light than might be expected penetrates the first layer of palisade cells because of the sieve effect and light channeling.

The sieve effect is due to the fact that chlorophyll is not uniformly distributed throughout cells but instead is confined to the chloroplasts. This packaging of chlorophyll results in shading between the chlorophyll molecules and creates gaps between the chloroplasts, where light is not absorbed—hence the reference to a sieve. Because of the sieve effect, the total absorption of light by a given amount of chlorophyll in a palisade cell is less than the light absorbed by the same amount of chlorophyll in a solution.

Light channeling occurs when some of the incident light is propagated through the central vacuole of the palisade cells and through the air spaces between the cells, an arrangement that facilitates the transmission of light into the leaf interior.

Below the palisade layers is the spongy mesophyll, where the cells are very irregular in shape and are surrounded by large air spaces. The large air spaces generate many interfaces between air and water that reflect and refract the light, thereby randomizing its direction of travel. This phenomenon is called *light scattering*. Light scattering is especially important in leaves because the multiple reflections between cell–air interfaces greatly increase the length of the path over which photons travel, thereby increasing the probability for absorption. In fact, photon path lengths within leaves are commonly four times or more longer than the thickness of the leaf. Thus the palisade cell properties that allow light to pass through, and the spongy mesophyll cell properties that are conducive to light scattering, result in more uniform light absorption throughout the leaf.

Some environments, such as deserts, have so much light that it is potentially harmful to leaves. In these environments leaves often have special anatomic features, such as hairs, salt glands, and epicuticular wax that increase the reflection of light from the leaf surface, thereby reducing light absorption. Such adaptations can decrease light absorption by as much as 40%, minimizing heating and other problems associated with the absorption of too much light.

Control Light Absorption

Chloroplast movement is widespread among algae, mosses, and leaves of higher plants. If chloroplast orientation and location are controlled, leaves can regulate how much of the incident light is absorbed. Under low light, chloroplasts gather at the cell surfaces parallel to the plane of the leaf so that they are aligned perpendicularly to the incident light—a position that maximizes absorption of light.

Under high light, the chloroplasts move to the cell surfaces that are parallel to the incident light, thus avoiding excess absorption of light. Such chloroplast rearrangement can decrease the amount of light absorbed by the leaf by about 15%. Chloroplast movement in leaves is a typical blue-light response. Blue light also controls chloroplast orientation in many of the lower plants, but in some algae, chloroplast movement is controlled by phytochrome.

In leaves, chloroplasts move along actin microfilaments in the cytoplasm, and calcium regulates their movement. Leaves have the highest light absorption when the leaf blade, or lamina, is perpendicular to the incident light. Some plants control light absorption by solar tracking; that is, their leaves continuously adjust the orientation of their laminae such that they remain perpendicular to the sun's rays. Alfalfa, cotton, soybean, bean, lupine, and some wild species of the mallow family (Malvaceae) are examples of the numerous plant species that are capable of solar tracking.

Solar-tracking leaves keep a nearly vertical position at sunrise, facing the eastern horizon, where the sun will rise. The leaf blades then lock on to the rising sun and follow its movement across the sky with an accuracy of ±15° until sunset, when the laminae are nearly vertical, facing the west, where the sun will set. During the night the leaf takes a horizontal position and reorients just before dawn so that it faces the eastern horizon in anticipation of another sunrise. Leaves track the sun only on clear days, and they stop when a cloud obscures the sun.

In the case of intermittent cloud cover, some leaves can reorient as rapidly as 90° per hour and thus can catch up to the new solar position when the sun emerges from behind a cloud. Solar tracking is another blue-light response, and the sensing of blue light in solar-tracking leaves occurs in specialized regions. In species of *Lavatera* (Malvaceae), the photosensitive region is located in or near the major leaf veins. In lupines, (*Lupinus*, Fabaceae), leaves consist of five or more leaflets, and the photosensitive region is located in the basal part of each leaflet lamina. In many cases, leaf orientation is controlled by a specialized organ called the pulvinus (plural *pulvini*), found at the junction between the blade and petiole.

The pulvinus contains motor cells that change their osmotic potential and generate mechanical forces that determine laminar orientation. In other plants, leaf orientation is controlled by small mechanical changes along the length of the petiole and by movements of the younger parts of the stem. Some solar-tracking plants can also move their leaves such that they avoid full exposure to sunlight, thus minimizing heating and water loss. Building on the term heliotropism (bending toward the sun), which is often used to describe sun-induced leaf movements, these sunavoiding leaves are called *paraheliotropic*, and leaves that maximize light interception by solar tracking are called *diaheliotropic*.

Some plant species can display diaheliotropic movements when they are well watered and paraheliotropic movements when they experience water stress. Since full sunlight usually exceeds the amount of light that can be utilized for photosynthesis, what advantage is gained by solar tracking? By keeping leaves perpendicular to the sun, solar-tracking plants maintain maximum photosynthetic rates throughout the day, including early morning and late afternoon. Moreover, air temperature is lower during the early morning and late afternoon, so water stress is lower. Solar tracking therefore gives an advantage to plants that grow in arid regions.

ENERGY EFFICIENCY OF PHOTOSYNTHESIS

The physics of energy capture for plants and photovoltaics is similar, but they have to do different things. The photovoltaic system is designed by engineers basically to make a profit. It just has to put electrons down a wire to power our instruments. In contrast, photosynthetic systems have to work within living things. There are things more important to living things than energy efficiency. It has to protect its DNA, for example. If you make a photosynthetic system that produces toxic by-products that destroy or damage the DNA, that wouldn't work. It has to be conservative to preserve the future.

The energy efficiency of photosynthesis is the ratio of the energy stored to the energy of light absorbed. The chemical energy stored is the difference between that contained in gaseous oxygen and organic compound products and the energy of water, carbon dioxide, and other reactants. The amount of energy stored can only be estimated because many products are formed, and these vary with the plant species and environmental conditions. If the equation for glucose formation given earlier is used to approximate the actual storage process, the production of one mole (*i.e.*, 6.02×10^{23} molecules; abbreviated *N*) of oxygen and one-sixth mole of glucose results in the storage of about 117 kilocalories (kcal) of chemical energy. This amount must then be compared to the energy of light absorbed to produce one mole of oxygen in order to calculate the efficiency of photosynthesis.

Light can be described as a wave of particles known as photons; these are units of energy, or light quanta. The quantity *N* photons is called an einstein. The energy of light varies inversely with the length of the photon waves; that is, the shorter the wavelength, the greater the energy content. The energy (e) of a photon is given by the equation $e = hc/\lambda$, where c is the velocity of light, h is Planck's constant, and λ is the light wavelength. The energy (E) of an einstein is $E = Ne = Nhc/\lambda = 28{,}600/\lambda$, when E is in kilocalories and λ is given in nanometres (nm; 1 nm = 10^{-9} metres). An einstein of red light with a wavelength of 680 nm has an energy of about 42 kcal. Blue light has a shorter wavelength and therefore more energy than red light. Regardless of whether the light is blue or red, however, the same number of einsteins are required for photosynthesis per mole of oxygen formed. The part of the solar spectrum used by plants has an estimated mean wavelength of 570 nanometres; therefore, the energy of light used during photosynthesis is approximately 28,600/570, or 50 kilocalories per einstein.

In order to compute the amount of light energy involved in photosynthesis, one other value is needed: the number of einsteins absorbed per mole of oxygen evolved. This is called the quantum requirement. The minimum quantum requirement for photosynthesis under optimal conditions is about nine. Thus the energy used is 9×50, or 450 kilocalories per mole of oxygen evolved. Therefore, the estimated maximum energy efficiency of photosynthesis is the energy stored per mole of oxygen evolved—117 kilocalories—divided by 450; that is, 117/450, or 26 per cent.

The actual percentage of solar energy stored by plants is much less than the maximum energy efficiency of photosynthesis. An agricultural crop in which the biomass (total dry weight) stores as much as 1 per cent of total solar energy received on an annual area-wide basis is exceptional, although a few cases of higher yields (perhaps as much as 3.5 per cent in sugarcane) are reported. There are several reasons for this difference between the predicted maximum efficiency of photosynthesis and the actual energy stored in biomass. First, more than half of the incident sunlight is composed of wavelengths too long to be absorbed, while some of the remainder is reflected or lost to the leaves.

Consequently, plants can at best absorb only about 34 per cent of the incident sunlight. Second, plants must carry out a variety of physiological processes in such nonphotosynthetic tissues as roots and stems; these processes, as well as cellular respiration in all parts of the plant, use up stored energy. Third, rates of photosynthesis in bright sunlight sometimes exceed the needs of the plants, resulting in the formation of excess sugars and starch. When this happens, the regulatory mechanisms of the plant slow down the process of photosynthesis, allowing more absorbed sunlight to go unused. Fourth, in many plants, energy is wasted by the process of photorespiration. Finally, the growing season may last only a few months of the year; sunlight received during other seasons is not used. Furthermore, it should be noted that if only agricultural products (*e.g.*, seeds, fruits, and tubers, rather than total biomass) are considered as the end product of the energy conversion process of photosynthesis, the efficiency falls even further.

CHLOROPLASTS, THE PHOTOSYNTHETIC UNITS OF GREEN PLANTS

The process of plant photosynthesis takes place entirely within the chloroplasts. Detailed studies of the role of these organelles date from the work of the British biochemist Robert Hill. About 1940 Hill discovered that green particles obtained from broken cells could produce oxygen from water in the presence of light and a chemical compound, such as ferric oxalate, able to serve as an electron acceptor. This process is known as the Hill reaction. During the 1950s Daniel Arnon and other American biochemists prepared plant cell fragments in which not only the Hill reaction but also the synthesis of the energy-storage compound ATP occurred. In addition, the coenzyme NADP was used as the final acceptor of electrons, replacing the nonphysiological electron acceptors used by Hill. His procedures were refined further so that individual small pieces of isolated chloroplast membranes, or lamellae, could perform the Hill reaction.

These small pieces of lamellae were then fragmented into pieces so small that they performed only the light reactions of the photosynthetic process. It is now possible also to isolate the entire chloroplast so that it can carry out the complete process of photosynthesis, from light absorption, oxygen formation, and the reduction of carbon dioxide to the formation of glucose and other products.

Structural Features

The intricate structural organization of the photosynthetic apparatus is essential for the efficient performance of the complex process of photosynthesis. The chloroplast is enclosed in a double outer membrane, and its size approximates a spheroid about 2,500 nanometres thick and 5,000 nanometres long. Some single-celled algae have one chloroplast that occupies more than half the cell volume. Leaf cells of higher plants contain many chloroplasts, each approximately the size of the one in some algal cells.

When thin sections of a chloroplast are examined under the electron microscope, several features are apparent. Chief among these are the intricate internal membranes (*i.e.*, the lamellae) and the stroma, a colourless matrix in which the lamellae are embedded. Also visible are starch granules, which appear as dense bodies.

The stroma is basically a solution of enzymes and small molecules. The dark reactions occur in the stroma, the soluble enzymes of which catalyze the conversion of carbon dioxide and minerals to carbohydrates and other organic compounds. The capacity for carbon fixation and reduction is lost if the outer membrane of the chloroplast is broken, allowing the stroma enzymes to leak out.

A single lamella, which contains all the photosynthetic pigments, is approximately 10–15 nanometres thick. The lamellae exist in more-or-less flat sheets, a few of which extend through much of the length of the chloroplast. Examination of cross sections of lamellae under the electron microscope shows that their edges are joined to form closed hollow disks that are called thylakoids ("saclike"). The chloroplasts of most higher plants have regions, called grana, in which the thylakoids are very tightly stacked. When viewed by electron microscopy at an oblique angle, the grana appear as stacks of disks. When viewed in cross section, it is apparent that some thylakoids extend from one grana through the stroma into other grana. The thin aqueous spaces inside the thylakoids are believed to be connected with each other via these stroma thylakoids. These thylakoid spaces are isolated from the stroma spaces by the relatively impermeable lamellae.

The light reactions occur exclusively in the thylakoids. The complex structural organization of lamellae is required for proper thylakoid function; intact thylakoids apparently are necessary for the formation of ATP. Thylakoids that have been broken down to smaller units can no longer form ATP, even when the conversion of light into chemical energy occurs during electron transport in these units. Such lamellar fragments can carry out the Hill reaction, with the transfer of electrons from water to $NADP^+$.

Chemical composition of lamellae » Lipids

Lamellae consist of about equal amounts of lipids and proteins. About one-fourth of the lipid portion of the lamellae consists of pigments and coenzymes; the remainder consists of various lipids, including polar compounds such as phospholipids and galactolipids. These polar lipid molecules have "head" groups that attract water (*i.e.*, are hydrophilic) and fatty acid "tails" that are oil soluble and repel water (*i.e.*, are hydrophobic). When polar lipids are placed in an aqueous environment, they can line up with the fatty acid tails side by side.

A second layer of phospholipids forms tail-to-tail with the first, establishing a lipid bilayer in which the hydrophilic heads are in contact with the aqueous solution on each side of the bilayer. Sandwiched between the heads are the hydrophobic tails, creating a hydrophobic environment from which water is excluded. This lipid bilayer is an essential feature of all biological membranes .

The hydrophobic parts of proteins and lipid-soluble cofactors and pigments are dissolved or embedded in the lipid bilayer. Lamellar membranes can function as electrical insulating material and permit a charge, or potential difference, to develop across the membrane. Such a charge can be a source of chemical or electrical energy.

Approximately one-fifth of the lamellar lipids are chlorophyll molecules; one type, chlorophyll *a*, is more abundant than the second type, chlorophyll *b*. The chlorophyll molecules are specifically bound to small protein molecules. Most of these chlorophyll-proteins are "light-harvesting" pigments. These absorb light and pass its energy on to special chlorophyll *a* molecules that are directly involved in the conversion of light energy to chemical energy. When one of these special chlorophyll *a* molecules is excited by light energy (as described later), it gives up an electron. There are two types of these special chlorophyll *a* molecules: one, called P_{680}, has an absorption spectrum that peaks at 684 nanometres; the other, called P_{700}, shows an absorption peak at 700 nanometres.

Although chlorophylls are the main light-absorbing molecules in green plants, other pigments such as carotenes and carotenoids (which are responsible for the yellow-orange colour of carrots) also can absorb light and may supplement chlorophyll as the light-absorbing molecules in some plant cells. The light energy absorbed by these pigments must be passed to chlorophyll before conversion to chemical energy can occur.

Chemical Composition of Lamellae: Proteins

Many of the lamellar proteins are components of the chlorophyll–protein complexes described above. Other proteins include enzymes and protein-containing coenzymes. Enzymes are required as organic catalysts for specific reactions within the lamellae. Protein coenzymes, also called cofactors, include important electron carrier molecules called cytochromes, which are iron-containing pigments with the pigment portions attached to protein molecules. During electron transfer, an electron is accepted by an iron atom in the pigment portion of a cytochrome molecule, which thus is reduced; then the electron is transferred to the iron atom in the next cytochrome carrier in the electron transfer chain, thus oxidizing the first cytochrome and reducing the next one in the chain.

In addition to the metal atoms found in the pigment portions of cytochrome molecules, metal atoms also are found in other protein molecules of the lamellae. In proteins with a total molecular weight of 900,000 (based on the weight of hydrogen as one), there are two atoms of manganese, 10 atoms of iron, and six atoms of copper. These metal atoms are required for the catalytic activity of some of the enzymes important in photosynthesis. The manganese atoms are involved in water-splitting and oxygen formation. Both copper- and iron-containing proteins function in electron transport between water and the final electron-acceptor molecule of the light stage of photosynthesis, an iron-containing protein called ferredoxin. Ferredoxin is a soluble component in the chloroplasts. In its reduced form, it gives electrons directly to the systems that reduce nitrate and sulfate and via NADPH to the system that reduces carbon dioxide. A copper-containing protein called plastocyanin (PC) carries electrons at one point in the electron transport chain. PC molecules are water soluble and can move through the inner space of the thylakoids, carrying electrons from one place to another.

Chemical Composition of Lamellae: Quinones

Small molecules called plastoquinones are found in substantial numbers in the lamellae. Like the cytochromes, quinones have important roles in carrying electrons between the components of the light reactions. Since they are lipid soluble, they can diffuse through the membrane. They can carry one or two electrons and, in their reduced form (with added electrons), they carry hydrogen atoms that can be released as hydrogen ions when the added electrons are passed on, for example, to a cytochrome.

Light absorption and energy transfer

The light energy absorbed by a chlorophyll molecule excites some electrons within the structure of the molecule to higher energy levels, or excited states. Light of shorter wavelength has more energy than light of longer wavelength , so that absorption of blue light creates an excited state of higher energy. A molecule raised to this higher energy state quickly gives up the "extra" energy as heat and falls to its lowest excited state. This lowest excited state is similar to that of a molecule that has just absorbed the longest wavelength light capable of exciting it. In the case of chlorophyll *a*, this lowest excited state corresponds to that of a molecule that has absorbed red light of about 680 nanometres. the return of a chlorophyll *a* molecule from its lowest excited state to its original low-energy state (ground state) requires the release of the extra energy of the excited state.

This can occur in one of several ways. In photosynthesis, most of this energy is conserved as chemical energy by the transfer of an electron from a special chlorophyll *a* molecule (P_{680} or P_{700}) to an electron acceptor. When this electron transfer is blocked by inhibitors, such as the herbicide dichlorophenylmethylurea (DCMU), or by low temperature, the energy can be released as red light. Such re-emission of light is called fluorescence. The examination of fluorescence from photosynthetic material in which electron transfer has been blocked has proved to be a useful tool for scientists studying the light reactions.

PLANTS ADAPT TO SUN AND SHADE

Some plants have enough developmental plasticity to adapt to a range of light regimes, growing as sun plants in sunny areas and as shade plants in shady habitats. Some shady habitats receive less than 1% of the PAR available in an exposed habitat. Leaves that are adapted to very sunny or very shady environments are often unable to survive in the other type of habitat.

Sun and shade leaves have some contrasting characteristics:

- *Shade leaves* have more total chlorophyll per reaction centre, have a higher ratio of chlorophyll *b* to chlorophyll *a*, and are usually thinner than sun leaves.
- *Sun leaves* have more rubisco, and a larger pool of xanthophyll cycle components than shade leaves.

Contrasting anatomic characteristics can also be found in leaves of the same plant that are exposed to different light regimes. Some anatomic differences between a leaf grown in the sun and a leaf grown in the shade. Sun-grown leaves are thicker and have longer palisade cells than leaves growing in the shade.

Even different parts of a single leaf show adaptations to their light microenvironment. Cells in the upper surface of the leaf, which are exposed to the highest prevailing photon flux, have characteristics of cells from leaves grown in full sunlight; cells in the lower surface of the leaf have characteristics of cells found in shade-grown leaves.

These morphological and biochemical modifications are associated with specific functions. Far-red light is absorbed primarily by PSI, and altering the ratio of PSI to PSII or changing the light-harvesting antennae associated with the photosystems makes it possible to maintain a better balance of energy flow through the two photosystems. These adaptations are found in nature; some shade plants show a 3:1 ratio of photosystem II to photosystem I reaction centers, compared with the 2:1 ratio found in sun plants. Other shade plants, rather than altering the ratio of PSI to PSII, add more antennae chlorophyll to PSII.

These adaptations appear to enhance light absorption and energy transfer in shady environments, where far-red light is more abundant. Sun and shade plants also differ in their respiration rates, and these differences alter the relationship between respiration and photosynthesis.

Plants Compete for Sunlight

Plants normally compete for sunlight. Held upright by stems and trunks, leaves configure a canopy that absorbs light and influences photosynthetic rates and growth beneath them. Leaves that are shaded by other leaves have much lower photosynthetic rates.

Some plants have very thick leaves that transmit little, if any, light. Other plants, such as those of the dandelion (*Taraxacum* sp.), have a rosette growth habit, in which leaves grow radially very close to each other and to the stem, thus preventing the growth of any leaves below them. Trees represent an outstanding adaptation for light interception.

The elaborate branching structure of trees vastly increases the interception of sunlight. Very little PAR penetrates the canopy of many forests; almost all of it is absorbed by leaves. Another feature of the shady habitat is sunflecks, patches of sunlight that pass through small gaps in the leaf canopy and move across shaded leaves as the sun moves. In a dense forest, sunflecks can change the photon flux impinging on a leaf in the forest floor more than tenfold within seconds.

For some of these leaves, a sunfleck contains nearly 50% of the total light energy available during the day, but this critical energy is available for only a few minutes in a very high dose. Sunflecks also play a role in the carbon metabolism of lower leaves in dense crops that are shaded by the upper leaves of the plant. Rapid responses of both the photosynthetic apparatus and the stomata to sunflecks have been of substantial interest to plant physiologists and ecologists because they represent unique physiological responses specialized for capturing a short burst of sunlight.

LEAVES AND LEAF STRUCTURE

Plants are the only photosynthetic organisms to have leaves (and not all plants have leaves). A leaf may be viewed as a solar collector crammed full of photosynthetic cells. The raw materials of photosynthesis, water and carbon dioxide, enter the cells of the leaf, and the products of photosynthesis, sugar and oxygen, leave the leaf.

Water enters the root and is transported up to the leaves through specialized plant cells known as xylem (pronounces zigh–lem). Land plants must guard against drying out (desiccation) and so have evolved specialized structures known as stomata to allow gas to enter and leave the leaf. Carbon dioxide cannot pass through the protective waxy layer covering the leaf (cuticle), but it can enter the leaf through an opening (the stoma; plural = stomata;

Greek for hole) flanked by two guard cells. Likewise, oxygen produced during photosynthesis can only pass out of the leaf through the opened stomata. Unfortunately for the plant, while these gases are moving between the inside and outside of the leaf, a great deal water is also lost. Cottonwood trees, for example, will lose 100 gallons of water per hour during hot desert days. Carbon dioxide enters single–celled and aquatic autotrophs through no specialized structures.

The Nature of Light

White light is separated into the different colours (=wavelengths) of light by passing it through a prism. Wavelength is defined as the distance from peak to peak (or trough to trough). The energy of is inversely porportional to the wavelength: longer wavelengths have less energy than do shorter ones.

The order of colours is determined by the wavelength of light. Visible light is one small part of the electromagnetic spectrum. The longer the wavelength of visible light, the more red the colour. Likewise the shorter wavelengths are towards the violet side of the spectrum. Wavelengths longer than red are referred to as infrared, while those shorter than violet are ultraviolet.

Light behaves both as a wave and a particle. Wave properties of light include the bending of the wave path when passing from one material (medium) into another (*i.e.* the prism, rainbows, pencil in a glass–of–water, etc.). The particle properties are demonstrated by the photoelectric effect. Zinc exposed to ultraviolet light becomes positively charged because light energy forces electrons from the zinc. These electrons can create an electrical current. Sodium, potassium and selenium have critical wavelengths in the visible light range. The critical wavelength is the maximum wavelength of light (visible or invisible) that creates a photoelectric effect.

Chlorophyll and Accessory Pigments

A pigment is any substance that absorbs light. The colour of the pigment comes from the wavelengths of light reflected (in other words, those not absorbed). Chlorophyll, the green pigment common to all photosynthetic cells, absorbs all wavelengths of visible light except green, which it reflects to be detected by our eyes.

Black pigments absorb all of the wavelengths that strike them. White pigments/lighter colours reflect all or almost all of the energy striking them. Pigments have their own characteristic absorption spectra, the absorption pattern of a given pigment.

Chlorophyll is a complex molecule. Several modifications of chlorophyll occur among plants and other photosynthetic organisms. All photosynthetic organisms (plants, certain protistans, prochlorobacteria, and cyanobacteria) have chlorophyll a. Accessory pigments absorb energy that chlorophyll a does not absorb. Accessory pigments include chlorophyll b (also c, d, and e in algae and protistans), xanthophylls, and carotenoids (such as beta–carotene). Chlorophyll a absorbs its energy from the Violet–Blue and Reddish orange–Red wavelengths, and little from the intermediate (Green–Yellow–Orange) wavelengths.

Carotenoids and chlorophyll b absorb some of the energy in the green wavelength. Why not so much in the orange and yellow wavelengths? Both chlorophylls also absorb in the orange–red end of the spectrum (with longer wavelengths and lower energy).

The origins of photosynthetic organisms in the sea may account for this. Shorter wavelengths (with more energy) do not penetrate much below 5 meters deep in sea water. The ability to absorb some energy from the longer (hence more penetrating) wavelengths might have been an advantage to early photosynthetic algae that were not able to be in the upper (photic) zone of the sea all the time.

The action spectrum of photosynthesis is the relative effectiveness of different wavelengths of light at generating electrons. If a pigment absorbs light energy, one of three things will occur. Energy is dissipated as heat. The energy may be emitted immediately as a longer wavelength, a phenomenon known as fluorescence. Energy may trigger a chemical reaction, as in photosynthesis.

Chlorophyll only triggers a chemical reaction when it is associated with proteins embedded in a membrane (as in a chloroplast) or the membrane infoldings found in photosynthetic prokaryotes such as cyanobacteria and prochlorobacteria.

Structure of the Chloroplastand Photosynthetic Membranes

The thylakoid is the structural unit of photosynthesis. Both photosynthetic prokaryotes and eukaryotes have these flattened sacs/vesicles containing photosynthetic chemicals. Only eukaryotes have chloroplasts with a surrounding membrane. Thylakoids are stacked like pancakes in stacks known collectively as grana. The areas between grana are referred to as stroma. While the mitochondrion has two membrane systems, the chloroplast has three, forming three compartments.

LIGHT-RESPONSE CURVES REVEAL PHOTO SYNTHETIC PROPERTIES

Measuring CO_2 fixation in intact leaves at increasing photon flux allows us to construct light-response curves that provide useful information about the photosynthetic properties of leaves. In the dark there is no photosynthetic carbon assimilation, and CO_2 is given off by the plant because of respiration. By convention, CO_2 assimilation is negative in this part of the light-response curve. As the photon flux increases, photosynthetic CO_2 assimilation increases until it equals CO_2 release by mitochondrial respiration. The point at which CO_2 uptake exactly balances CO_2 release is called the light compensation point.

The photon flux at which different leaves reach the light compensation point varies with species and developmental conditions. One of the more interesting differences is found between plants grown in full sunlight and those grown in the shade. Light compensation points of sun plants range from 10 to 20 μmol m^{-2} s^{-1}; corresponding values for shade plants are 1 to 5 μmol m^{-2} s^{-1}. The values for shade plants are lower because respiration rates in shade plants are very low, so little net photosynthesis suffices to bring the net rates of CO_2 exchange to zero. Low respiratory rates seem to represent a basic adaptation that allows shade plants to survive in light-limited environments. Increasing photon flux above the light compensation point results in a proportional increase in photosynthetic rate, yielding a linear relationship between photon flux and photosynthetic rate. Such a linear relationship comes about because photosynthesis is *light limited* at those levels of incident light, so more light stimulates more photosynthesis.

In this linear portion of the curve, the slope of the line reveals the maximum quantum yield of photosynthesis for the leaf. Recall that quantum yield is the relation between a given light-dependent product (in this case CO_2 assimilation) and the number of absorbed photons. Quantum yields vary from 0, where none of the light energy is used in photosynthesis, to 1, where all the absorbed light is used.

The quantum yield of photochemistry is about 0.95, and the quantum yield of oxygen evolution by isolated chloroplasts is about 0.1 (10 photons per molecule of O_2). In the intact leaf, measured quantum yields for CO_2 fixation vary between 0.04 and 0.06. Healthy leaves from many species of C_3 plants, kept under low O_2 concentrations that inhibit photorespiration, usually show a quantum yield of 0.1. In normal air, the quantum yield of C_3 plants is lower, typically 0.05.

Quantum yield varies with temperature and CO_2 concentration because of their effect on the ratio of the carboxylase and oxygenase reactions of rubisco. Below 30°C, quantum yields of C_3 plants are generally higher than those of C_4 plants; above 30°C, the situation is usually reversed. Despite their different growth habitats, sun and shade plants show similar quantum yields. At higher photon fluxes, the photosynthetic response to light starts to level off and reaches *saturation*. Once the saturation point is reached, further increases in photon flux no longer affect photosynthetic rates, indicating that factors other than incident light, such as electron transport rate, rubisco activity, or the metabolism of triose phosphates, have become limiting to photosynthesis.

After the saturation point, photosynthesis is commonly referred to as CO_2 *limited*, reflecting the inability of the Calvin cycle enzymes to keep pace with the absorbed light energy. Light saturation levels for shade plants are substantially lower than those for sun plants. These levels usually reflect the maximal photon flux to which the leaf was exposed during growth. The light-response curve of most leaves saturates between 500 and 1000 $\mu mol\ m^{-2}\ s^{-1}$, photon fluxes well below full sunlight (which is about 2000 $\mu mol\ m^{-2}\ s^{-1}$). Although individual leaves are rarely able to utilize full sunlight, whole plants usually consist of many leaves that shade each other.

For example, only a small fraction of a tree's leaves are exposed to full sun at any given time of the day. The rest of the leaves receive subsaturating photon fluxes in the form of small patches of light that pass through gaps in the leaf canopy or in the form of light transmitted through other leaves. Because the photosynthetic response of the intact plant is the sum of the photosynthetic activity of all the leaves, only rarely is photosynthesis saturated at the level of the whole plant.

Light-response curves of individual trees and of the forest canopy show that photosynthetic rate increases with photon flux and photosynthesis usually does not saturate, even in full sunlight. Along these lines, crop productivity is related to the total amount of light received during the growing season, and given enough water and nutrients, the more light a crop receives, the higher the biomass.

Leaves Must Dissipate Excess Light Energy

When exposed to excess light, leaves must dissipate the surplus absorbed light energy so that it does not harm the photosynthetic apparatus. There are several routes for energy dissipation involving *nonphotochemical quenching,* which is the quenching of chlorophyll fluorescence by mechanisms other than photochemistry. The most important example involves the transfer of absorbed light energy away from electron transport toward heat production. Although the molecular mechanisms are not yet fully understood, the xanthophyll cycle appears to be an important avenue for dissipation of excess light energy.

The Xanthophyll Cycle

RThe xanthophyll cycle, which comprises the three carotenoids violaxanthin, antheraxanthin, and zeaxanthin, is involved in the dissipation of excess light energy in the leaf. Under high light, violaxanthin is converted to antheraxanthin and then to zeaxanthin. Note that the two aromatic rings of violaxanthin have a bound oxygen atom in them, antheraxanthin has one, and zeaxanthin has none. Experiments have shown that zeaxanthin is the most effective of the three xanthophylls in heat dissipation, and antheraxanthin is only half as effective.

Whereas the levels of antheraxanthin remain relatively constant throughout the day, the zeaxanthin content increases at high irradiances and decreases at low irradiances. In leaves growing under full sunlight, zeaxanthin and antheraxanthin can make up 60% of the total xanthophyll cycle pool at maximal irradiance levels attained at midday.

In these conditions a substantial amount of excess light energy absorbed by the thylakoid membranes can be dissipated as heat, thus preventing damage to the photosynthetic machinery of the chloroplast. The fraction of light energy that is dissipated depends on irradiance, species, growth conditions, nutrient status, and ambient temperature.

The Xanthophyll Cycle in Sun and Shade Leaves

Leaves that grow in full sunlight contain a substantially larger xanthophyll pool than shade leaves, so they can dissipate higher amounts of excess light energy. Nevertheless, the xanthophyll cycle also operates in plants that grow in the low light of the forest understory, where they are only occasionally exposed to high light when sunlight passes through gaps in the overlying leaf canopy, forming sunflecks. Exposure to one sunfleck results in the conversion of much of the violaxanthin in the leaf to zeaxanthin.

In contrast to typical leaves, in which violaxanthin levels increase again when irradiances drop, the zeaxanthin formed in shade leaves of the forest understory is retained and protects the leaf against exposure to subsequent sunflecks. The xanthophyll cycle is also found in species such as conifers, the leaves of which remain green during winter, when photosynthetic rates are very low yet light absorption remains high. Contrary to the diurnal cycling of the xanthophyll pool observed in the summer, zeaxanthin levels remain high all day during the winter. Presumably this mechanism maximizes dissipation of light energy, thereby protecting the leaves against photooxidation during winter.

In addition to protecting the photosynthetic system against high light, the xanthophyll cycle may help protect against high temperatures. Chloroplasts are more tolerant of heat when they accumulate zeaxanthin. Thus, plants may employ more than one biochemical mechanism to guard against the deleterious effect of excess heat.

Leaves Must Dissipate Vast Quantities of Heat

The heat load on a leaf exposed to full sunlight is very high. In fact, a leaf with an effective thickness of water of 300 ìm would warm up by 100°C every minute if all available solar energy were absorbed and no heat were lost. However, this enormous heat load is dissipated by the emission of longwave radiation, by sensible (*i.e.*, perceptible) heat loss, and by evaporative (or latent) heat loss:

- Air circulation around the leaf removes heat from the leaf surfaces if the temperature of the leaf is higher than that of the air; this phenomenon is called sensible heat loss.
- Evaporative heat loss occurs because the evaporation of water requires energy. Thus as water evaporates from a leaf, it withdraws heat from the leaf and cools it. The human body is cooled by the same principle, through perspiration.

Sensible heat loss and evaporative heat loss are the most important processes in the regulation of leaf temperature, and the ratio of the two is called the Bowen ratio:

$$\text{Bowen Ratio} = \frac{\text{Sensible heat loss}}{\text{Evaporative heat loss}}$$

In well-watered crops, transpiration, and hence water evaporation from the leaf, is high, so the Bowen ratio is low. On the other hand, when evaporative cooling is limited, the Bowen ratio is large. For example, in some cacti, stomata closure prevents evaporative cooling; all the heat is dissipated by sensible heat loss, and the Bowen ratio is infinite. Plants with very high

Bowen ratios conserve water but have to endure very high leaf temperatures in order to maintain a sufficient temperature gradient between the leaf and the air. Slow growth is usually correlated with these adaptations.

Isoprene Synthesis Helps Leaves Cope with Heat

We have seen how the xanthophyll cycle can protect against high light, but how do chloroplasts cope with the high leaf temperatures that usually accompany high light? Isoprene synthesis appears to confer stability to photosynthetic membranes at high light and temperatures. Many plants, including American oak (*Quercus* sp.), aspen (*Populus* sp.), and kudzu (*Pueraria lobata*) emit gaseous five-carbon molecules such as isoprene. On a global scale, these emissions amount to 5×10^{14} g released to the atmosphere each year. These gaseous hydrocarbons are responsible for the pine scent (α- and β-pinene) in coniferous forests and can form a blue haze above forests on hot days.

Because isoprene and related hydrocarbons play an important role in atmospheric chemistry, they have attracted much attention from atmospheric scientists. Isoprene emission from leaves can constitute a significant fraction of the carbon assimilated in photosynthesis. For example, up to 2% of the carbon fixed by photosynthesis in aspen and oak leaves at 30°C is released as isoprene. Sun leaves synthesize more isoprene than shade leaves, and synthesis is proportional to leaf temperature and water stress.

Evidence that isoprene confers stability to photosynthetic membranes under high temperatures comes from three types of experimental results:

1. Whereas preventing isoprene emission with an inhibitor increases susceptibility to damage by heat, adding isoprene to plants that do not produce isoprene confers heat stability.
2. Mutant plants unable to emit isoprene are more easily damaged by high temperatures than are wild-type plants.
3. Isoprene is rapidly synthesized enzymatically in response to elevated leaf temperatures.

Absorption of Too Much Light Can Lead to Photoinhibition

When leaves are exposed to more light than they can utilize, the reaction centre of PSII is inactivated and damaged, in a phenomenon called photoinhibition. The characteristics of photoinhibition in the intact leaf depend on the amount of light to which the plant is exposed, and two types of photoinhibition are identified: dynamic photoinhibition and chronic photoinhibition. Under moderate excess light, dynamic photoinhibition is observed. Quantum efficiency decreases, but the maximum photosynthetic rate remains unchanged.

Dynamic photoinhibition is caused by the diversion of absorbed light energy toward heat dissipation—hence the decrease in quantum efficiency. This decrease is often temporary, and quantum efficiency can return to its initial higher value when photon flux decreases below saturation levels. Chronic photoinhibition results from exposure to high levels of excess light that damage the photosynthetic system and decrease both quantum efficiency and maximum photosynthetic rate.

Chronic photoinhibition is associated with damage and replacement of the D1 protein from the reaction centre of PSII. In contrast to dynamic photoinhibition, these effects are relatively long-lasting, persisting for weeks or months. Early researchers of photoinhibition interpreted all decreases in quantum efficiency as damage to the photosynthetic apparatus. It is now recognized that short-term decreases in quantum efficiency seem to reflect protective mechanisms, whereas chronic photoinhibition represents actual damage to the chloroplast resulting from excess light, or a failure of the protective mechanisms. How significant is photoinhibition in nature? Dynamic photoinhibition appears to occur normally at midday, when leaves are exposed to maximum amounts of light and there is a corresponding reduction in carbon fixation. Photoinhibition is more pronounced at low temperatures, and it becomes chronic under more extreme climatic conditions.

Studies of natural willow populations, and crops of *Brassica napus* (oilseed rape) and *Zea mays* (maize), have shown that the cumulative effects of a daily depression in photosynthetic rates caused by photoinhibition decrease biomass by 10% at the end of the growing season. This may not seem a particularly large effect, but it could be significant in natural plant populations competing for limited resources—conditions under which any reduction in carbon allocated to reproduction can adversely affect reproductive success and survival.

PROCESS STAGES OF PHOTOSYNTHESIS

Photosynthesis is a two stage process. The first process is the Light Dependent Process (Light Reactions), requires the direct energy of light to make energy carrier molecules that are used in the second process. The Light Independent Process (or Dark Reactions) occurs when the products of the Light Reaction are used to form C–C covalent bonds of carbohydrates. The Dark Reactions can usually occur in the dark, if the energy carriers from the light process are present. Recent evidence suggests that a major enzyme of the Dark Reaction is indirectly stimulated by light, thus the term Dark Reaction is somewhat of a misnomer. The Light Reactions occur in the grana and the Dark Reactions take place in the stroma of the chloroplasts.

LIGHT REACTIONS

In the Light Dependent Processes (Light Reactions) light strikes chlorophyll a in such a way as to excite electrons to a higher energy state. In a series of reactions the energy is converted (along an electron transport process) into ATP and NADPH. Water is split in the process, releasing oxygen as a by–product of the reaction. The ATP and NADPH are used to make C–C bonds in the Light Independent Process (Dark Reactions). In the Light Independent Process, carbon dioxide from the atmosphere (or water for aquatic/marine organisms) is captured and modified by the addition of Hydrogen to form carbohydrates (general formula of carbohydrates is $[CH_2O]_n$). The incorporation of carbon dioxide into organic compounds is known as carbon fixation.

The energy for this comes from the first phase of the photosynthetic process. Living systems cannot directly utilize light energy, but can, through a complicated series of reactions, convert it into C–C bond energy that can be released by glycolysis and other metabolic processes. Photosystems are arrangements of chlorophyll and other pigments packed into thylakoids. Many Prokaryotes have only one photosystem, Photosystem II (so numbered because, while it was most likely the first to evolve, it was the second one discovered). Eukaryotes have Photosystem II plus Photosystem I. Photosystem I uses chlorophyll a, in the form referred to as P700. Photosystem II uses a form of chlorophyll a known as P680. Both "active" forms of chlorophyll a function in photosynthesis due to their association with proteins in the thylakoid membrane.

Photophosphorylation is the process of converting energy from a light–excited electron into the pyrophosphate bond of an ADP molecule. This occurs when the electrons from water are excited by the light in the presence of P680. The energy transfer is similar to the chemiosmotic electron transport occurring in the mitochondria. Light energy causes the removal of an electron from a molecule of P680 that is part of Photosystem II. The P680 requires an electron, which is taken from a water molecule, breaking the water into H^+ ions and O^{-2} ions. These O^{-2} ions combine to form the diatomic O_2 that is released. The electron is "boosted" to a higher energy state and attached to a primary electron acceptor, which begins a series of redox reactions, passing the electron through a series of electron carriers, eventually attaching it to a molecule in Photosystem I. Light acts on a molecule of P700 in Photosystem I, causing an electron to be "boosted" to a still higher potential.

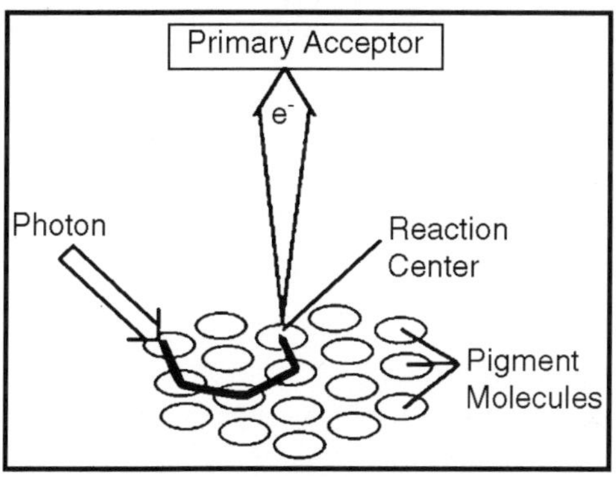

Fig. Action of a Photosystem

The electron is attached to a different primary electron acceptor (that is a different molecule from the one associated with Photosystem II). The electron is passed again through a series of redox reactions, eventually being attached to $NADP^+$ and H^+ to form NADPH, an energy carrier needed in the Light Independent Reaction. The electron from Photosystem II replaces the excited electron in the P700 molecule.

There is thus a continuous flow of electrons from water to NADPH. This energy is used in Carbon Fixation. Cyclic Electron Flow occurs in some eukaryotes and primitive photosynthetic bacteria. No NADPH is produced, only ATP. This occurs when cells may require additional ATP, or when there is no $NADP^+$ to reduce to NADPH. In Photosystem II, the pumping to H ions into the thylakoid and the conversion of ADP + P into ATP is driven by electron gradients established in the thylakoid membrane.

The above diagrams present the "old" view of photophosphorylation. We now know where the process occurs in the chloroplast, and can link that to chemiosmotic synthesis of ATP.

Halobacteria, which grow in extremely salty water, are facultative aerobes, they can grow when oxygen is absent. Purple pigments, known as retinal (a pigment also found in the human eye) act similar to chlorophyll. The complex of retinal and membrane proteins is known as bacteriorhodopsin, which generates electrons which establish a proton gradient that powers an ADP–ATP pump, generating ATP from sunlight without chlorophyll. This supports the theory that chemiosmotic processes are universal in their ability to generate ATP.

DARK REACTION

Carbon–Fixing Reactions are also known as the Dark Reactions (or Light Independent Reactions). Carbon dioxide enters single–celled and aquatic autotrophs through no specialized structures, diffusing into the cells. Land plants must guard against drying out (desiccation) and so have evolved specialized structures known as stomata to allow gas to enter and leave the leaf. The Calvin Cycle occurs in the stroma of chloroplasts (where would it occur in a prokaryote?). Carbon dioxide is captured by the chemical ribulose biphosphate (RuBP). RuBP is a 5–C chemical. Six molecules of carbon dioxide enter the Calvin Cycle, eventually producing one molecule of glucose. The first stable product of the Calvin Cycle is phosphoglycerate (PGA), a 3–C chemical.

The energy from ATP and NADPH energy carriers generated by the photo-systems is used to attach phosphates to (phosphorylate) the PGA. Eventually there are 12 molecules of glyceraldehyde phosphate (also known as phosphoglyceraldehyde or PGAL, a 3–C), two of which are removed from the cycle to make a glucose.

The remaining PGAL molecules are converted by ATP energy to reform 6 RuBP molecules, and thus start the cycle again. Remember the complexity of life, each reaction in this process, as in Kreb's Cycle, is catalyzed by a different reaction–specific enzyme.

C–4 Pathway

Some plants have developed a preliminary step to the Calvin Cycle (which is also referred to as a C–3 pathway), this preamble step is known as C–4. While most C–fixation begins with RuBP, C–4 begins with a new molecule, phosphoen-olpyruvate (PEP), a 3–C chemical that is converted into oxaloacetic acid (OAA, a 4–C chemical) when carbon dioxide is combined with PEP.

The OAA is converted to Malic Acid and then transported from the mesophyll cell into the bundle–sheath cell, where OAA is broken down into PEP plus carbon dioxide.

The carbon dioxide then enters the Calvin Cycle, with PEP returning to the mesophyll cell. The resulting sugars are now adjacent to the leaf veins and can readily be transported throughout the plant. The capture of carbon dioxide by PEP is mediated by the enzyme PEP carboxylase, which has a stronger affinity for carbon dioxide than does RuBP carboxylase When carbon dioxide levels decline below the threshold for RuBP carboxylase, RuBP is catalyzed with oxygen instead of carbon dioxide.

The product of that reaction forms glycolic acid, a chemical that can be broken down by photorespiration, producing neither NADH nor ATP, in effect dismantling the Calvin Cycle. C–4 plants, which often grow close together, have had to adjust to decreased levels of carbon dioxide by artificially raising the carbon dioxide concentration in certain cells to prevent photorespiration.

C–4 plants evolved in the tropics and are adapted to higher temperatures than are the C–3 plants found at higher latitudes. Common C–4 plants include crabgrass, corn, and sugar cane. Note that OAA and Malic Acid also have functions in other processes, thus the chemicals would have been present in all plants, leading scientists to hypothesize that C–4 mechanisms evolved several times independently in response to a similar environmental condition, a type of evolution known as convergent evolution.

Carbon Cycle

Plants may be viewed as carbon sinks, removing carbon dioxide from the atmosphere and oceans by fixing it into organic chemicals. Plants also produce some carbon dioxide by their respiration, but this is quickly used by photosynthesis. Plants also convert energy from light into chemical energy of C–C covalent bonds. Animals are carbon dioxide producers that derive their energy from carbohydrates and other chemicals produced by plants by the process of photosynthesis.

The balance between the plant carbon dioxide removal and animal carbon dioxide generation is equalized also by the formation of carbonates in the oceans. This removes excess carbon dioxide from the air and water (both of which are in equilibrium with regard to carbon dioxide). Fossil fuels, such as petroleum and coal, as well as more recent fuels such as peat and wood generate carbon dioxide when burned. Fossil fuels are formed ultimately by organic processes, and represent also a tremendous carbon sink. Human activity has greatly increased the concentration of carbon dioxide in air.

This increase has led to global warming, an increase in temperatures around the world, the Greenhouse Effect. The increase in carbon dioxide and other pollutants in the air has also led to acid rain, where water falls through polluted air and chemically combines with carbon dioxide, nitrous oxides, and sulfur oxides, producing rainfall with pH as low as 4. This results in fish kills and changes in soil pH which can alter the natural vegetation and uses of the land. The Global Warming problem can lead to melting of the ice caps in Greenland and Antarctica, raising sea–level as much as 120 meters. Changes in sea–level and temperature would affect climate changes, altering belts of grain production and rainfall patterns.

3

Cell Biology of Plant

PLANT CELLS

So far we have considered only animal cells and it is of interest that Hackett in the *International Review of Cytology*, has pointed out that in plant cells glycolysis involves the plastids, the soluble fraction of cells and possibly the nucleus as well as the mitochondrial enzymes. He points out that many enzymes which are involved in the Krebs cycle are not exclusive to the mitochondria and that hydrogen transfer is not confined to these organelles.

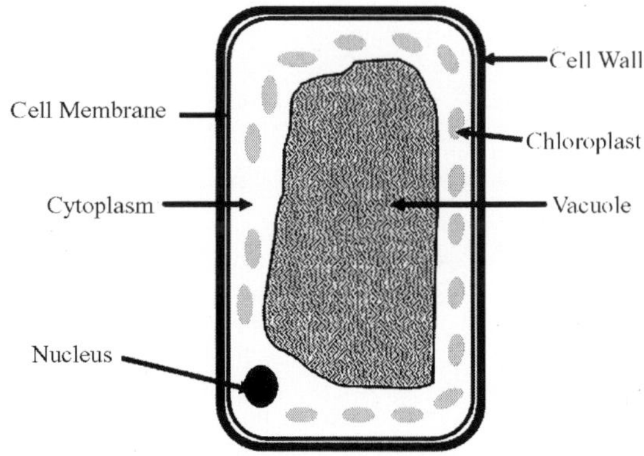

Fig. Plant Cell.

Mitochondria he says react with the nucleus in the process of phosphorylation, they react with the chloroplasts in photosynthesis and they react with the microsomes in protein synthesis in the endoplasmic reticulum. He believes that a close relation between the cell membrane and the mitochondria may play an active part in the movement of substances into the cell or possibly in the growth of the cell wall of plants.

Furthermore, he reminds us that the real unit is the cell itself, its various parts work as an integrated unit and one should only break them down for the purpose of trying to analyse the various processes in which they participate.

PLANT CELL STRUCTURE

To depict the structure of a plant cell, an epidermal cell of an onion will be used as an example. The epidermis is the final tissue that covers all organs above ground. The cells of the onion epidermis are common specimens on the first day of a German basic botanical course. Since they contain no chlorophyll, they are actually no "typical" plant cells. The picture above shows an onion's epidermal cells. They are elongated and the ratio of length to width can vary strongly. Each cell is enclosed by a wall. In the region of the cell poles and where three cells adjoin, large intercellular spaces can be observed.

Elsewhere a pectin-containing middle lamina cements neighbouring cells together like bricks. The cell wall is perforated at regular intervalls, so that adjoining cells are in contact. The holes of the perforation are called simple pits and the plasma cords that run through them plasmodesmata. The surface of the epidermal cells seems to be folded, an effect that is caused by the water-repellent, waxy cuticle.

PROTOPLASM, CYTOPLASM AND CYTOSOL: THE CELL'S CONTENT

The "living" content of a cell, the protoplasm, is surrounded by a membrane called plasma membrane or plasmalemma. The protoplasm is usually next to the cell wall, so that the plasmalemma can hardly be seen. To display it, the cells are transferred into a high salt or sugar solution. As a result the protoplasm shrinks and detaches itself from the wall. The process is reversible and is called plasmolysis.

This behaviour is due to the properties of the membrane and the plasma. It is reviewed in more detail elsewhere. A substance that causes plasmolysis is called plasmolyticum and - depending on its chemical composition (potassium ions orcalcium ions, for example)—the protoplasm takes on different shapes. The plasmolyticum has accordingly an influence on the properties of the membrane. The properties of the plasma membrane differ from that of the tonoplast. The tonoplast is the membrane that surrounds the vacuole. The difference is especially striking if cells with a coloured vacuole content are used. Often the vacuole is criss-crossed by numerous plasma cords. The plasma cannot therefore not simply be viewed as a solution that is influenced by the rules of hydrodynamics alone. Rather, it contains viscous, structure-determining components, whose chemical, physicochemical and structural properties have only been recognised recently and in fragments.

Cytoplasm and Caryoplasm

The nucleus is a rather conspicuous part of nearly every living plant cell. Its structure separated from the rest of the cell by the nuclear envelope, a membrane system that consists of two discrete membranes as can be seen on electromicroscopic images. The nuclear content is called the caryoplasm while that of the rest of the cell is called cytoplasm. But these terms are only valid at certain stages of a cell's life cycle. In the course of cell-division and mitosis, the nuclear envelope disintegrates and the nucleus is replaced by the chromosomes.

It makes consequently no sense to speak of caryo- and cytoplasm during these stages. The nucleus of plant cells is usually of a round or elliptic appearance, sometimes it is also shaped like a spindle. One nucleus per cell is the rule, but cells with two or more nuclei are no rare exception. The cells of certain algae of the genus *Chladophora* have many nuclei, they are polyenergid. The nucleoli that can often be perceived after staining are substructures of the nucleus. They, too, disintegrate during cell division and mitosis and do not reshape before a new nucleus has been formed.

Plastids

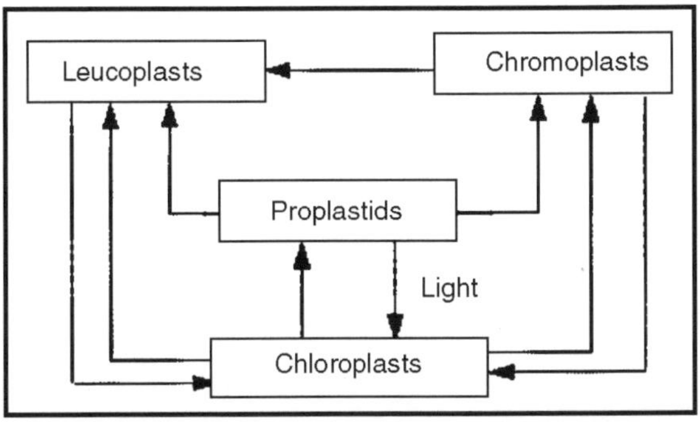

Plastids are organelles that occur only in plants. Their most prominent members are the chloroplasts. Others plastids are the coloured chromoplasts and the colourless leucoplasts as well as their proplastids. Proplastidsare vestigial bodies that are generated during germ cell development due to degeneration of plastids, for example. They may differentiate into complete plastids during the development of the plant embryo. Their ripening into chloroplasts occurs usually only after light exposure.

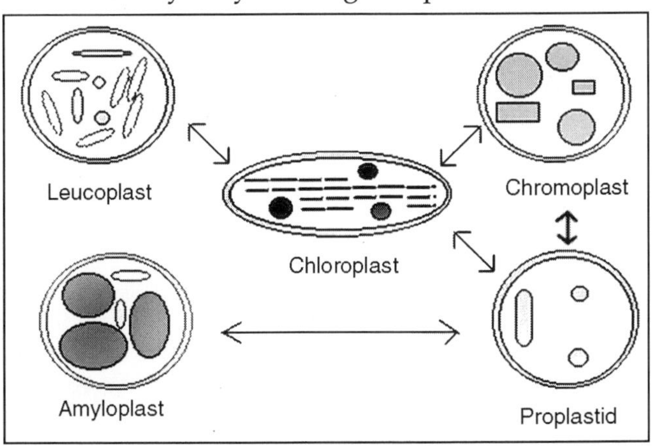

Chloroplasts contain the green plant colour chlorophyll. They are the places where photosynthesis takes place. Chloroplasts enable the plants to convert solar energy into chemical energy. Because of this process, plants are called primary producers. The existence of consumers, like most animals, depends on them. Chloroplasts occur in most cell types, but only in organs above ground.

They can be especially well observed in tissues consisting of a single layer as in the leaflike structures of some mosses (*Funaria hygrometrica* or *Mnium hornum*) or in the water plant *Vallisneria*). Here they are rather large and of a lens-shaped appearance. During daytime, when the light is diffuse, they occur mainly at the upper and lower surface of the chloroplast. They appear to be round under top view. If exposed to strong light, they gather in parallel to the lateral sides of the cell which gives them an elleptic appearance upon top view. Chloroplasts are the site of starch production and -storage. Starch can easily be detected with the aid of potassium iodide (LUGOL's reagent). The starch-iodine complex is deeply blue-violet.

Starch production during photosynthesis can be made visible by placing a mask at a leaf that covers it partially while leaving some places exposed to sunlight. After one day of exposure, the leaf is first bleached to get rid of other pigments and afterwards treated with potassium iodide. An image is gained that is the exact replication of the mask and at the same time represents starch synthesis in the leaf. This experiment has first been done by J. v SACHS, probably the most outstanding plant physiologist of the 20th century. He thought that starch was the primary product of photosynthesis. This assumption proved wrong. It is well-known today that the first products of photosynthesis are monomeric sugars (glucose and others) and that only part of them is used for starch production.

The structures of the chloroplasts of higher cells resemble largely that of mosses. Their average diameter is 4 - 8 mm, an average cells contains 10 through 50. Their chlorophyll is unevenly distributed. At high resolutions chlorophyll-rich and chlorophyll-poor areas can be distinguished. This is due to the inner structure of the chloroplast: it is organised into grana (chlorophyll-rich) and stroma (chlorophyll-poor).

Upon stimulation with short-waved light (blue or violet) the chlorophyll emits an intensive red autofluorescence that looks especially impressive in a fluorescence microscope. The differences between grana and stroma become very obvious. The uniformity of the chloroplasts of all higher plants points out that the optimal form has been found rather early in evolution and has not been changed since.

This is different with algae. The chloroplasts of green algae (Chlorophyceens) are very varied in shape. Many species have just one chloroplast that covers nearly the whole space of the cell's interior. It is screw-like in *Spirogyra*-species, star-shaped in *Zygnema* and *Zygnemopsis* and netlike in *Oedogonium*.

The disc-shaped chloroplast of *Mougeotia* can be viewed either from above or in profile depending on the amount of light used. Its rotation is a well-analysed example of an induced chloroplast movement. The chloroplasts of many species of algae contain often well-visible pyrenoids, structures, that produce and structure starch.

Chromoplasts are red, orange or yellow plastids. The colour is usually the result of yellow xantophyll and yellow to red carotinoids. Both compounds do also exist in chloroplasts, but are concealed by chlorophyll. Chlorophyll is broken down much faster than carotinoids as can be observed in the coloured leaves in autumn. Fluid transitions between chromo- and chloroplasts exist, just as between chromo- and leucoplasts. Typical chromoplasts cause the orange colour of the carrot, the red colour of the ripe pimento and tomato as well as the colour of numerous flowers. Carotinoids are not very water-soluble and do therefore often crystallise within the chromoplasts.

Their crystals can be disc-shaped, needle-like, jagged or sickle-like. In many cases, flower and leaf colours are caused by the coloured content of the vacuole. The colour of the vacuole and that of the plastids may lead to a mixed colour.

The leaves of the copper beech, where the vacuole's content is red and that of the chloroplasts is green are a typical example. The plastids of the red and brown algae are traditionally counted among the chromoplasts although they contain chlorophyll. The green colour is concealed by the red phycoerythrin (Rhodophyceae) or the brown fucoxanthin (Phaeophyta).

Leucoplasts are common, colourless plastids. They develop from proplastids, but form no homogeneous group of their own. A part of them can differentiate into chloroplasts or chromoplasts at light exposure, while this is not due for others. The guard cells, for example, contain leucoplasts, that are permanently exposed to light without developing into chloroplasts. Leucoplasts do also occur within colourless leaves (variegated leaves) or plant parts.

There exists a number of examples which show that they developed from chloroplasts that lost their ability to produce chlorophyll. There are even species, like *Neottia*, an orchid that cannot produce chlorophyll at all and are thus dependent on a parasitic or saprophytic lifestyle (saprophy is the feeding from dead organic material). A second class of leucoplasts occurs within the non-green tissues of otherwise green plants. It is especially common within roots. Though these leucoplasts are capable to become green, they do usually not since they are not exposed to light.

The leucoplasts of the calyptra (a calyptra is any hood or cap of cells protecting a plant part) contain starch and are therefore counted among the amyloplasts (starch-containing leucoplasts). They have, as is explained later, the function of statolithes, that have an important part in the perception of gravity (geotropism).

Starch

We got to know starch in the section above as a content of chloroplasts and leucoplasts (amyloplasts). It is produced by the polymerisation of glucose residues, which again are products of photosynthesis. Since the plant is able to transport sugars from leaf to root or from leaf to seed and fruit, starch production can also take place in these organs. Different species produce starch grains of different shape. Since the shape of starch grains informs about their origin, they are helpful in the identification of seeds and other starch-containing plant parts.

The following numbers show the variations in their diameters. Starch grains from potato tubers have a diameter of 70 - 100 μm, that of the endosperm of wheat 30 - 45 μm and that of corn endosperm 12 - 18 μm. Their shape reflects their development. The starch molecules are long-stretched and only sparsely branched. They are deposited within the plastids and their development begins at a so-called formation centre from where it proceeds radially. Layer follows layer and the thickness of one layer is dependent on the average molecular length.

A starch grain is therefore organised like a crystal (semi-crystalline). This can be shown very impressively with a polarisation microscope. Within the moistened specimen can a layering be viewed that is dependent on the water content of the single molecular parts.

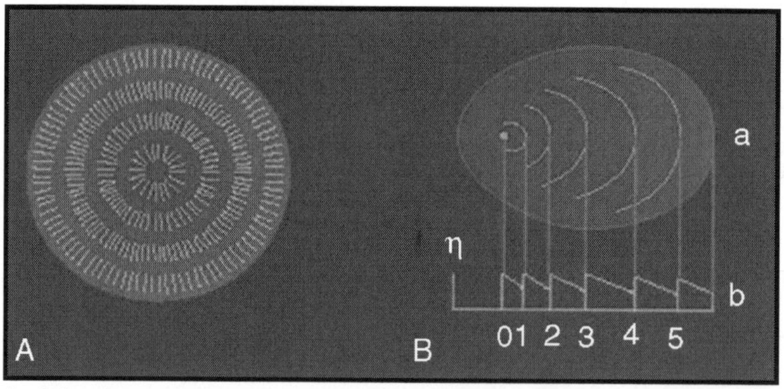

A. Model of a starch grain structure. The single lines symbolise starch molecules. They are arranged in a radial pattern. B. Layering of the starch grains, a. Formation centre and layer borders, b. Diagram of the refraction conditions. The ordinate shows the refraction index. (according to A. FREY-WYSSLING, 1938) The denser the molecular packaging, the less water is deposited. Layers with less water content refract light stronger than those with much water. After drying of the preparation no layering can be perceived any more. Depending the central or peripheral placing of the formation centre, starch grains with either concentric or eccentric layering develop. The starch grains of graminaceous plants (wheat, corn, etc.) are usually concentric, while those of the potato are always eccentric.

Sometimes plastids with two to three formation centres occur in graminaceous plants. This leads to the development of several starch grains. During growth common outer layers may be formed. Compound starch grains are typical for oat, they are built from a large number of smaller grains.

Starch grains in bean seeds (*Phaseolus vulgaris*) are very big, their shape is round or oval, the spacing of the layers is very regular. Their centers can easily be hollowed out by addition of water displaying radial ruptures in microscopic images. In the sap of *Euphorbia splendens* dumb-bell shaped starch grains can be found.

CRYSTALLINE INCLUSIONS WITHIN CELLS

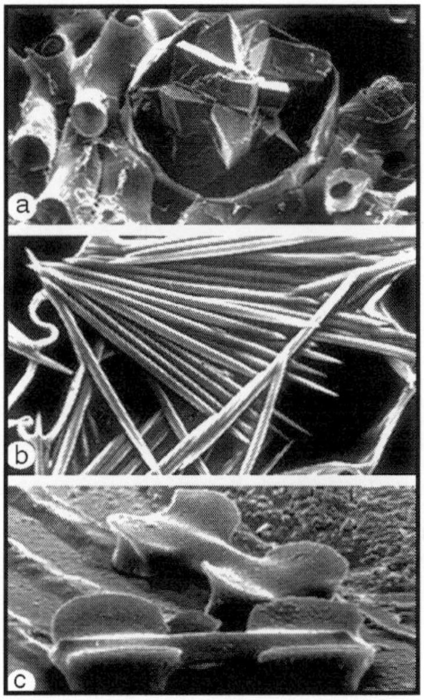

Many plant cells contain crystalline inclusions of different chemical composition and shape. Crystalline aggregations are called druses, bundles of needle-shaped crystals are termed raphids. Scanning electron microscopic images: Top picture: Calcium oxalate druse in the mesophyll cells of an oleander leaf *(Nerium oleander)*. Typical druse shape of dicots. Middle picture: Calcium oxalate needles (raphids) of a vanilla root (Orchidaceae). Typical raphid bundle of monocots. Lowest picture: Silicate bodies of silicate cells in the epidermis of *Schizachyrium sanguineum* (a gramineaen species of the old world tropics). Characteristic mineralisation of a gramineaen cell. [W. BARTHLOTT, MARTENS, 1979 (c), W. BARTHLOTT, unpublished (a, b)]

THE CELL WALL

Except for very few examples, plant cells are surrounded by a cellulose containing cell wall. It is flexible and distortable during growth, but loses its ability for distortion after growth has stopped, while a limited flexibility remains. Because of these changes, it is distinguished between primary and secondary cell walls. As we will see when talking about electron microscopic pictures of the cell wall, both forms differ mainly in the arrangement of their cellulose microfibrils. While they are unorganised within the amorphous matrix of the primary cell wall, they are organised into several ordered layers that are arranged one on top of the other at right angles in the secondary cell wall. The secondary cell walls of many cells, especially those of vascular tissues, are incrusted with strengthening material.

Two important ones are:

1. Lignin, the ground substance of wood and
2. Suberin, the ground substance of cork

PLANT CELL WALLS

Higher plants resist the forces of gravity and powerful lateral forces through the cumulative strength of the walls that surround individual cells. These walls consist mainly of cellulose, noncellulosic polysaccharides and lignin, in proportions that depend upon the specific functions of the cell and its stage of development. Spatially and temporally controlled heterogeneity in the physicochemical properties of wall polysaccharides is observed at the tissue and individual cell levels, and emerging *in situ* technologies are providing evidence that this heterogeneity also occurs across a single cell wall. We consider the origins of cell wall heterogeneity and identify contributing factors that are inherent in the molecular mechanisms of polysaccharide biosynthesis and are crucial for the changing biological functions of the wall during growth and development. We propose several key questions to be addressed in cell wall biology, together with an alternative two-phase model for the assembly of noncellulosic polysaccharides in plants.

Heterogeneity in the polysaccharide constituents of plant cell walls has been recognized since the earliest chemical analyses of isolated wall preparations. Clear and dramatic differences in average wall compositions are commonly observed, and actual differences are likely to be even greater than reported, because a good deal of heterogeneity can be disguised by the averaging effects of extracting constituent polysaccharides from plant or wall preparations that contain components from several different cell types. These observations are driving the development of a wide array of *in situ* methods for the definition of local wall composition, including methods based on enzymatic hydrolysis and mass spectroscopy, specific monoclonal antibodies, atomic force microscopy, Raman spectroscopy and real-time examination of cellular organelles.

In considering this heterogeneity, we shall see that certain aspects of the molecular mechanisms that mediate polysaccharide synthesis are inherently imprecise. In some cases the plant cell is able to control this imprecision and produces wall polysaccharides that are relatively homogeneous with respect to chemical structure and physicochemical properties. In other cases, the cell can generate polysaccharide products that are themselves chemically and physicochemically heterogeneous, a situation in which 'built-in' heterogeneity is a prerequisite for cell wall function. In another example, a synthetic system with the same inherent limitations in precision builds polysaccharides that have both nonrandom and random structural features. The unusual and perhaps counterintuitive mechanism, through which an enzyme system can precisely generate random or other statistically defined structural features, is a characteristic of plant cell wall polysaccharide biosynthesis that appears to allow for both evolutionary and environmental diversity. And as if this were not complex enough already, there are postsynthetic modifications that can even further alter the chemical and physicochemical properties of wall polysaccharides.

DIVERSE FUNCTIONS OF CELL WALLS IN PLANTS

Plants do not have specialized skeletal systems of the type found in animals; instead, the considerable strength that enables large plants such as the giant *Sequoia* redwood tree to withstand the forces of gravity is provided by the cumulative strength of the walls that surround individual cells. But the load-bearing walls also have to be sufficiently flexible to withstand the large tensile and compressive forces that occur as, for example, the plant sways in the breeze.

The wall is subjected to further forces from within, where osmotic pressure represents another threat to cell and plant integrity. In young cells, the walls must be porous, to allow water, low–molecular weight nutrients and hormones to diffuse freely between cells and tissues, whereas in specialized vascular tissues required for long-distance movement of water and nutrients through the plant, the walls must be impermeable. In most tissues, cell walls are also important for intercellular adhesion. These various functional requirements are accommodated through the formation of reinforced gel structures consisting of cellulosic microfibrils, which have a high tensile strength, embedded in a gel-like matrix phase that consists predominantly of noncellulosic polysaccharides. This matrix provides flexibility and some mechanical support for the cell, but its inherent porosity also allows water and other small molecules to diffuse through the wall.

As cells of higher plants stop growing and increasingly assume a structural role, their walls thicken, cellulose microfibrils become laminated to increase wall strength and the walls often become encrusted with lignin, which is a large polyphenolic molecule that renders the wall resistant to compressive forces and restricts the passage of small molecules. At the same time, the wall will 'lock in' the final shape of the cell, and many different cell shapes can be detected in a single section of a plant organ. Smaller amounts of structural and enzymatic proteins and various proteoglycans and glycoproteins are also found in the wall, but these will not be discussed in detail here.

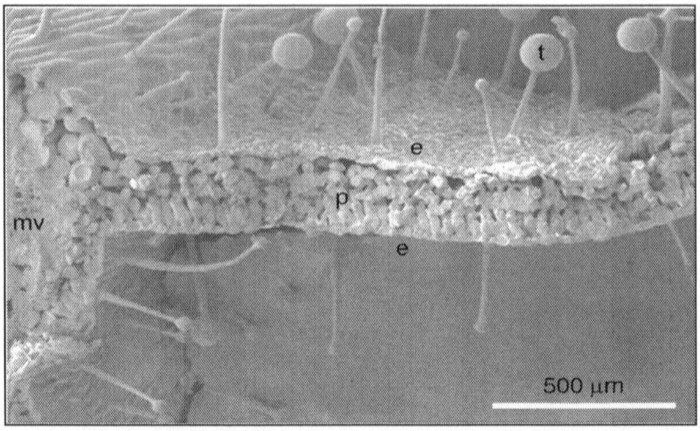

Fig. Heterogeneity in Cell Shape in a Leaf from *Nicotiana Benthamiana*.

In this mature leaf, which has stopped expanding, there are parenchyma cells (p) that have thin walls, fibre cells that have thickened walls, xylem cells that are no longer living and need a wall impervious to the water they transport, and trichome cells (t) that are greatly elongated and require a stiff wall to support them above the surface of the leaf. The cell walls surrounding these cell types play a major role in 'locking in' the final shape of the cell. The midvein (mv) and epidermal layers (e) are also indicated.

In addition to their crucial structural role, cell walls are likely to represent a significant source of metabolizable energy for some plants and may be subject to continual turnover in certain tissues as a result of this function. For example, the walls of the storage cells of seeds are not load bearing but in some cases are extraordinarily thick.

Wall polysaccharides in these seeds act as an alternative to starch as a storage macromolecule. Examples include the galactomannans of seeds like fenugreek, carob, ivory nut and guar, as well as pectins in lupin seeds and xyloglucans in nasturtium and tamarind seeds. Grains of many grasses, particularly the economically important cereals such as wheat, maize, rice and barley, contain an endosperm that is rich in starch but also comprises a significant amount of cell wall material that can be used as a source of energy by the germinated grain. Indeed, it is estimated that up to 18% of metabolizable glucose in the germinated barley grain is released from the (1,3;1,4)-α-D-glucans of the starchy endosperm cell wall. A similar energy reserve function has been proposed for cell wall (1,3;1,4)-α-D-glucans in vegetative tissues of the grasses.

Finally, fragments of noncellulosic polysaccharides of cell walls have been implicated in growth regulation and in signaling pathways associated with a loss of cell wall integrity. Various oligogalacturonan fragments from pectic polysaccharides are bound by receptor-like kinases and can initiate a range of cellular changes, including defence responses that are executed following pathogen attack. Indeed, one might reasonably ask whether the evolution of the extreme complexity in pectic polysaccharides might have been driven by the potential of these polysaccharides to release an array of different signaling oligosaccharides. If this were true, an obvious next question would be whether other wall polysaccharides, such as the heteroxylans, undertake a similar function, especially in the *Poaceae,* where levels of pectic polysaccharides in the wall are very low.

Heterogeneity in cell Wall Polysaccharide Structures

Cellulose is the common component of most plant cell walls, and microfibrils of cellulose constitute the reinforcing rods of the wall. This macromolecule consists of unbranched, unsubstituted (1,4)-α-D-glucan chains that are able to form noncovalent microfibrillar complexes through extensive inter- and intramolecular hydrogen bonding and hydrophobic interactions. There is a good deal of heterogeneity in the chain length of cellulose, and its degree of crystallinity can also vary, but the fundamental chemical structure is essentially identical in the walls of different plants, tissues and organs.

In contrast, the noncellulosic polysaccharides that compose the gel-like matrix of walls are structurally more complex and show considerable heterogeneity, from the tissue to the molecular level. Across the plant kingdom, a diverse array of polysaccharides is responsible for the formation of the wall matrix, but observed differences in the extant chemistries of the polysaccharides belie the diversity of the evolutionary histories that have converged to a common physicochemical outcome, namely the formation of a chemically heterogeneous but functionally competent gel-like component in the wall. In other words, the required physicochemical properties of wall polysaccharides can be met through a range of chemical strategies.

The backbone structures of these polysaccharides are based on (1,4)-β-linked monosaccharides in the case of cellulose, xyloglucan, (1,3;1,4)-β-glucan, heteroxylan and heteromannan, whereas the backbone of the galacturonans that are major constituents of pectic polysaccharides are based on (1,4)-α-linked chains of galacturonosyl residues. Extensive chain aggregation of the type found in cellulose is sterically hindered in the noncellulosic wall polysaccharides mainly through the addition of short oligosaccharides, monosaccharides or acetyl groups.

However, in the (1,3;1,4)-β-glucans, main chain aggregation is prevented through the irregular conformation of the polysaccharide. Noncellulosic polysaccharides of walls from dicotyledonous plants are rich in pectic polysaccharides and xyloglucans, whereas walls in the *Poaceae* family of monocotyledons, which includes commercially important cereals and grasses, have relatively small amounts of pectic polysaccharides and xyloglucans and commensurately higher levels of heteroxylans and (1,3;1,4)-β-D-glucans. In the endosperm of legumes, walls often contain relatively high levels of mannans, galactomannans and galactoglucomannans.

Differences are also detected in the compositions of the 'primary' walls that are laid down during cell expansion and the 'secondary' walls that are deposited once cell expansion has ceased, although in practice it is often difficult to define when primary wall formation is replaced by secondary wall formation. Cellulose and lignin levels are typically higher in secondary walls than in primary walls. Compositional and physical differences have also been widely reported between primary and secondary walls; noncellulosic polysaccharides of secondary walls generally have lower degrees of branching or main chain substitution and are considered to be less soluble in aqueous media.

Thus, physical and chemical heterogeneity in plant cell walls at the species and developmental levels is extensive, but newer technologies have revealed additional levels of wall diversity. Monoclonal antibodies directed against cell wall polysaccharides specifically recognize epitopes that compose small regions of the polysaccharide and hence can distinguish different fine structural features in the polysaccharides.

When sections of organs such as leaves and roots are probed with monoclonal antibodies directed against epitopes present on wall polysaccharides such as xyloglucans or pectins, binding varies by orders of magnitude between different cells in the section.

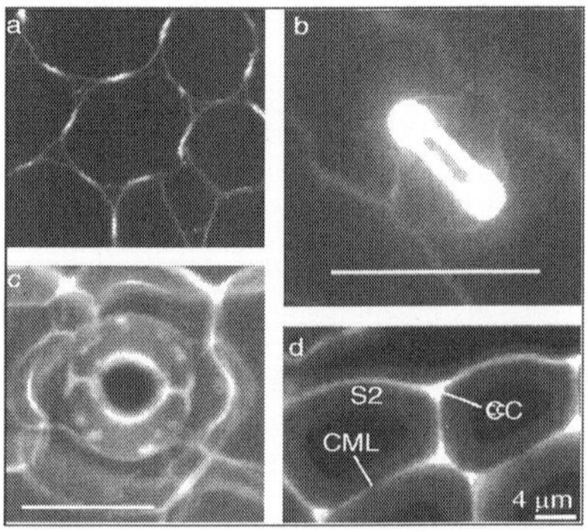

Fig. Heterogeneity in the Compositions of Cell Walls.

(a) Immunofluorescence microscopy and antibodies with different specificities for arabinan polysaccharides and their substructures were used to examine arabinan heterogeneity in stem tissues of *Arabidopsis*. In the pith parenchyma, the LM16 epitope was restricted to regions of adhered cell walls and was absent from cell wall regions surrounding intercellular air spaces.

(b) Distinctive UV autofluorescence associated with guard cells in maize. The data showed that guard cells had different amounts of phenolic esters of pectin than did epidermal cells.

(c) Heterogeneity in the patterns, colour and intensity of autofluorescence in the monocot *Commelina communis* has also been demonstrated.

(d) Raman images of a cross section of poplar wood prepared by integrating signals over defined wavenumber areas that showed constituents such as lignin and carbohydrate. The cell corner (CC) and the compound middle lamella (CML) adjacent to primary walls have a higher intensity of the lignin signal than in the S2 layer.

Molecular Mechanisms for Wall Polysaccharide Biosynthesis

The generally accepted mechanism of wall polysaccharide synthesis involves enzymes that add glycosyl residues from activated sugar nucleotide donors to the elongating polymeric chain. Two major classes of these glycosyltransferases have been recognized. One group, which is often referred to as the polysaccharide synthase group, is believed to catalyze the iterative addition of glycosyl residues to the end of the elongating polysaccharide chain, without releasing the chain between successive glycosylation events. It has been generally assumed that the incoming glycosyl residue is transferred from the sugar nucleotide donor substrate to the nonreducing end of the polysaccharide, although it has also been suggested that plant cell wall polysaccharide biosynthesis might involve addition of glycosyl residues to the reducing end of the elongating chain.

The iterative synthase enzymes are assumed to catalyze the polymerization not only of cellulose but also of the main chain of noncellulosic wall polysaccharides; these enzymes include the (1,4)-βmannan synthase that polymerizes the (1,4)-β-mannan backbone of mannans, galactomannans and galactoglucomannans and the (1,4)-β-glucan synthase that synthesizes the backbone of xyloglucans. These enzymes are encoded by members of the cellulose synthase (*CesA*) and cellulose synthase-like (*Csl*) gene superfamily and are members of the GT2 group of glycosyltransferases. The GAUT enzymes are also believed to be iterative glycosyltransferases that synthesize the (1,4)-β-galacturonan backbone of pectic polysaccharides, but the GAUT galacturonyl transferases belong to the GT8 group of enzymes, and their structures are likely to be quite distinct from the GT2 group involved in the synthesis of β-linked polysaccharide chains.

The second group of glycosyltransferases is believed to add single glycosyl substituents to the main chain residues or to add short oligosaccharide chains that are appended to the main chain through single catalytic encounters with the nascent polysaccharide chain. Examples of this group of glycosyltransferases include the arabinosyl transferases that add á-arabinofuranosyl or á-glucuronyl residues to the (1,4)-â-xylan backbone of arabinoxylans, the xylosyl transferases that add (1,6)-β-xylosyl residues to the (1,4)-β-glucan backbone of nascent xyloglucan molecules and the galactosyl transferases that add (1,6)-β-galactopyranosyl residues to the backbone of (1,4)-β-mannans or (1,4)-β-glucomannans.

Despite the best efforts of biochemists over many decades, there are still a number of fundamental questions about the biosynthetic process that we have not answered. The mechanism of chain initiation is not known, and potential primer molecules for wall polysaccharide biosynthesis have not been convincingly identified. Similarly, the mechanisms controlling chain termination and thus the overall length of the polysaccharide chain have not been elucidated. Do polysaccharide chains simply 'fall off' the synthase enzymes when they get to a certain length or when the enzyme itself begins to be turned over?

Do the hydrolytic enzymes that have been increasingly recognized in tissues in which net synthesis of wall polysaccharides is occurring hydrolyze the elongating chain and release the nascent polysaccharide from the synthase enzyme? Might decreasing concentrations of glycosyl donor substrates lead to termination of the chain? It was recently suggested that a terminal oligosaccharide detected at the reducing end of certain heteroxylans might be involved in chain termination. Whichever of these or other possibilities might apply, plant cell wall polysaccharide synthesis does not appear to be guided by a template, and it is difficult to envision how chain termination could be anything but an imprecise process.

It follows that the resultant polysaccharide chain length will be heterogeneous, as is usually observed. This is not to say, however, that plant cells do not have some control over wall polysaccharide chain length, because clear differences are apparent, and in some cases chain lengths fall within reasonably narrow limits.

A second potential level of imprecision is afforded by the glycosyltransferases that add the various glycosyl units to the backbone of the wall polysaccharide. We do not know whether the addition of the substituent is a precise process or not, insofar as there are examples of relatively regular spacing of substituents along the chain and other examples in which substituents are, in statistical terms, asymmetrically distributed. In the former case, one might expect that the backbone synthase and the substituting glycosyltransferase would have to operate as a highly coordinated enzyme complex.

On the other hand, if the main chain synthase and glycosyltransferase were operating independently, it is again difficult to see how main chain substitution could be a precise process with respect to the extent and spacing of added substituents. Equally plausible is the possibility that an enzyme complex could mediate the generation of a polysaccharide in which every main chain residue was substituted or in which every second or third main chain residue was substituted. This would result in a relatively homogeneous polysaccharide, but postsynthetic removal of substituents at regular intervals, as has been described for arabinoxylans, or asymmetrical postsynthetic modifications such as methyl esterification, as has been described for some pectic polysaccharides, could introduce heterogeneity into a polysaccharide that was initially homogeneous with respect to its fine structure.

Against this background of heterogeneity that will arise generally as a result of biosynthetic mechanisms, the specific types of heterogeneity actually observed in the major groups of noncellulosic wall polysaccharides are summarized in the sections below.

Alternative Chemistries for Noncellulosic Polysaccharides

Two major chemical strategies have been used by plants to match noncellulosic polysaccharide structures to their functional requirements as components of the gel-like matrix phase of walls. Structurally regular polysaccharides with extended conformations of the type commonly found in cellulose and related wall polysaccharides will tend to align and aggregate in aqueous media; the aggregates will be stabilized through the formation of intermolecular hydrogen bonding and hydrophobic interactions between sugar rings.

This will reduce their solubility and enhance their ability to form gels. So, how do plants introduce limited and possibly controllable irregularity into the noncellulosic wall polysaccharides to limit this tendency to aggregate? In most instances the structurally regular and conformationally extended backbone or main chain that would normally align and aggregate into fibrillar or insoluble complexes is prevented from doing so through the addition of single monosaccharide substituents or short oligosaccharides along the otherwise regular backbone chain. During gel formation in the wall, the presence of some structurally regular 'junction zones' along the polysaccharide chain facilitates localized intermolecular interactions, which, in combination with 'solubilizing' noninteracting regions of the chain, allows the polysaccharide to form a gel.

An example of this chemical strategy is embodied in the xyloglucans, which consist of a (1,4)-β-glucan main chain. These 'cellulosic' backbone chains do not aggregate because they are substituted with short oligosaccharide chains at the C(O)6 of the main chain (1,4)-β-glucosyl residues. Similarly, the arabinoxylans of the *Poaceae* consist of a (1,4)-β-xylan main chain that would normally be insoluble and fibrillar in nature but is instead substituted with α-arabinosyl residues.

There are also regions of the (1,4)-β-xylan backbone that carry no substituents; these will therefore be capable of intermolecular interactions that will result in limited junction zone formation and will allow gel formation in the wall. The chemical structures of galactomannans and pectic polysaccharides reflect the same strategy.

The second, quite different strategy to achieve wall polysaccharides with extended conformations but with limited tendencies to align and aggregate is exemplified by the relatively recently evolved (1,3;1,4)-β-glucans of the *Poaceae*. Like the xyloglucans, their structure is based upon a conformationally regular 'cellulosic' (1,4)-β-glucan backbone. However, in this alternative strategy, intermolecular aggregation is limited through the introduction of irregularly spaced (1,3)-β-linkages, which result in molecular 'kinks' in the otherwise 'cellulosic' backbone. Provided the (1,3)-β-linkages are spaced at irregular intervals, the asymmetric polysaccharide chains will not aggregate over extended regions.

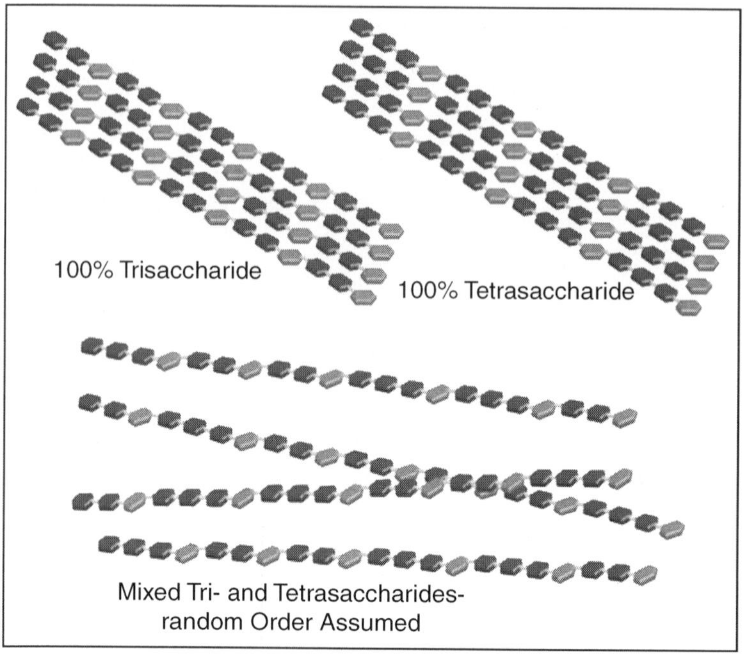

Fig. The Effects of (1,3;1,4)-β-glucan fine structure on aggregation and solubility.

The (1,3;1,4)-β-glucans of the *Poaceae* have structures based upon a conformationally regular 'cellulosic' (1,4)-β-glucan backbone. Intermolecular aggregation is limited through the introduction of irregularly spaced (1,3)-β-linkages, which result in molecular 'kinks' in the (1,4)-β-glucan backbone. The (1,3)-linked â-glucosyl residues are shown in red, whereas the (1,4)-â-glucosyl residues are shown in blue. Relatively high or relatively low ratios of cellotriosyl/cellotetraosyl units indicate a predominance of cellotriosyl or cellotetraosyl residues, respectively. If one or another of the oligosaccharide units predominates, the overall polysaccharide chain becomes more regular in shape and hence less soluble because of chain alignment. Random arrangement of a mixture of the trisaccharide and the tetrasaccharide units means that the chains will not aggregate over extended regions and can remain in aqueous solution, as depicted in the lower diagram. Such differences in physicochemical properties could be used to the plant's advantage in meeting different functional requirements in different cells or at different stages of development.

Heterogeneity in Pectic Polysaccharides

Pectins are a major component of the primary cell wall, particularly of dicotyledonous plants. Pectins contribute to the mechanical strength, porosity, adhesion and stiffness of the cell wall and have been implicated in many processes, including wall slippage, extension and intercellular signaling. In addition, pectic polysaccharides are a source of signaling molecules. Pectins are extremely complex molecules that include a number of structural classes, such as homogalacturonan, xylogalacturonan, apiogalacturonan and rhamnogalacturonan. The galacturonan chains can chelate calcium ions, apiosyl residues can coordinate borate ions, and there is good evidence that covalent linkages form between different structural classes of the pectic polysaccharide group. Methylated and acetylated residues are often found on pectin and indeed on several other polysaccharides and can profoundly influence their structures and physicochemical properties.

Considering this complexity, it is not unexpected that the level of heterogeneity in the pectic polysaccharides is high. Here we will explain just one example of how structural heterogeneity is generated and how this alters the physicochemical and functional properties of the cell wall matrix. A common mechanism facilitating local changes in the properties of pectic polysaccharides is the degree to which their polygalacturonate residues are methyl esterified. This involves heterogeneity both within a chain and between chains and can now be characterized analytically. Methyl esterification will neutralize the charge on galacturonyl residues and thereby abolish the ability of the negatively charged galacturonyl residues to interact with calcium ions. Specific monoclonal antibodies, which can differentially detect pectins with particular methylation profiles, show spatial heterogeneity around a single higher-plant cell with respect to the pattern of pectin esterification, which may be linked to cell adhesion.

There are a number of pectin methyl esterases that selectively remove methyl groups from homogalacturonan in either a blockwise or a nonblockwise fashion. Where methyl groups are removed in a blockwise manner, an exposed region of the polysaccharide that is able to form calcium-mediated cross-links is created; these cross-links contribute to the 'stickiness' required for cell adhesion. Where the removal of ester groups is nonblockwise, the homogalacturonan appears to be less 'sticky', and such junction zones cannot form so readily. As one might expect, this particular form of homogalacturonan is found in regions where cells separate.

Heterogeneity in Heteroxylans

Heteroxylans have a (1,4)-β-xylan backbone that can be substituted with single á-arabinofuranosyl, á-glucuronosyl or 4-methyl-á-glucuronosyl residues. These residues can be linked to the $C(O)_3$ or $C(O)_2$ atoms of the backbone xylopyranosyl residues, or in some cases the xylopyranosyl residues can be substituted at both $C(O)_2$ and $C(O)_3$. Short oligosaccharide substituents are also found on some heteroxylans and, in the grasses in particular, á-arabinofuranosyl substituents can be esterified with hydroxycinnamic acids such as ferulic acid and *p*-coumaric acid.

Two carbohydrate-binding modules and two monoclonal antibodies have been used to locate specific types of xylans both within intact cell walls and in solubilized wall material. Both classes of probe revealed distinct types of heteroxylans, which differ in both structure and charge and are present in both primary and secondary walls. Similarly, in the endosperm of the grasses, wall arabinoxylans digested with (1,4)-β-xylan endohydrolase release a population of oligomeric arabinoxylosides that provide an enzymatic fingerprint of the polysaccharide's chemical structure. Examination of the arabinoxylan fingerprint from wheat endosperm walls showed variation in the structure and amount of this polysaccharide both spatially and temporally through grain development. The results were confirmed through Fourier transform infrared microspectroscopy (FT-IR), which showed that arabinoxylans in developing grain contained more backbone xylosyl residues disubstituted with arabinosyl residues than did arabinoxylans in mature grain and that the structure of the arabinoxylan was dependent upon its position in either prismatic or central cells.

Heterogeneity in Xyloglucans

Xyloglucans are thought to be particularly important in the primary walls of many plants because of their ability to noncovalently cross-link cellulose microfibrils. Their structure is based on a cellulosic (1,4)-β-glucan backbone substituted at $C(O)_6$ with á-linked xylosyl residues, typically at about 70% of the backbone glucosyl residues. However, further substitution of xylosyl residues is required to prevent xyloglucan self-association, which interferes with binding to cellulose. A typical basic substructure involves xylosyl substituents on three consecutive glucosyl backbone residues, followed by an unsubstituted glucosyl residue. The availability of endoglucanase enzymes that depolymerize the xyloglucan chain at unsubstituted glucosyl residues allows the isolation of oligosaccharides for detailed analysis. A wide range of carbohydrate and selective acetyl substituents on the basic xyloglucan structure have been characterized in this way.

Most studies of partial glucanase digests are consistent with an essentially random arrangement of oligosaccharide substructures, but this may not always be the case. One promising new approach to characterizing the heterogeneity of xyloglucans is the use of mass spectrometry to define glucanase-derived oligosaccharides on very small sample sizes. The selectivity of specific endoglucanase digestion, coupled with the sensitivity of mass spectrometric detection, allows not only the profiling of small amounts of isolated xyloglucan polymers but also the profiling of xyloglucans present within specific cell types or in isolated organelle fractions.

Furthermore, *in situ* analysis of xyloglucan oligosaccharide substructures in walls can be carried out at high resolution by aiming the ionization laser of a MALDI-TOF mass spectrometer at desired locations within enzyme-treated tissues. Although mass spectral methods depend to some extent on prior knowledge of xyloglucan structure and cannot be easily applied to polysaccharides such as arabinoxylans, which consist only of pentose sugars, the method has potential to open up the dissection of location-specific heterogeneity in xyloglucan structures, especially if the method can be complemented with antibodies directed against an array of xyloglucan substructures.

Heterogeneity in (1,3;1,4)-β-glucans

The (1,3;1,4)-β-glucans, which are abundant in walls of the *Poaceae*, represent one example in which heterogeneity in fine structure appears to be essential for function in the wall and is therefore built into the (1,3;1,4)-β-glucan during biosynthesis. Detailed structural analyses of the (1,3;1,4)-β-glucans demonstrate that randomness can be recognized at one level, whereas other structural features of the molecule are nonrandom. How can this be?

The water-soluble (1,3;1,4)-β-glucan from barley endosperm consists of about 72% (1,4)-β-glucosyl residues and 28% (1,3)-β-glucosyl residues. The (1,3)-β-glucosyl residues are always flanked on both their reducing end and nonreducing end sides by (1,4)-β-glucosyl residues; two or more adjacent (1,3)-β-glucosyl residues are not found. In this respect the fine structure is nonrandom, because if the (1,3)-β-glucosyl residues and (1,4)-β-glucosyl residues were randomly distributed, on statistical grounds one would expect to find significant numbers of two or more adjacent (1,3)-β-glucosyl residues.

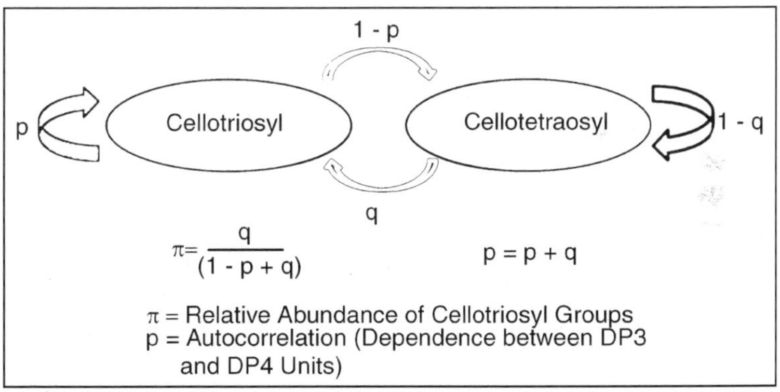

Fig. Fine Structural Analysis of the (1,3;1,4)-β-glucan from Barley.

The (1,4)-β-linkages of most (1,3;1,4)-β-glucans are arranged predominantly in groups of two or three adjacent linkages. In the water-soluble (1,3;1,4)-β-glucan from barley grain, this arrangement accounts for about 90% of the polysaccharide. The remaining 10% consists of longer blocks of from 4 up to 12◎–15 adjacent (1,4)-β-glucosyl residues. If we focus on the groups of two and three adjacent (1,4)-β-glucosyl linkages, (1,3;1,4)-β-glucans can be viewed as (1,3)-β-linked copolymers of cellotriosyl and cellotetraosyl residues in which the ratio of cellotriosyl (DP3) to cellotetraosyl (DP4) units varies from 1.5 to 4.5, depending on the source of the (1,3;1,4)-β-glucan. Markov chain analyses were used to investigate the arrangement of the cellotriosyl (DP3) to cellotetraosyl (DP4) units and showed that the cellotriosyl (DP3) to cellotetraosyl (DP4) units were arranged at random along the (1,3;1,4)-β-glucan chain.

The structure of a water-soluble barley (1,3;1,4)-β-glucan was investigated by treating it as a Markov chain. The polysaccharide consists mostly of cellotriosyl and cellotetraosyl units linked by (1,3)-β-linkages. This analysis was used to explore whether the addition of a particular unit oligosaccharide structure to the elongating chain depended on the adjacent unit oligosaccharide in the chain. Through controlled enzymatic hydrolysis, such that most of the products represented two units of the polysaccharide chain, it was possible to quantitate the amount of hexasaccharide (which consisted of two cellotriosyl units), heptasaccharide (which consisted of one cellotriosyl unit and one cellotetraosyl unit) and octasaccharide (which consisted of two cellotetraosyl units).

Applying this quantitative measure of near-neighbour analysis in the Markov chain would give a value for ρ of about 1.0 for blocks of adjacent cellotriosyl units and adjacent cellotetraosyl units, with values near –1.0 if the polysaccharide consisted of alternating cellotriosyl and cellotetraosyl units and values close to zero if the cellotriosyl and cellotetraosyl units were randomly arranged. The experimentally determined values for ρ for two different but related (1,3;1,4)-β-glucan preparations were –0.003 and 0.050. The data therefore indicated that the (1,3)-β-linked cellotriosyl and cellotetraosyl units were arranged essentially at random along the polysaccharide chain. p, the probability that a cellotriosyl unit is followed by another cellotriosyl unit. q, the probability that a cellotetraosyl unit is followed by a cellotriosyl unit.

The heterogeneity in fine structure imparted by this chemistry carries important consequences for the physicochemical and hence functional properties of the (1,3;1,4)-β-glucan. Thus, mid-level ratios of cellotriosyl/cellotetraosyl units, say 1.5:2.5, would meet the functional requirements of a porous, gel-like wall matrix, whereas much higher or lower ratios would characterize a more conformationally regular, less soluble (1,3;1,4)-β-glucan that would have increased capacity to aggregate with other (1,3;1,4)-β-glucan molecules and possibly with cellulose and other wall polysaccharides such as heteroxylans. The wide adoption of (1,3;1,4)-â-glucans in the *Poaceae* might be attributable to the ability of these species to synthesize (1,3;1,4)-β-glucans with mid-range ratios of cellotriosyl/cellotetraosyl units.

Biological models for the synthesis of the (1,3;1,4)-β-glucans must account for their fine structural features, three of which challenge traditional cellular and biochemical views of how wall polysaccharides are synthesized.

These are:

i. The occurrence of single, isolated (1,3)-β-glucosyl residues amongst the predominating (1,4)-β-glucosyl residues,

ii. The random arrangement of cellotriosyl units and cellotetraosyl units along the chain
iii. The incorporation of longer blocks of up to 12–15 adjacent (1,4)-β glucosyl residues in the molecule.

A two-phase assembly system has been proposed to account for the structural features of (1,3;1,4)-β-glucans. As illustrated diagrammatically in Figure, the first phase involves the assembly of a population of cellodextrins, which are linked to a Golgi membrane associated macromolecule. It is envisioned that a population of (1,4)-β-oligoglucosides is synthesized by the cellulose synthase–like CslF or CslH enzymes, some of which are located in the Golgi.

After transfer of the membrane-bound (1,4)-β-oligoglucosides to the plasma membrane, the second phase of the model suggests that the (1,4)-β-oligoglucosides are taken at random by another enzyme, possibly a callose synthase, a xyloglucan endotransglycosylase or even another CslF isoenzyme, and linked through (1,3)-β-linkages. The model is consistent with the cellular location of the final (1,3;1,4)-β-glucan polysaccharide. During extensive immunochemical examination of (1,3;1,4)-β-glucan synthesis in a range of different cell types, the polysaccharide could not be detected in the Golgi, even where its concentration in the wall was very high. The Golgi-located cellodextrins proposed to represent the first phase of (1,3;1,4)-β-glucan synthesis would not be detected by the (1,3;1,4)-β-glucan antibodies used. However, another group has recently disagreed and has concluded that some polymeric (1,3;1,4)-β-glucan is located in Golgi-derived vesicles of maize coleoptiles. We predict that these cell biological aspects of cell wall polysaccharide synthesis will remain the topic of debate for some time.

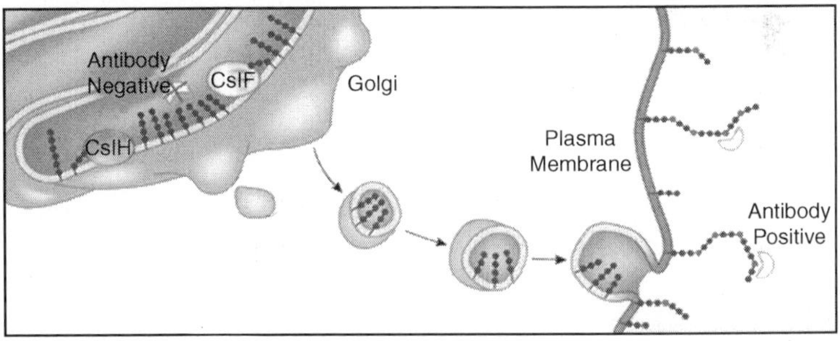

Fig. An Alternative, Two-phase Mechanism for Noncellulosic Polysaccharide Assembly.

Noncellulosic polysaccharides of the matrix phase in plant cell walls are generally considered to be synthesized in their entirety in the Golgi apparatus, from which they are transported via Golgi-derived vesicles to the plasma membrane. Fusion of the vesicles with the plasma membrane results in the deposition of the polysaccharides into the periplasmic space and their eventual incorporation into the wall. In an alternative theory, it has been proposed that oligosaccharides might be assembled in the Golgi, where they remain attached to a recyclable membrane lipid (or protein). This would be consistent with early but not generally accepted indications that membrane-associated sitosterol or other sterols might act as intermediates in cellulose synthesis. The oligosaccharides would subsequently be polymerized by a plasma membrane–bound transferase that would deposit the nascent polysaccharide chain directly into the wall.

Callose synthase and xyloglucan endotransglycosylases were suggested as candidate oligosaccharide polymerases. In this diagram, a model for the potential assembly of (1,3;1,4)-β-glucans is presented, through which a population of different-length cellodextrins are synthesized in the Golgi by CslF and CslH enzymes. These cellodextrins, which the (1,3;1,4)-β-glucan antibody does not recognize, would then be transported to the plasma membrane via Golgi-derived vesicles. Upon fusion with the plasma membrane, the cellodextrins are released to allow the second stage os assembly to occur where they are joined by (1,3)-β-linkages. This yields the final polysaccharide now recognized by the (1,3;1,4)-β-glucan antibodies. It has been postulated that an assembly system similar to this might also apply to other noncellulosic polysaccharides of plant cell wall.

Although the theory of two-phase assembly of wall polysaccharides in the Golgi and at the plasma membrane was developed initially to account for experimental data related not only to the cell biology of (1,3;1,4)-β-glucan biosynthesis but also to the inbuilt heterogeneity in their fine structure, the model might be equally applicable to the synthesis of other wall polysaccharides. Xyloglucans have a strong repetitive element in their structures and might be considered a polymer of xylosylated tetraglucoside units. Pectic polysaccharides contain some repeating elements, and extracellular arabinogalactan proteins are believed to be made up of repeating oligosaccharide units.

Heterogeneity in Mannans

The cell wall polymer family for which fine structure has been most precisely defined and reconciled with biosynthesis processes is the galactomannans, members of which have a (1,4)-β-mannan backbone along which some but not all 6-C(OH) positions are substituted with á-galactose. Mannose/galactose ratios vary from about 1:1 to as high as 20:1. On the basis of enzyme digestion and quantitative analysis of digestion-resistant oligosaccharides, it has been shown that second-order Markov chain statistics are sufficient to describe the fine structure of a range of galactomannans. For a processive biosynthesis mechanism in which galactosyl transferase activity is coupled with backbone mannan synthase, a second-order Markov chain is consistent with the galactomannan subsite recognition sequence of the galactosyltransferase encompassing at least three backbone mannosyl residues—that is, the site of substitution and the two preceding residues towards the reducing end of the growing galactomannan chain. Datasets from three different plant species generated three distinctly different sets of probabilities and thus different galactosyl-substitution patterns. Subsequently, heterologously expressed galactosyl transferases in transgenic tobacco resulted in the production of galactomannan with Markov probability values characteristic of the original source of the galactosyl transferase. The Markov statistics describing polymer structure also allow a precise prediction of the probability of unsubstituted mannan segments of defined length, and these have been shown to be of consequence in the formation of interpolymeric junction zones that can form the basis for three-dimensional gel-like matrices of the type thought to be present within plant cell walls.

A Potential Role for Hydrolases in Creating Heterogeneity

An emerging theme in cell wall biology is the presence of polysaccharide hydrolases in tissues in which walls are being deposited and there is net synthesis of constituent wall polysaccharides. One example is the presence of (1,4)-β-glucan endohydrolases (cellulases) such as Korrigan in tissues in which cellulose synthesis is occurring. The role of the cellulase has not been defined unequivocally, but it may be involved in the release of nascent cellulosic chains from the cellulose synthase enzyme complex, or it might be involved in the trimming of noncrystalline cellulose from the surface of microfibrils. The latter possibility is consistent with observations that cellulose crystallinity is decreased in *Arabidopsis* lines that carry a silencing mutation in the *kor* gene.

Molecular heterogeneity in heteroxylans might be generated by partial hydrolysis of the nascent polysaccharide rather than during the biosynthetic process itself. Newly synthesized arabinoxylans in maize and barley coleoptiles are usually highly substituted. Indeed, most if not all the xylosyl residues of the (1,4)-β-xylan backbone are initially substituted with arabinosyl residues. However, as the wall matures the degree of substitution decreases, and this has been attributed to the activity of arabinoxylan arabinofuranohydrolases, which can remove arabinosyl residues from the (1,4)-β-xylan backbone.

Irrefutable emerging evidence points to high levels of chemical, and hence structural, heterogeneity in the polysaccharide constituents of plant cell walls. The heterogeneity can be measured between walls in different parts of the same plant, between walls in a single tissue section and between regions within a single cell wall. In addressing the 'big questions' in plant cell wall biology in the immediate future, we will need to ensure that our explanations adequately rationalize heterogeneity as a crucial component of wall structure and function. Of the major unanswered questions facing cell wall biologists, the unequivocal identification of enzymes involved in the biosynthesis of the backbones of xyloglucans and heteroxylans is particularly pressing. Given the evidence for the participation of multienzyme complexes in cellulose biosynthesis, it will be important to investigate the highly likely possibility that the synthesis of noncellulosic wall polysaccharides also requires the coordinated action of several enzymes and other proteins. The latter question becomes even more intriguing in the case of substituted wall polysaccharides, in which enzymes involved in the synthesis of the backbone might form a complex with enzymes that simultaneously add monosaccharide or oligosaccharide substituents to the backbone. Might hydrolases also form a part of such complexes?

The recent observation that different isoforms of a single enzyme family can direct the synthesis of wall polysaccharides with different structures and thus different physicochemical properties raises the question of what the molecular mechanisms used by enzymes to synthesize wall polysaccharides in plants are. To thoroughly and accurately describe these mechanisms, three-dimensional structures of the synthases will be required, but at this stage there are no crystal structures of plant polysaccharide synthases of the GT2 group in the public databases.

The conventional view of the cellular steps involved in noncellulosic wall polysaccharide assembly has also been challenged, although proposed alternative pathways are supported largely by indirect data. It is likely that new technologies will enable us to investigate alternative models of wall polysaccharide assembly at the cellular level in the immediate future; the results of these investigations will be potentially transformational in terms of the way we view the cell biology of wall metabolism in plants.

Similarly, emerging *in situ* methods will enable us to define in more detail the nature of polysaccharide-polysaccharide, polysaccharide-protein and polysaccharide-lignin interactions in the wall, to define whether these interactions involve covalent or noncovalent linkages and to identify the enzyme systems responsible for mediating the interactions. It has been known for some years that oligosaccharides released from wall polysaccharides by the action of microbial enzymes can act as signal molecules and, in particular, signal a loss of cell wall integrity that elicits a range of plant defence responses to pathogen attack.

Given the role of the cell walls of both the microorganism and the plant as contact points during their first encounter, the question of cell wall integrity and oligosaccharide signaling will continue to be of importance in describing how plants resist biotic stress.

The structural complexity and heterogeneity of pectins are consistent with a role for these polysaccharides in these processes, but are there other polysaccharides in the walls of the *Poaceae*, where pectin levels in walls are often very low, that assume a similar role?

Finally, it is important to acknowledge the recent appearance of two major research drivers that we can predict will be focused in the immediate future on the key questions outlined above. First, massive investments are being directed by both the private and public sectors towards the use of crop and plant residues, which consist in large part of cell wall material, as biomass sources for bioethanol and other biofuels production. Second, plant cell walls are attracting intense interest as a major source of dietary fibre in human foods, as increased fibre intake has been shown to significantly reduce the risks of such widespread and serious diet-related conditions as type 2 diabetes, cardiovascular disease, colorectal cancer and diverticular disease. We see a precise definition of heterogeneity and the underlying mechanisms for achieving the broad range of physicochemical properties as central for the ongoing biotechnological use of plant cell walls and their constituent polysaccharides.

MITOCHONDRIA: POWER PLANT OF CELL

For thousands of years, men puzzled over the question of where body heat comes from. This heat was recognized as nearly synonymous with life itself–but what produced it? The source of the energy of man or beast was equally mysterious. In recent decades it has become clear that both heat and energy can be traced to some sausage-shaped organelles in our cells, the mitochondria. Of course the ultimate source of energy for all plants and animals is sunlight. But the sun's energy can be harnessed by plants, through photosynthesis and stored in molecules of carbohydrates.

When animals eat these carbohydrates and break them down to carbon dioxide and water, with the help of oxygen and an arsenal of enzymes, large amounts of energy become available. Animals immediately convert this energy into molecules of high-energy ATP (adenosine triphosphate) -- the universal currency of energy in living things. Excluding only the very first stages in carbohydrate breakdown, which are called glycolysis, the entire, complicated process of energy transfer to ATP takes place within the mitochondria.

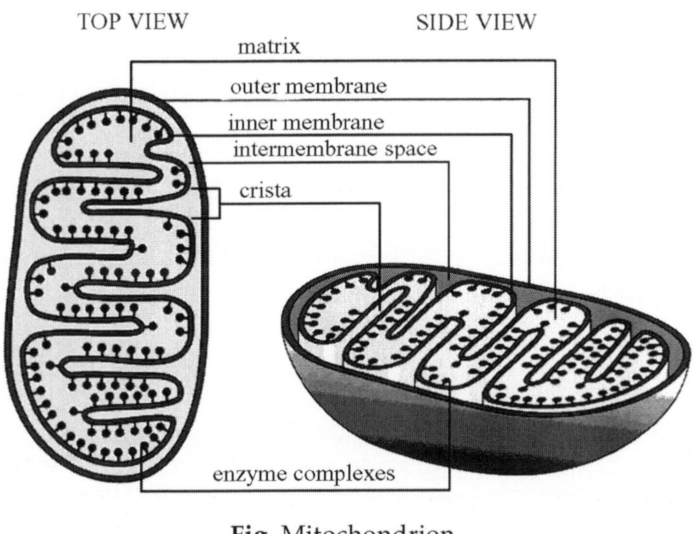

Fig. Mitochondrion.

Mitochondria are the largest organelles in the cell, after the nucleus, yet some cells have more than a thousand of them. Thin and long, they vary in diameter from 0.5 to 1 micrometer and in length up to 7 micrometers. Their thread-like outlines can be seen with a good light microscope.

Although they were first observed in 1850, it took a century to visualize their internal structure and understand their function. Actually the story goes back even further, to the 18th century, when Lavoisier showed that animals require oxygen for respiration. This led chemists to study cellular respiration, the process by which living cells use oxygen to release the chemical energy stored in foodstuffs.

Early in the 20th century biochemists discovered that these reactions fall into two major groups:

1 The carbon pathway, where a series of chemical reactions, each requiring a specific enzyme, breaks down the carbohydrates into carbon dioxide and hydrogen;
2 The hydrogen pathway, which transfers the hydrogen to oxyen in stages, forming water and releasing energy.

The whole process is organized much like an assembly line. In the hydrogen pathway, the hydrogen's electrons pass through an "electron transport chain" made up of enzymes that act as carriers and as the electrons move from enzyme to enzyme, they give up part of their energy, which is stored in molecules of ATP.

The ironcontaining enzymes, cytochrome a, b and c, carry out the final stages of the process. All in all, three molecules of ATP are formed for every atom of oxygen that is used up in respiration. For a long time, biochemists studied these reactions in cell extracts without worrying about what parts of the cells might be involved.

One scientist, B.F. Kingsbury of Cornell University, did suggest that mitochondria might be the site of cellular respiration back in 1912, but this was ignored. In the 1940's, using an ultracentrifuge, Albert Claude and his associates isolated various cell particles including mitochondria. Shortly afterwards, in 1948, Albert Lehninger of Johns Hopkins and Eugene Kennedy of Harvard showed that the reactions leading to the synthesis of ATP occurred in the mitochondria. In the early 1950's, Palade and the Swedish scientist Fritiof Sjostrand reported that mitochondria are bounded by a membrane and that they have a system of parallel, regularly spaced inner ridges which were named cristae.

It is now clear that there are, in fact, two membranes around the mitochondria of plants, protozoa, molluscs, insects and man: an outer membrane, set off by a fluidfilled gap; and an inner membrane which is folded inward at various points to increase its surface, forming the cristae, so that the enzyme molecules of the electron transport chain, which are attached to this inner surface in specific sequences, can be packed more densely side by side in the mitochondria.

This general design for mitochondria seems to have existed unchanged, through the evolution of plants and animals, for more than 1.2 billion years. Though mitochondria are generally visualized as ovalshaped, they can change shape quite drastically. They swell and contract in response to various hormones and drugs. Even in very low concentrations, Lehninger found, the thyroid hormone thyroxin will make mitochondria swell and will simultaneously "uncouple" the electron transport chain from the synthesis of ATP, so that even though carbohydrates may be metabolized and oxygen consumed, no ATP will be manufactured.

ATP, on the other hand, makes mitochondria contract. This swelling and contraction appear related to the movement of water in, out and through cells, which is of particular importance in the kidney. The more scientists learn about mitochondria, however, the more their curiosity is aroused, for these are without a doubt the strangest of the cell's organelles.

To begin with, mitochondria contain their own hereditary material (DNA), which resembles that of bacteria far more than that of animal cell nuclei. Because of this similarity, many scientists believe that mitochondria are derived from bacteria that infected a primitive cell and evolved into a useful symbiotic relationship with it. As Lewis Thomas puts it, "A good case can be made for our nonexistence as entities.

We are shared, rented, occupied. At the interior of our cells, driving them, providing the oxidative energy that sends us out for the improvement of each shining day, are the mitochondria and in a strict sense they are not ours. They turn out to be little separate creatures, the colonial posterity of migrant procaryotes, probably primitive bacteria that swam into ancestral precursors of our eucaryotic cells and stayed there.

Ever since, they have maintained themselves and their ways, replicating in their own fashion, privately, with their own DNA and RNA quite different from ours. They are as much symbionts as the rhizobial bacteria in the roots of beans. Without them, we would not move a muscle, drum a finger, think a thought. My mitochondria comprise a very large proportion of me. Looked at in this way, I could be taken for very large, motile colony of respiring bacteria, operating a complex system of nuclei, microtubules and neurons for the pleasure and sustenance of their families and running, at the moment, a typewriter, continues Thomas somewhat facetiously.I am obliged to do a great deal of essential work for my mitochondria. My nuclei code out the outer membranes of each and a good many of the enzymes attached to the cristae must be synthesized by me. Each of them, by all accounts, makes only enough of its own materials to get along on and the rest must come from me. And I am the one who has to do the worrying.

During the past few years, scientists have begun to learn which proteins and enzymes of the mitochondria are synthesized according to directions in the mitochondrial DNA and which are controlled by the genes in the nucleus. This information would be of considerable importance for the design of antibiotics and drugs, as well as for the understanding of genetic diseases. For example, researchers have found that chloramphenicol–a powerful antibiotic which can cause a fatal anemia–interferes with the synthesis of mitochondrial DNA, while other antibiotics do not. Mitochondria are just as important to the life of each cell as the heart is to the human body.

At least 95 percent of our cells' energy comes from mitochondria. (Only the mammalian red blood cells, which have no nuclei, get along without mitochondria; they obtain whatever energy they need from the partial breakdown of glucose in their cytoplasm). Growing cells continuously synthesize nucleic acids, proteins and other components, for which they need the energy that is stored in molecules of ATP.

Muscle cells which carry out mechanical work, plant cells which pump water against the force of gravity and many other cells also require a continuous source of energy. Besides supplying this energy, mitochondria apparently help to control the concentration of water, calcium and other ions in the cytoplasm.

They are also deeply involved in the breakdown and recycling of sugars, fatty acids and amino acids to recover their energy, as well as in the formation of urea as a waste product, to be excreted through the kidney. These strange organelles may thus hold clues to a variety of seemingly unrelated diseases. They have been linked to some liver diseases (*e.g.*, viral hepatitis, obstructive jaundice and cirrhosis of the liver); some muscle diseases (*e.g.*, familial myotrophic dystrophy); and some kidney diseases.

In all these conditions, pathologists have seen large amounts of materials pile up inside the mitochondria as "inclusions" of varying shapes and sizes, presumably waiting to be processed. What these inclusions consist of remains a mystery, but if scientists could find out, they might discover which specific components (for instance, enzymes) are involved in each disease. In addition, several calcium disorders may be related to mitochondrial defects. For example, abnormal concentrations of calcium (possibly caused by malfunctioning mitochondria) may lead to arthritis, bursitis, or arterial disease.

While all these leads are promising, the greatest spinoff from research on mitochondria may be in the area of energy–not only for the cell, but for general use by mankind. As the oil crisis has made us realise, we presently depend on a limited supply of fossil fuels–coal and oil which are of biological origin but cannot be replenished.

On the other hand, nature contains a far greater resource: a molecular know-how and technology gained over millennia, by which biological systems manufacture ATP from solar energy.

This technology awaits us. It exists in very similar, if not identical forms in mitochondria and in their mirror image, the chloroplasts, which are found in green plants, where they do the work of photosynthesis, converting the sun's energy into carbohydrates.

Like mitochondria, chloroplasts have an outer and an inner membrane. They have some of their own DNA. They, too, may have originated as some kind of infection, probably with blue-green algae. And it is on chloroplasts that we ultimately depend for nearly all our energy–that of our bodies as well as that which is stored in deposits of oil and coal. If scientists ever unravel the details of the process by which chloroplasts store solar energy in carbohydrates and mitochondria release it, we may be able to produce unlimited quantities of usable energy.

MICROTUBULES, THE CELL'S PHYSICAL PROPS

All the organelles described so far are bound by membranes. For a long time, the rest of the cytoplasm–a liquid called the cell sap, or ground substance–appeared totally unstructured. But as scientists learned to use newer and gentler fixatives to prepare cells for electron miscroscopy, more and more tiny structures materialized in this soup. The most prominent of the new structures are two organelles which provide an intricate system of physical support for the cell– the equivalent of buttresses or skeletal bones– as well as a contractile mechanism for cell movement: microtubules and microfilaments.

Neither is bound by a membrane. Microtubules were first noticed in the middle 1950's, but they were seen only rarely until the development of glutaraldehyde as a gentle fixative in 1963. Extremely thin cylinders, about 200 to 300 angstroms in diameter and of variable length, microtubules are constructed chiefly of proteins called tubulins. In 1967, Edwin Taylor and his associates at the University of Chicago discovered that colchicine–a chemical used to arrest cell division–did its work by binding to the protein tubulin.

This pointed to the role of microtubules in mitosis and helped to answer a question which had worried generations of biologists: How do chromosomes sort themselves out into two sets and separate during cell division? The details of mitosis would have been very hard to explain without understanding the chemistry of the spindle fibres which organize this separation. In fact, if there were no microtubules, scientists would have had to invent them.Something of this sort was needed to account not only for mitosis, but for the surprisingly firm structure of such seemingly vulnerable fibres as the long, thin axons of nerve cells, as well as for the unique structure of some red blood cells, which are held in a disc-like shape by hoops composed of microtubules. Microtubules are also involved in assisting the transport of substances in and out of the cell and in the motion of cells themselves.

They make up the cilia and flagella, whip-like filaments which project from the surface of cells and move rhythmically. Large numbers of cilia are found on cells that line the respiratory system, for instance, where they help to sweep out dust and debris. Both cilia and flagella play an important role in human reproduction: the coordinated beating of cilia in the oviduct produces a sort of current which draws the female's egg into the uterus, while the rapidly thrashing tail of sperm actually is a flagellum.

PLANT CELL MEMBRANE

The membrane that surrounds a cell is made up of proteins and lipids. Depending on the membrane's location and role in the body, lipids can make up anywhere from 20 to 80 per cent of the membrane, with the remainder being proteins. Cholesterol, which is not found in plant cells, is a type of lipid that helps stiffen the membrane.

You may not remember it, but you crossed a membrane to get in here. Every cell is contained within a membrane punctuated with special gates, channels, and pumps. These gadgets let in—or force out—selected molecules. Their purpose is to carefully protect the cell's internal environment, a thick brew (called the cytosol) of salts, nutrients, and proteins that accounts for about 50 per cent of the cell's volume (organelles make up the rest).

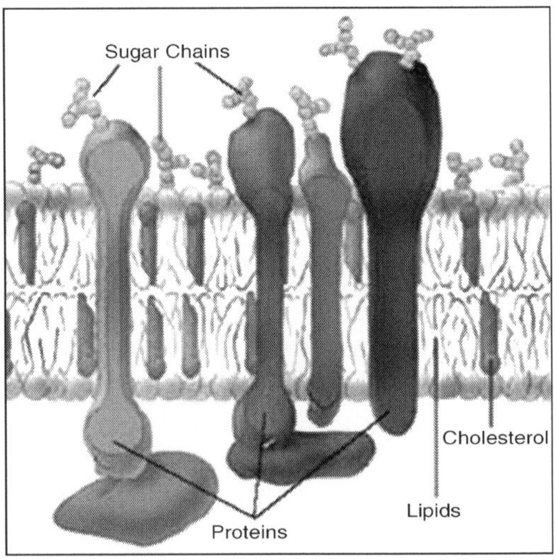

The cell's outer membrane is made up of a mix of proteins and lipids (fats). Lipids give membranes their flexibility. Proteins transmit chemical messages into the cell, and they also monitor and maintain the cell's chemical climate. On the outside of cell membranes, attached to some of the proteins and lipids, are chains of sugar molecules that help each cell type do its job. If you tried to bounce on the cell's outer surface as you did on the nuclear membrane, all these sugar molecules and protruding proteins would make it rather tricky (and sticky).

ENDOPLASMIC RETICULUM

If you peer over the side of the nucleus, you'll notice groups of enormous, interconnected sacs snuggling close by. Each sac is only a few inches across but can extend to lengths of 100 feet or more. This network of sacs, the endoplasmic reticulum (ER), often makes up more than 10 per cent of a cell's total volume.

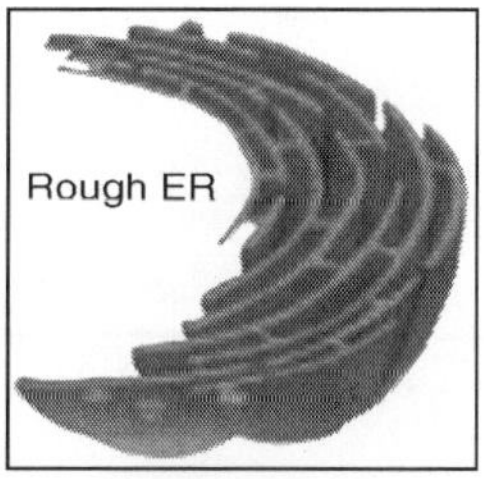

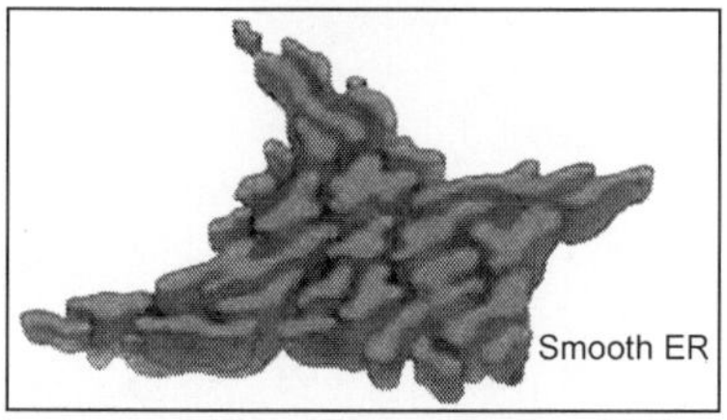

The endoplasmic reticulum comes in two types: Rough ER is covered with ribosomes and prepares newly made proteins; smooth ER specializes in making lipids and breaking down toxic molecules.Take a closer look, and you'll see that the sacs are covered with bumps about 2 inches wide. Those bumps, called ribosome's, are sophisticated molecular machines made up of more than 70 proteins and 4 strands of RNA, a chemical relative of DNA. Ribosomes have a critical job: assembling all the cell's proteins. Without ribosomes, life as we know it would cease to exist. To make a protein, ribosomes weld together chemical building blocks one by one. As naked, infant protein chains begin to curl out of ribosomes, they thread directly into the ER. There, hard–working enzymes clothe them with specialized strands of sugars.

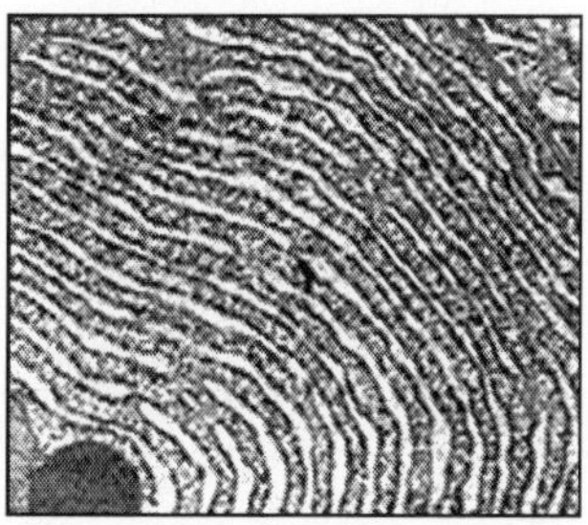

Fig. Rough ER

Now, climb off the nucleus and out onto the ER. As you venture farther from the nucleus, you'll notice the ribosomes start to thin out. Be careful! Those ribosomes serve as nice hand—and footholds now. But as they become scarce or disappear, you could slide into the smooth ER, unable to climb out.

In addition to having few or no ribosomes, the smooth ER has a different shape and function than the ribosome–studded rough ER. A labyrinth of branched tubules, the smooth ER specializes in synthesizing lipids and also contains enzymes that break down harmful substances. Most cell types have very little smooth ER, but some cells—like those in the liver, which are responsible for neutralizing toxins—contain lots of it.

Next, look out into the cytosol. Do you see some free–floating ribosomes? The proteins made on those ribosomes stay in the cytosol. In contrast, proteins made on the rough ER's ribosomes end up in other organelles or are sent out of the cell to function elsewhere in the body. A few examples of proteins that leave the cell (called secreted proteins) are antibodies, insulin, digestive enzymes, and many hormones.

RX: RIBOSOME BLOCKERS

In a dramatic technical feat, scientists obtained the first structural snapshot of an entire ribosome in 1999. This more recent image captures a bacterial ribosome in the act of making a protein (the long, straight spiral in the lightest shade of blue). It also shows that—unlike typical cellular machines, which are clusters of proteins (shown here as purple ribbons)–ribosomes are composed mostly of RNA (the large, light blue and grey loopy ladders). Detailed studies of ribosomal structures could lead to improved antibiotic medicines.

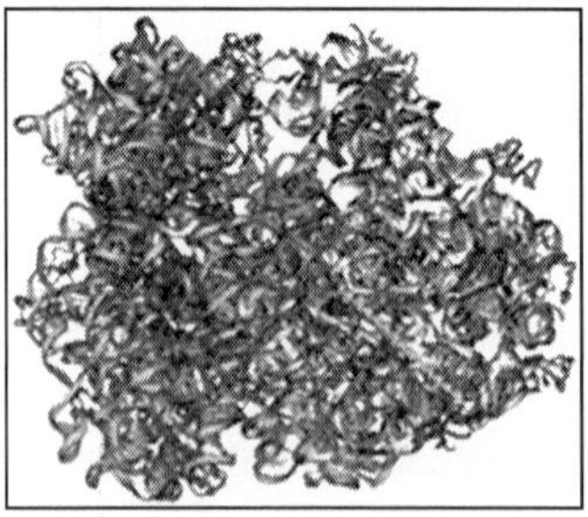

All cellular organisms, including bacteria, have ribosomes. And all ribosomes are composed of proteins and ribosomal RNA. But the precise shapes of these biological machines differ in several very specific ways between humans and bacteria. That's a good thing for researchers trying to develop bacteria–killing medicines called antibiotics because it means that scientists may be able to devise therapies that knock out bacterial ribosomes (and the bacteria along with them) without affecting the human hosts.

Several antibiotic medicines currently on the market work by inhibiting the ribosomes of bacteria that cause infections. Because many microorganisms have developed resistance to these medicines, we urgently need new antibiotics to replace those that are no longer effective in fighting disease. Using sophisticated imaging techniques like X–ray crystallography, researchers have snapped molecular pictures of antibiotics in the act of grabbing onto a bacterial ribosome.

Studying these three-dimensional images in detail gives scientists new ideas about how to custom design molecules that grip bacterial ribosomes even more strongly. Such molecules may lead to the development of new and more effective antibiotic drugs.

GOLGI

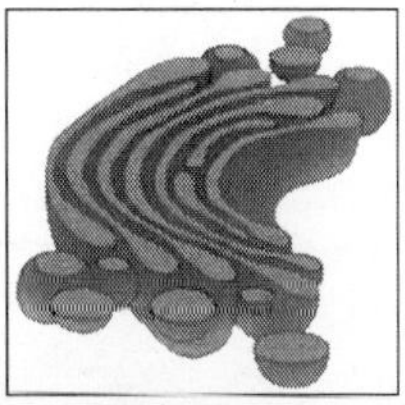

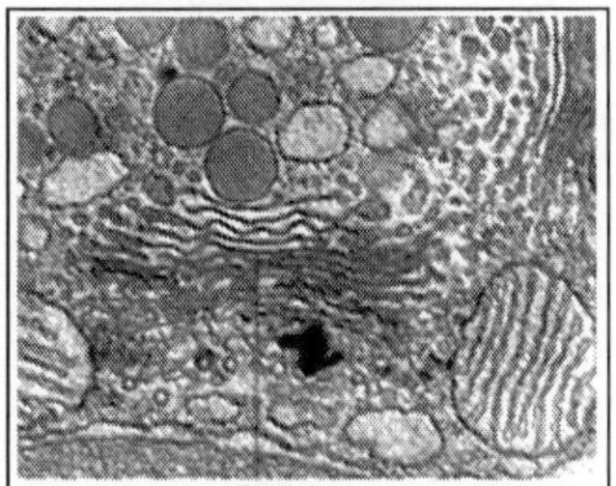

Golgi

TINA CARVALHO

Now, let's slog through the cytosol a bit. Notice that stack of a half dozen flattened balloons, each a few inches across and about 2 feet long?

That's the Golgi complex, also called the Golgi apparatus or, simply, the Golgi. Like an upscale gift shop that monograms, wraps, and mails its merchandise, the Golgi receives newly made proteins and lipids from the ER, puts the finishing touches on them, addresses them, and sends them to their final destinations. One of the places these molecules can end up is in lysosomes.

LYSOSOMES

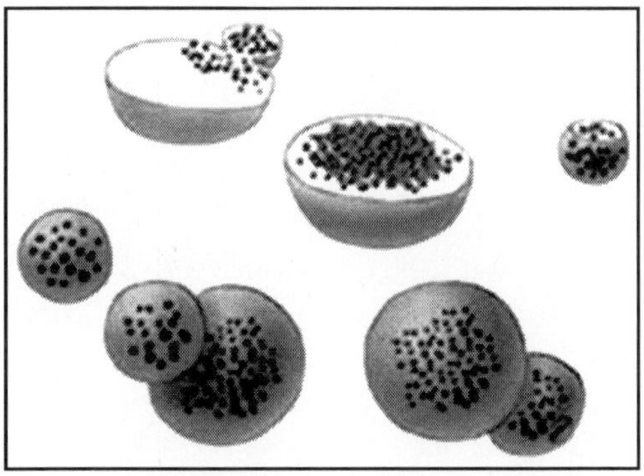

See that bubble about 10 feet across? That's a lysosome. Let's go—I think you'll like this. Perhaps even more than other organelles, lysosomes can vary widely in size—from 5 inches to 30 feet across.

Go ahead, put your ear next to it. Hear the sizzling and gurgling? That's the sound of powerful enzymes and acids chewing to bits anything that ends up inside.

But materials aren't just melted into oblivion in the lysosome. Instead, they are precisely chipped into their component parts, almost all of which the cell recycles as nutrients or building blocks. Lysosomes also act as cellular garbage trucks, hauling away unusable waste and dumping it outside the cell. From there, the body has various ways of getting rid of it.

CELLULAR POWER OF PLANTS BIOLOGY

Blink. Breathe. Wiggle your toes. These subtle movements—as well as the many chemical reactions that take place inside organelles—require vast amounts of cellular energy. The main energy source in your body is a small molecule called ATP, for adenosine triphosphate.

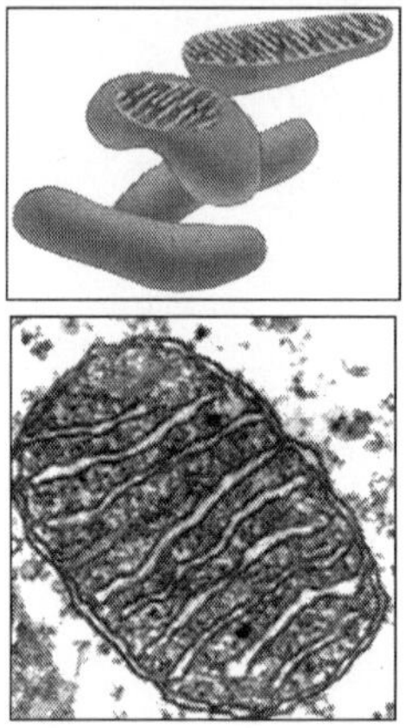

Fig. Mitochondrion

ATP is made in organelles called mitochondria. Let's see if we can find some. They look like blimps about as long as pickup trucks but somewhat narrower. Oh, a few of them are over there. As we get nearer, you may hear a low whirring or humming sound, similar to that made by a power station. It's no coincidence. Just as power plants convert energy from fossil fuels or hydroelectric dams into electricity, mitochondria convert energy from your food into ATP.

Like all other organelles, mitochondria are encased in an outer membrane. But they also have an inner membrane. Remarkably, this inner membrane is four or five times larger than the outer membrane. So, to fit inside the organelle, it doubles over in many places, extending long, fingerlike folds into the center of the organelle. These folds serve an important function: They dramatically increase the surface area available to the cell machinery that makes ATP. In other words, they vastly increase the ATP–production capacity of mitochondria. The mazelike space inside mitochondria is filled with a strong brew of hundreds of enzymes, DNA (mitochondria are the only organelles to have their own genetic material), special mitochondrial ribosomes, and other molecules necessary to turn on mitochondrial genes.

	Actual Size (Average)	Perceived Size when Magnified 3 Million Times
Cell diameter	30 micrometers	300 feet
Nucleus diameter	5 micrometers	50 feet
Mitochondrion length	Typically 1–2 micrometers, but can be up to 7 micrometers long	18 feet
Lysosome diameter	50–3,000 nanometers	5 inches to 30 feet
Ribosome diameter	20–30 nanometers	2–3 inches
Microtubule width	25 nanometers	3 inches
Intermediate filament width	10 nanometers	1.2 inches
Actin filament width	5–9 nanometers	0.5–1 inch

CYTOSKELETON: THE CELL'S SKELETON

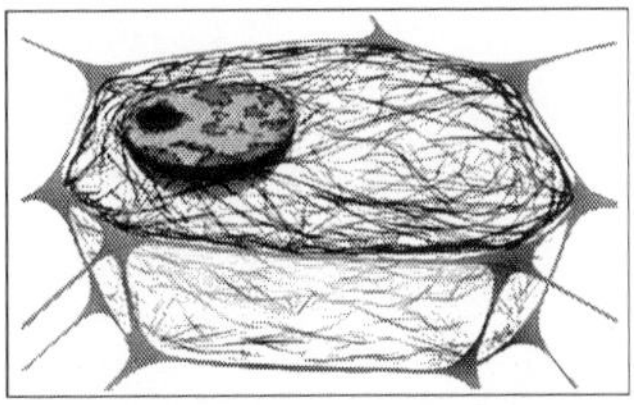

Now, about all those pipes, ropes, and rods you've been bumping into. Together, they are called the cytoskeleton—the cell's skeleton. Like the bony skeletons that give us stability, the cytoskeleton gives our cells shape, strength, and the ability to move, but it does much more than that.

Think about your own cells for a moment. Right now, some of your cells are splitting in half, moving, or changing shape. If you are a man, your sperm use long tails called flagella to swim. If you are a woman, hairlike fibres called cilia sweep newly released eggs from your ovaries into your uterus. And all that is thanks to the cytoskeleton. As you can see, the cytoskeleton is incredibly versatile. It is made up of three types of fibres that constantly shrink and grow to meet the needs of the cell: microtubules, intermediate filaments, and actin filaments. Each type of fibre looks, feels, and functions differently.

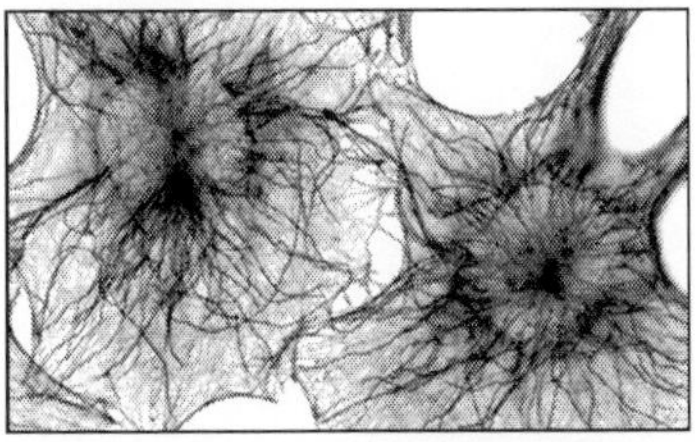

The 3-inch-wide flexible pipes you just banged your head on are called microtubules. Made of the strong protein tubulin, microtubules are the heavy lifters of the cytoskeleton. They do the tough physical labour of separating duplicate chromosomes when cells copy themselves and serve as sturdy railway tracks on which countless molecules and materials shuttle to and fro. They also hold the ER and Golgi neatly in stacks and form the main component of flagella and cilia. Grab one of those inch–thick ropes. Yeah, you can swing on it—it won't snap.

These strands, called intermediate filaments, are unusual because they vary greatly according to their location and function in the body. For example, some intermediate filaments form tough coverings, such as in nails, hair, and the outer layer of skin (not to mention animal claws and scales). Others are found in nerve cells, muscle cells, the heart, and internal organs. In each of these tissues, the filaments are made of different proteins. So if doctors analyse intermediate filaments in tumors, they can determine the origin of—and possible treatments for—some kinds of cancer.

See that bundle of long rods near the edge of the cell? You can touch it, but don't try to bend the rods. They shatter easily. These rods, slightly thinner than intermediate filaments, are actin filaments. They are made up of two chains of the protein actin twisted together. Although actin filaments are the most brittle of the cytoskeletal fibres, they are also the most versatile in terms of the shapes they can take. They can gather together into bundles, weblike networks, or even three–dimensional gels. They shorten or lengthen to allow cells to move and change shape. Together with a protein partner called myosin, actin filaments make possible the muscle contractions necessary for everything from your action on a sports field to the automatic beating of your heart.

CELL GROWTH AND CELL DIVISION

Growth is defined as an irreversible increase in mass that is typically associated with an increase in volume. Plant cell growth is associated with meristems and must be carefully regulated in order for organogensis and histogenesis to occur in the appropriate patterns. The plant regulates growth by regulating the extensibility of its cell walls.

A cell that has nonextensible cells walls can take up some water, but eventually the physical pressure of the water inside the cell pressing out on the cell wall (the turgor pressure) prevents the entry of additional water and any further change in volume. In contrast, a cell that has extensible cell walls can take up a substantial volume of water and thus increase in size.

Turgor pressure that would otherwise prevent water entry momentarily decreases because the walls keep stretching.

Typically cell growth occurs in small increments:

- Wall extensibility increases, reducing turgor pressure;
- Reduced turgor pressure allows water to enter the cell, increasing cell volume;
- Wall extensibility decreases, allowing the cell to build up turgor and preventing further water entry

- The cell undergoes a cycle of synthesis of cytoplasmic and wall components, adding to the cell's mass. This cycle of incremental growth is repeated many times until the cell reaches its final size.

The plant hormones auxin and gibberellin are produced in the vicinity of the apical meristems and usually act in concert to induce cell growth. Both hormones regulate wall extensibility, but carry out this function in different ways. Auxin induces the activity of cell membrane H^+ adenosine triphosphatase (ATPase) molecules.

Proton (H^+) extrusion lowers the pH of the cell wall, thus activating the cell wall enzyme expansin. Expansin cleaves the hydrogen bonds between two cell wall components:

The cellulose microfibrils and the hemicellulose molecules that link adjacent cellulose microfibrils. Breakage of these bonds allows these structural wall components to reposition themselves farther apart, increasing wall extensibility.

Gibberellic acid, on the other hand, stimulates the activity of another cell wall enzyme called xyloglucan endotrans-glycosylase (XET). Xyloglucans are a type of hemicellulose that is cleaved by the XET enzyme.

The hormones auxin and gibberellin each promote cell expansion by loosening the bonds between adjacent cell wall molecules. Each hormone acts on a different molecular target. Hemicellulose molecules also allows the cellulose microfibrils to move farther apart, increasing wall extensibility. Cell division and cell growth are often tightly linked.

When the rate of cell division is balanced by cell growth, as in the apical meristems, average cell size does not increase. As the meristem grows away from earlier formed cells, the ratio of growth to division increases, resulting in overall cell enlargement. As the tissues mature further, cell division ceases completely, giving rise to zones of pure cell enlargement where most of the visible growth of the plant occurs.

This relationship between division and growth, coupled with observations of the predictable planes of cell division during histogenesis, indicates that cell division is carefully regulated during plant development. Molecules called cyclin-dependent kinases (CDKs) are key regulators of cell cycling (including cell division) in plants. CDKs are activated by association with a regulatory subunit called a cyclin and by phosphorylation and dephosphorylation events.

The plant hormone cytokinin appears to regulate the cell cycle by interacting with the CDKs. Cytokinins enhance the synthesis of the cyclin subunits that are required for the cell to enter the deoxyribonucleic acid (DNA) synthesis phase of the cell cycle.

Cytokinins also enhance the CDK dephosphorylation step that is required for the cell to progress into mitosis. Both of these processes are inhibited by the hormone abscisic acid; thus a "developmental tug-of-war" occurs between a division-enhancing hormone and a division-suppressing hormone. The delicate balance between them determines the rate of cell division and this type of interaction is probably typical of the hormonal regulation of many aspects of plant development.

Differentiation

Differentiation is the process whereby cells, tissues, and organs become different from each other and from their precursors. The concept can be applied to organogenesis since cotyledons, foliage leaves, sepals, and petals may all develop from similar appearing precursors, the leaf primordia.

As these organs mature, they become different from each other in size, shape, and the development of distinctive cell types. For instance, the epidermis tissue of petals is sharply differentiated from that of cotyledons, foliage leaves and sepals that are photosynthetic organs. Correlated with a photosynthetic function, the epidermis of these organs is made up of flat, transparent cells that allow the penetration of light into internal tissues.

Specialized guard cells that allow CO_2 to enter the leaf are also present. In contrast, the epidermal cells of petals contain brightly coloured carotenoid or anthocyanin pigments. These cells also have a papillate shape that imparts a velvetlike sheen to the petal surface. Since petals carry on minimal photosynthesis, they often lack guard cells. The process of differentiation is best understood on a cellular level. For instance, guard cells are highly specialized epidermal cells. Early in the development of a leaf, protodermal precursor cells undergo a distinctive pattern of cell divisions. At first the cell divisions are asymmetric, producing one large and one small derivative. The large derivative stops dividing and differentiates as an unspecialized epidermal cell, while the small derivative undergoes another asymmetric division. At an unknown stop signal, the small derivative undergoes a symmetric division, giving rise to two equal sized cells that become the guard cells. Unlike their plain neighbours, these cells develop a distinctive kidney shape, unevenly thickened cell walls, large, conspicuous chloroplasts, and finally form a pore (the stomatal aperture) between them.

UNIQUENESS OF PLANT DEVELOPMENT

Although plants share many features of development with animals such as apical-basal polarity, regulation of the balance between cell growth and cell division, formation of distinctive patterns of organs, cells and tissues, and differentiation, some aspects of development are unique to plants.

Among these are:

- The formation and maintenance of the perpetually embryonic regions, the apical meristems. The meristems have an indeterminate growth pattern that result in the occurrence of growth, organogenesis, and histogenesis throughout the life of the plant.
- Plant cells have rigid cell walls that prevent cell movement. Thus organogenesis and histogenesis must occur through differential growth and regulation of the planes of cell division. Cell-cell

communication is important in plant development, but cell recognition is likely less important than it is in animals since plant cells keep the same neighbours throughout their life.
- Plant cells are totipotent; that is, able to differentiate as a different cell type if given the appropriate stimulus. Totipotency is likely a reflection of the plant's sedentary lifestyle. Plants can't escape predators and other kinds of damage, but they can readily repair wounds and reconnect vascular strands by differentiating the appropriate cell types.

FERTILIZATION

Fertilization, the process in which gametes—a male's sperm and a female's egg or ovum—fuse together, producing a single cell that develops into an adult organism. Fertilization occurs in both plants and animals that reproduce sexually—that is, when a male and a female are needed to produce an offspring.

For information on plant fertilization. Fertilization is a precise period in the reproductive process. It begins when the sperm contacts the outer surface of the egg and it ends when the sperm's nucleus fuses with the egg's nucleus. Fertilization is not instantaneous—it may take 30 minutes in sea urchins and up to several hours in mammals. After nuclear fusion, the fertilized egg is called a zygote.

When the zygote divides to a two-cell stage, it is called an embryo. Fertilization is necessary to produce a single cell that contains a full complement of genes. When a cell undergoes meiosis, gametes are formed—a sperm cell or an egg cell. Each gamete contains only half the genetic material of the original cell. During sperm and egg fusion in fertilization, the full amount of genetic material is restored: half contributed by the male parent and half contributed by the female. In humans, there are 46 chromosomes (carriers of genetic material) in each human body cell—except in the sperm and egg, which each have 23 chromosomes.

As soon as fertilization is complete, the zygote that is formed has a complete set of 46 chromosomes containing genetic information from both parents. The fertilization process also activates cell division. Without activation from the sperm, an egg typically remains dormant and soon dies. In general, it is fertilization that sets the egg on an irreversible pathway of cell division and embryo development.

The Fertilization Process

Fertilization is complete when the sperm's nucleus fuses with the egg's nucleus. Researchers have identified several specific steps in this process. The first step is the sperm approaching the egg. In some organisms, sperm just swim randomly toward the egg (or eggs). In others, the eggs secrete a chemical substance that attracts the sperm toward the eggs.

In one species of sea urchin (an aquatic animal often used in fertilization research), the sperm swim toward a small protein molecule in the egg's protective outer layer, or surface coat. In humans there is evidence that sperm are attracted to the fluid surrounding the egg.

The second step of fertilization is the attachment of several sperm to the egg's surface scoat. All animal eggs have surface coats, which are variously named the vitelline envelope (in abalone and frogs) or the zona pellucida (in mammals).

This attachment step may last for just a few seconds or for several minutes. The third step is a complex process in which the sperm penetrate the egg's surface coat. The head, or front end, of the sperm of almost all animals except fish contains an acrosome, a membrane-enclosed compartment. The acrosome releases proteins that dissolve the surface coat of an egg of the same species. In mammals, a molecule of the egg's surface coat triggers the sperm's acrosome to explosively release its contents onto the surface coat, where the proteins dissolve a tiny hole.

A single sperm is then able to make a slitlike channel in the surface coat, through which it swims to reach the egg's cell membrane. In fish eggs that do not have acrosomes, specialized channels, called micropyles, enable a single sperm to swim down through the egg's surface coat to reach the cell membrane.

When more than one sperm enters the egg, the resulting zygote typically develops abnormally. The next step in fertilization—the fusion of sperm and egg cell membranes—is poorly understood. When the membranes fuse, a single sperm and the egg become one cell. This process takes only seconds, and it is directly observable by researchers.

Specific proteins on the surface of the sperm appear to induce this fusion process, but the exact mechanism is not yet known. After fusion of the cell membranes the sperm is motionless. The egg extends cytoplasmic fingers to surround the sperm and pull it into the egg's cytoplasm.

Filaments called microtubules begin to grow from the inner surface of the egg cell's membrane inward toward the cell's centre, resembling spokes of a bicycle wheel growing from the rim inward toward the wheel's hub. As the microtubules grow, the sperm and egg nuclei are pushed toward the egg's centre. Finally, in a process that is also poorly understood, the egg and sperm nuclear envelopes (outer membranes) fuse, permitting the chromosomes from the egg and sperm to mix within a common space. A zygote is formed, and development of an embryo begins.

Types of Fertilization

Two types of fertilization occur in animals: external and internal. In external fertilization the egg and sperm come together outside of the parents' bodies. Animals such as sea urchins, starfish, clams, mussels, frogs, corals, and many fish reproduce in this way. The gametes are released, or spawned, by the adults into the ocean or a pond. Fertilization takes place in this watery environment, where embryos start to develop. A disadvantage to external fertilization is that the meeting of egg and sperm is somewhat left to chance.

Swift water currents, water temperature changes, predators, and a variety of other interruptions can prevent fertilization from occurring. A number of adaptations help ensure that offspring will successfully be produced. The most important adaptation is the production of litreally millions of sperm and eggs—if even a tiny fraction of these gametes survive to become zygotes, many offspring will still result. Males and females also use behavioural clues, chemical signals, or other stimuli to coordinate spawning so that sperm and eggs appear in the water at the same time and in the same place.

In animals that use external fertilization, there is no parental care for the developing embryos. Instead, the eggs of these animals contain a food supply in the form of a yolk that nourishes the embryos until they hatch and are able to feed on their own. Internal fertilization takes place inside the female's body.

The male typically has a penis or other structure that delivers sperm into the female's reproductive tract. All mammals, reptiles, and birds as well as some invertebrates, including snails, worms, and insects, use internal fertilization. Internal fertilization does not necessarily require that the developing embryo remains inside the female's body.

In honey bees, the queen bee deposits the fertilized eggs into special compartments in the honeycomb. These compartments are supplied with food resources for the young bees to use as they develop. Various adaptations have evolved in the reproductive process of internal-fertilizing organisms.

Because the sperm and egg are always protected inside the male's and female's bodies—and are deliberately placed into close contact during mating—relatively few sperm and eggs are produced. Many animals in this group provide extensive parental care of their young.

In most mammals, including humans, two specialized structures in the female's body further help to protect and nourish the developing embryo. One is the uterus, which is the cushioned chamber where the embryo matures before birth; the other is the placenta, which is a blood-rich organ that supplies nutrients to the embryo and also removes its wastes.

Research Issues

Although reproduction is well studied in many kinds of organisms, fertilization is one of the least understood of all fundamental biological processes. Our knowledge of this fascinating topic has been vastly improved by many recent discoveries. Researchers have discovered how to clone the genes that direct the fertilization process. Yet many important questions still remain.

Scientists are actively trying to determine issues such as how sperm and egg cells recognize that they are from the same species; what molecules sperm use to attach to egg coats; and how signals on the sperm's surface are relayed inside to trigger the acrosome reaction. With continued study, answers to these questions will one day be known.

GAMETE

Gamete, sexual reproductive cell that fuses with another sexual cell in the process of fertilization. The cell resulting from the union of two gametes is called a zygote; the zygote usually undergoes a series of cell divisions until it develops into a complete organism. Gametes, also called germ cells, vary widely in structure. The simplest sexual organisms are isogamous, that is, they produce a single kind of gamete.

The identical gametes unite in pairs to produce zygotes. Although all isogametes are apparently alike in structure, they are thought to be different in physiological constitution, because gametes from the same individual do not successfully unite. The simplest isogametes, those of lower fungi such as molds, are small cells that grow on the ends of body filaments and become detached when mature.

Other lower organisms, such as lower algae and protozoa, have gametes, which are formed by division of the protoplasm of single cells. All higher plants are heterogamous, that is, they produce two kinds of gametes. The female gamete is called the egg; the male gamete is called the sperm.

The organ of gamete production in plants is called a gametangium. All animals and animal-like lower organisms that reproduce sexually, except a few protozoans, are also heterogamous. The male gametes are called spermatozoa; female gametes, ova or eggs. The gamete-producing organs of animals are called gonads.

The formation of gametes in the gonads of animals is called gametogenesis. By this process the number of chromosomes in the sex cells is reduced in number from diploid to haploid, which is half the number of chromosomes in the normal body cells of the species. The diploid number of human chromosomes. When a human sex cell divides to form two gametes, each gamete receives only half, or 23, of the normal complement of chromosomes. This type of cell division is called meiosis. The normal total of chromosomes is restored in fertilization when two gametes fuse, each contributing half of the chromosomes required by the zygote.

TOOLS FOR STUDYING IN PLANT CELLS

Cell biologists would love to do what you just did—shrink down and actually see, touch, and hear the inner workings of cells. Because that's impossible, they've developed an ever–growing collection of approaches to study cellular innards from the outside. Among them are biochemistry, physical analysis, microscopy, computer analysis, and molecular genetics.

Using these techniques, researchers can exhaustively inventory the individual molecular bits and pieces that make up cells, eavesdrop on cellular communication, and spy on cells as they adapt to changing environments. Together, the approaches provide vivid details about how cells work together in the body's organs and tissues. We'll start by discussing the traditional tools of the trade—microscopes—then touch on the new frontiers of quantum dots and computational biology.

LIGHT MICROSCOPES: THE FIRST WINDOWS INTO CELLS

Scientists first saw cells by using traditional light microscopes. In fact, it was, looking through a microscope at a thin slice of cork, who coined the word "cell." He chose the word to describe the boxlike holes in the plant cells because they reminded him of the cells of a monastery.Scientists gradually got better at grinding glass into lenses and at whipping up chemicals to selectively stain cellular parts so they could see them better. By the late 1800s, biologists already had identified some of the largest organelles (the nucleus, mitochondria, and Golgi).Researchers using high–tech light microscopes and glowing molecular labels can now watch biological processes in real time. The scientists start by chemically attaching a fluorescent dye or protein to a molecule that interests them.

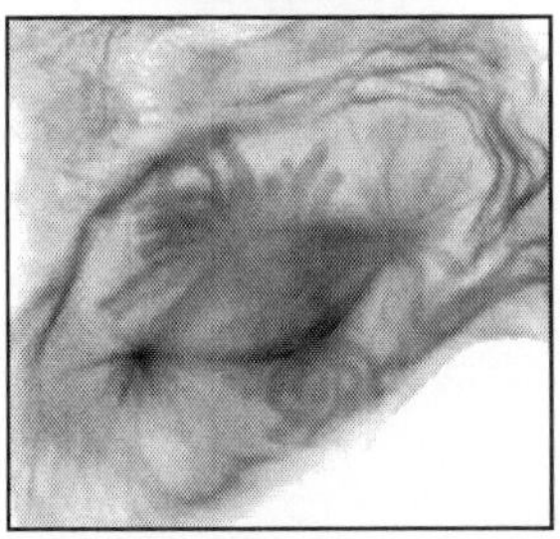

The coloured glow then allows the scientists to locate the molecules in living cells and to track processes—such as cell movement, division, or infection—that involve the molecules.Fluorescent labels come in many colours, including brilliant red, magenta, yellow, green, and blue. By using a collection of them at the same time, researchers can label multiple structures inside a cell and can track several processes at once. The technicolour result provides great insight into living cells—and is stunning cellular art.

ELECTRON MICROSCOPES: THE MOST POWERFUL OF ALL

In the 1930s, scientists developed a new type of microscope, an electron microscope that allowed them to see beyond what some ever dreamed possible. The revolutionary concept behind the machine grew out of physicists' insights into the nature of electrons.

As its name implies, the electron microscope depends not on light, but on electrons. The microscopes accelerate electrons in a vacuum, shoot them out of an electron gun, and focus them with doughnut–shaped magnets. As the electrons bombard the sample, they are absorbed or scattered by different cell parts, forming an image on a detection plate.

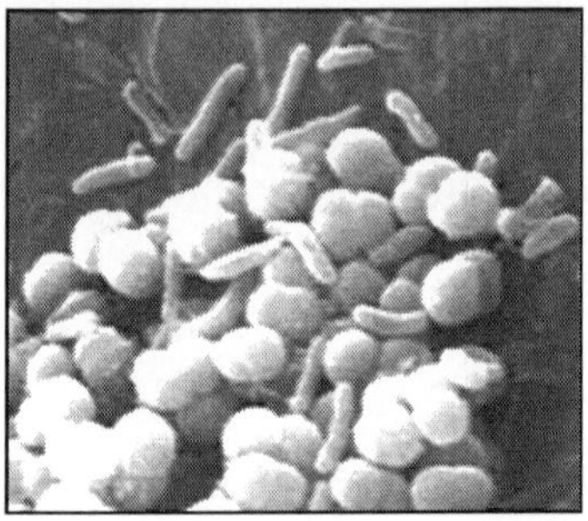

Although electron microscopes enable scientists to see things hundreds of times smaller than anything visible through light microscopes, they have a serious drawback: They can't be used to study living cells. Biological tissues don't survive the technique's harsh chemicals, deadly vacuum, and powerful blast of electrons.Electron microscopes come in two main flavours: transmission and scanning. Some transmission electron microscopes can magnify objects up to 1 million times, enabling scientists to see viruses and even some large molecules.

To obtain this level of detail, however, the samples usually must be sliced so thin that they yield only flat, two–dimensional images. Photos from transmission electron microscopes are typically viewed in black and white.Scanning electron microscopes cannot magnify samples as powerfully as transmission scopes, but they allow scientists to study the often intricate surface features of larger samples.

This provides a window to see up close the three–dimensional terrain of intact cells, material surfaces, microscopic organisms, and insects. Scientists sometimes use computer drawing programmes to highlight parts of these images with colour.

STUDYING SINGLE MOLECULES: CONNECTING THE QUANTUM DOTS

Dyes called quantum dots can simultaneously reveal the fine details of many cell structures. Someday, the technique may be used for speedy disease diagnosis, DNA testing, or analysis of biological samples.

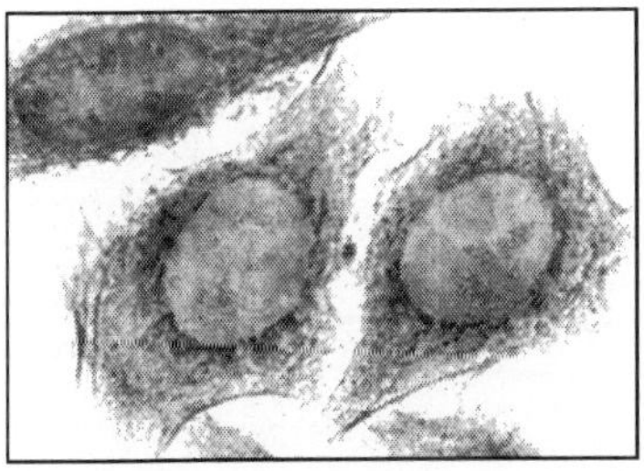

Whether they use microscopes, genetic methods, or any other technique to observe specific molecules, scientists typically flag every molecule of a certain type, then study these molecules as a group.

It's rather like trying to understand a profession—say, teaching, architecture, or medicine—by tagging and observing all the workers in that profession simultaneously. Although these global approaches have taught us a lot, many scientists long to examine individual molecules in real time—the equivalent of following individual teachers as they go about their daily routines.

Now, new techniques are beginning to allow scientists to do just that. One technology, called quantum dots, uses microscopic semiconductor crystals to label specific proteins and genes. The crystals, each far less than a millionth of an inch in diameter, radiate brilliant colours when exposed to ultraviolet light.

Dots of slightly different sizes glow in different fluorescent colours—larger dots shine red, while smaller dots shine blue, with a rainbow of colours in between. Researchers can create up to 40,000 different labels by mixing quantum dots of different colours and intensities as an artist would mix paint. In addition to coming in a vast array of colours, the dots also are brighter and more versatile than more traditional fluorescent dyes: They can be used to visualize individual molecules or, like the older labeling techniques, to visualize every molecule of a given type.

Quantum dots promise to advance not only cell biology but also a host of other areas. Someday, the technology may allow doctors to rapidly analyse thousands of genes and proteins from cancer patients and tailor treatments to each person's molecular profile. These bright dots also could help improve the speed, accuracy, and affordability of diagnostic tests for everything from HIV infection to allergies. And, when hitched to medicines, quantum dots might deliver a specific dose of a drug directly to a certain type of cell.

COMPUTERS CLARIFY COMPLEXITY

Say you're hungry and stranded in a blizzard: If you eat before you seek shelter, you might freeze to death, but if you don't eat first, you might not have the strength to get yourself out of the storm. That's analogous to the decisions cells have to make every day to survive.

For years, scientists have examined cell behaviours—like the response to cold or hunger—one at a time. And even that they did bit by bit, labouriously hammering out the specific roles of certain molecules. This approach made it difficult or impossible to study the relative contributions of—and the interplay between—genes that share responsibility for cell behaviours, such as the 100 or so genes involved in the control of blood pressure.

Now, computers are allowing scientists to examine many factors involved in cellular behaviours and decisions all at the same time. The field of computational biology blossomed with the advent of high-end computers. For example, sequencing the 3.2 billion base pairs of the human genome, which was completed in 2003, depended on computers advanced enough to tackle the challenge.Now, state-of-the-art equipment and a wealth of biological data from genome projects and other technologies are opening up many new research opportunities in computer analysis and modeling. So, much as microscopes and biochemical techniques revolutionized cell biology centuries ago, computers promise to advance the field just as dramatically in this new century.

SCIENCE SCHISMS

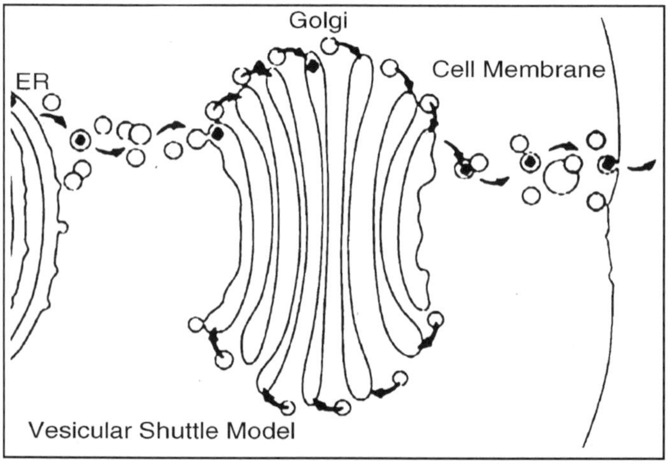

In living cells, material moves both forward (above arrows in figure) and backward (below arrows in figure) from each tip of the Golgi. For simplicity, we have illustrated forward movement only on the top and backward movement only on the bottom of the Golgi cartoon.One great thing about science is that you're allowed to argue about your work.To gain new information, scientists ask a lot of questions. Often, the answers spur more questions.

The never–ending circle not only feeds curiosity; it also can lead to important and sometimes unexpected breakthroughs. Sometimes, scientists studying the same topic but using different experimental approaches come up with different conclusions. Take the Golgi, for example. Think it's noncontroversial? The details of how this organelle forms inside your cells have kept two camps of researchers in a lively battle.

On one side of the debate is Graham Warren of Yale University School of Medicine in New Haven, Connecticut, who argues that the Golgi is an architectural structure that cannot be made from scratch. He believes that newly made proteins are packaged in the rough ER and are sent for further processing to a pre–existing structure (the Golgi) that is made up of different compartments. This is called the vesicular shuttle model.

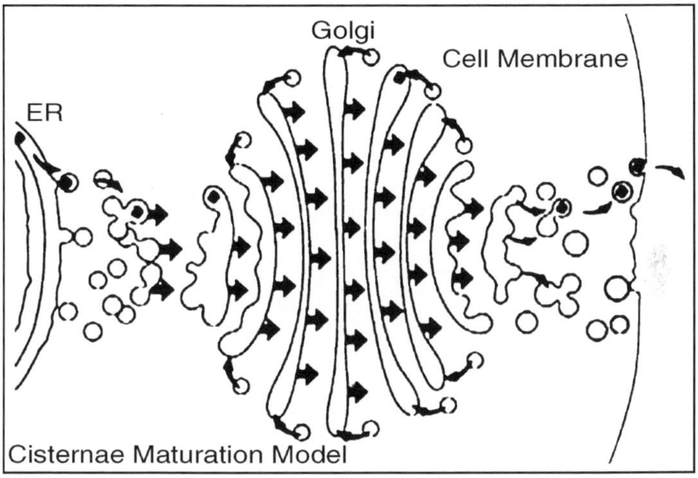

On the other side is Jennifer Lippincott–Schwartz of the National Institute of Child Health and Human Development (part of the National Institutes of Health) in Bethesda, Maryland. She says that the Golgi makes itself from scratch. According to her theory, packages of processing enzymes and newly made proteins that originate in the ER fuse together to form the Golgi. As the proteins are processed and mature, they create the next Golgi compartment. Intriguing new data suggest that perhaps neither model is completely correct. This will likely lead to yet another model. You may not see what all the fuss is about, but the differing Golgi theories say very different things about how cells function. Understanding basic cellular processes, such as how the Golgi works, ultimately can have a profound impact on the development of methods to diagnose, treat, and prevent diseases that involve those processes.

4

Management of Viral and Bacterial Plant Pathogens

BIOLOGICAL VIRUSES

The prime directive of all organisms is to reproduce and survive, which is also the case for viruses, which in most cases are considered a nuisance to humans.

VIRUSES

Viruses possess both living and non-living characteristics. The unique characteristic that differentiates viruses from other organisms is the fact that they require other organisms to host themselves in order to survive, hence they are deemed *obligate parasites*.

Viruses can be spread in the following exemplar ways:

- *Airborne*: Viruses that infect their hosts from the open air
- *Blood Borne*: Transmission of the virus between organisms when infected blood enters an organisms circulatory system
- *Contamination*: Caused from the consumption of materials by organisms such as water and food which have viruses within

Therefore viruses have many means of getting transmitted from one organism to another.

CELL ASSIMILATION BY A VIRUS

Viruses are tiny micro-organisms, and due to their size and simplicity, they are unable to replicate independently. Therefore, when a virus is situated in a host, it requires the means to reproduce before it dies out without producing more viruses. This is done by altering the genetic make up of a cell to start coding for materials required to make more viruses. By altering the cell instructions, more viruses can be produced which in turn, can affect more cells and continue their existence as a species. The following is a step by step guide of how an example bacteriophage (a virus that infects bacteria) takes control of its host cell and reproduces itself.

- The virus approaches the bacteria and attaches itself to the cell membrane
- The tail gives the virus the means to thrust its genetic information into the bacteria
- Nucleotides from the host are 'stolen' in order for the virus to create copies of itself
- The viral DNA alters the genetic coding of the host cell to create protein coats for the newly create viral DNA strands
- The viral DNA enters its DNA coat
- The cell is swollen with many copies of the original virus and bursts, allowing the viruses to attach themselves to other nearby cells
- The process begins all over again with many more viruses attacking the hosts' cells

Without a means of defence, the host that is under attack from the virus would soon die.

BIOLOGICAL CELL DEFENCE

Organisms must find a means of defence against antigens such a viruses described on the previous page. If this was not the case, bacteria, fungi and viruses would replicate out of control inside other organisms which would most likely already be extinct.

Therefore organisms employ many types of defence to stop this happening. Means of defence can be categorised into first and second lines of defence, with the first line usually having direct contact with the external environment.

First Lines of Defence

- *Skin* is an excellent line of defence because it provides an almost impenetrable biological barrier protecting the internal environment.
- *Lysozyme* is an enzyme found in tears and saliva that has powerful digestive capabilities, and can break down foreign agents to a harmless status before they enter the body.
- The clotting of blood near open wounds prevents an open space for antigens to easily enter the organism by coagulating the blood.
- *Mucus* and *cilia* found in the nose and throat can catch foreign agents entering these open cavities then sweep them outside via coughing, sneezing and vomiting.
- The *cell wall* of plants consists of fibrous proteins which provide a barrier to potential parasites (antigens).

If these first lines of defence fail, then there are further defences found within the body to ensure that the foreign agent is eliminated.

Second Lines of Defence

Second lines of defence deal with antigens that have bypassed the first lines of defence and still remain a threat to the infected organism. *Interferons* are a family of proteins that are released by a cell that is under attack by an antigen. These interferons attach themselves to *receptors* on the plasma membrane of other cells, effectively instructing it of the previous cells' situation.

This tells these neighbouring cells that an antigen is nearby and instructs them to begin coding for antiviral proteins, which upon action, defend the cell by shutting it down. In light of this, any invading antigen will not be able to replicated its DNA (or mRNA) and *protein coat* inside the cell, effectively preventing the spread of it in the organism. These antiviral proteins provide the organism with protection against a wide range of viruses.

This action brought about by interferon is a defensive measure, while *white blood cells* in the second line of defence in animals can provide a means of attacking these antigens. One method of attacking antigens is by a method called *phagocytosis,* where the contents of the antigen are broken down by molecules called phagocytes. These phagocytes contain digestive enzymes in their lysosomes (an organelle in phagocytes) such as lysozyme. White blood cells such as a *neutrophil* or a monocyte are capable of undergoing phagocytosis, which is illustrated below.

- The bacterium inside the cell gives out chemical messages that are picked up by the phagocyte.
- The bacteria targets the cell as a possible host and moves towards it.
- The cell is prepared for this and the bacterium becomes trapped in a vacuole that forms around it.
- The bacterium is a sitting duck that is harmless at present.
- The lysosomes detect the bacterium and the digestive enzymes inside them begin to break the bacterium down.
- The remnants of the lysosome and bacterium materials are absorbed into the cytoplasm.

The above illustrates one method of ridding an organism of an internal threat caused by an antigen. This is a non-specific response to an antigen.

PASSIVE AND ACTIVE TYPES OF IMMUNITY

The previous page investigated the role of white blood cells in phagocytosis. *White blood cells* are also responsible for *antibody* formation. Certain antibodies are synthesised in response to the presence of certain *antigens*

Specific Immune Responses

Lymphocytes are a type of white blood cell capable of producing a *specific immune response* to unique antigens. Some of these lymphocytes are capable of entrapping antigens on their surface. When lymphocytes catch these antigens they can then begin to code for unique antibodies, structures that are capable of catching these antigens. The lymphocytes code for a particular antibody on response to a particular antigen. The antibody that is formed will be capable of catching free antigens therefore neutralising the threat as seen below.

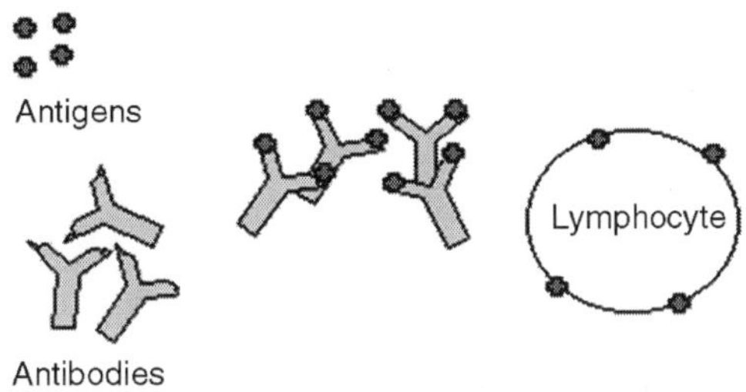

B lymphocytes (B Cells) produce free moving antibodies as above while *T lymphocytes* (T Cells) produce antibodies on their surface.

Types of Immunity

When attacked an organism has several means in which it can prepare to defend itself in event of attack.

- *Active Immunity*: *Vaccines* are used for health purposes to expose our bodies to a particular antigen. These antigens are usually killed or severely weakened to decrease their potency. After destroying these pathogens, the body stores some T cells as memory cells, due to the fact they code for a particular antigen and can be when needed. This memory in T cells can be a means of artificially acquiring immunity while a genuine attack by a pathogen is a naturally acquired type of immunity.
- *Passive Immunity*: This is where immunity to particular antigens as a result of genetic traits passed on from parents rendering the offspring immune to a particular pathogenic threat.

VIRUS DISEASES

About 1870, there occurred in the tobacco-growing region of Holland a disease of this plant which reached such damaging proportions that farmers sought the help of government scientists. Adolf Mayer, director of the agricultural experiment station at Wageningen, undertook a study of the disease about 1880. He published his report in Dutch in 1885 and in German in 1886. He was German-born and German-trained and was apparently acqtainted with the work of Koch. In searching for an explanation of the cause of the disease, he studied soil fertility, air-temperature changes, cultural methods, seed transmission, and nematodes but did not relate anyone of those factors to the etiology.

He described the disease, and, from the mosaic pattern common on leaves of affected plants, he suggested the name "mosaic," which has been retained until the present as the common name of the disease. Mayer discovered that when he macerated the tissue of a diseased leaf and injected the juice into the midrib of a healthy plant, the latter, after about 10 days, began to show typical symptoms. These appeared, not on the inoculated leaf, but on the youngest leaves of the growing plant, and as new leaves expanded, they were also diseased. This is the first known record of the experimental transmission of a virus disease of plant or animal. However, Mayer, impressed by the ease of transmission, endeavored to find out more about the infective agency.

The possibility of an enzyme was considered, but this seemed to him to be an untenable theory since, he removed the infectious agent when he filtered the juice through a double layer of filter paper. Enzyme precipitates with weak alcohol were noninfectious.

Continuous heat at 60 °C. aid not alter infectivity of the juice, but several hours at 80 °C destroyed it. He applied the techniques of Koch" and others in an attempt to isolate a bacterium which could be shown to be the causal !agent. All bacteria isolated were noninfectious, and cultures of a number-I of known bacteria, when used as inoculum, gave negative results. Mayer 1 closed the investigation, still believing that a bacterium which had escaped I him would probably be found to be the infective agent.

About the same time that Mayer was conducting his studies of tobacco mosaic, Erwin F. Smith undertook an investigation of the peach-yellows disease, which had long been destructive in the United States. He was unable to transmit the inciting entity by transfer of plant juice as Mayer had done. He succeeded in transmitting it only by transplanting a bud from a diseased tree to a healthy tree, whereupon the latter eventually became diseased. He received Mayer's paper before he published his final results. He was of the opinion that the two diseases had much in common in the nature of the causal agent.

Dmitria I vanowski was a Russian investigator at the Botanical Laboratory of the Academy of Science at St. Petersburg. The mosaic disease had become troublesome is the Crime, and he upertook look with it about 1890. He confirmed Mayer's result with transmission of the causal entity by means of expressed plant juice. He destroyed the infectivity of the juice with heat. He interpreted the causal entity as bacterium because of the lack of "fungi and other parasites."

He did I not find, as Mayer reported, that filtration of the plant juice through a double layer of filter paper removed the causal factor, and he pointed out i that Mayer was in error in assuming that this procedure would exclude I bacteria from the .filtrate anowski passed the infectious juice through a Chamberland filter candle and found the filtrate to retain infectivity. This fact is of great is torical interest because it is the first record of the I passage of either a plant or an animat virus through a bacteria-proof filter. This technique was so new that even I vanowski was skeptical, and he was apparently unaware of the great significance of the result. He thought there was a possibility that the bacteria of the tobacco plant in some way went through the filter and made the filtrate infectious, although when he passed liquids which were favourable media for bacterial growth through the candle, they remained sterile for several months. Another explanation of the results, which seemed plausible to him in light of prevailing opinions, was that, a toxin secreted by the infectious bacteria in the sap passed through the filter and made the filtrate infectious. The work of I vanowski thus added one hew fact to those proved by Mayer, namely, that the infectious entity in juices from infected tobacco plants passed through a Chamberland filter.

The next investigator to approach this problem was Martinus Willem Beijerinck, Dutch bacteriologist, who was a colleague of Mayer at Wageningen in the early eighties ana moved to Delft in 1887. He took up a study of tobacco mosaic about 1897. He found the juice of infected plants, when filtered through porcelain to be sterile but still infectious. He sought anaerobic bacteria in the filtrate without success. A minute amount of filtrate was sufficient to inoculate several plants, and he demonstrated that the contagious entity was increased in some way in the infected plant. He called it a "contagium vivum £luidum" and referred to it subsequently in the paper as a virus. Thus when Beijerinck published his results in 1898, he confirmed the findings of Mayer and of Ivanowski and claimed more emphatically than either of them that the causal entity was not a bacterium or any conceivable corpuscular material. He found that the virus would remain infectious in infected leaves which were dried down and kept in the herbarium for two years.

He regarded it, therefore, as capable of retaining its existence outside of the plant. He compared the tobacco virus with that of peach yellows and regarded the latter as a contagium capable of existing only in living plant tissue. In the same category as peach yellows he placed the plant variegations of Aubutilon and other ornamental plants which were transmitted to nonvariegated plants by grafting.

Erwin Baur published a paper on infectious variegation in Abutilon in 1904. Baur was a German botanist who later became prominent as a geneticist. He added little to the facts already known but confirmed the fact that. Abutilon variegation was due to a virus more like that of peach yellows than that of tobacco mosaic. He maintained that the virus could not be an organism, since, if a parasite had to depend upon man to make artificial grafts to perpetuate it, it would have long ago disappeared in nature. His argument is not sound in the light of present-day knowledge of insect transmission of.most viruses. At that time no one seems to have considered the possibility of insect transmission of plant Viruses.

The question as to whether the plant viruses were living or nonliving entities was debated for another 30 years. The events of those three decades will be discussed in the chapter on Virus Diseases. Suffice it to say here that at the beginning of the twentieth century the field of virus diseases of plants was just opening. Two major types that we recognize today had been studied, *i.e.*, tobacco mosaic and peach yellows. It was shown that the causal entities were transmissible and that they increased in the host. One was filterable through bacteria-proof filters. One was capable of existing for a time outside the living host; the other was not. Most of the present store of knowledge about plant viruses was still to be learned.

BACTERIAL DISEASES OF PLANTS

Woronin described bacteria in the root nodules of leguminous plants in 1866. Burrill of the University of Inois was the first to relate a bacterium to the cause of a plant disease. In reports extending from 1878 to 1884 he showed that the fire blight of pear was incited by bacterial organism. It is of interest to note that the first acterial disease of plants was demonstrated only two years after Koch had finally demonstrated the first bacterial disease of animals. In 1883 Wakker in Holland, published results of a thorough investigation of the yellow slime disease of hyacinth, which was shown to be incited by a bacterial organism.

Arthur, working at the New York Experiment Station, in articles published from 1885 to 1887, confirmed Burrill's work. He isolated the organism in pure culture, using the newly developed plate culture technique of Koch, and fully demonstrated the pathogenicity of the bacterium. Savastano demonstrated in 1887 that a bacterium incited the knot disease of olive. Erwin F. Smith began to study bacterial diseases of plants soon after 1890. His first attention was directed towards the bacterial wilt of cucurbits. His first report was in 1893, and a major I paper was published in 1895. He took up in-succession the brown rot of solanaceous plants and the black rot of crucifers, the latter of which had been first described by Painmel in 1893.

It is of significance that the leadership and most of the- important work on bacterial diseases of plants from 1878 to 1900 and even later was found I in the United States. In fact there was a distinct air of skepticism on I the part of European botanists concerning the importance of bacteria as I plant pathogens. When De Bary published his "Comparative Morphology" in 1884, he was aware of Wakker's work on the hyacinth I disease but apparently did not know of Burrill's work on fire blight. In, the section on bacteria he devoted one short paragraph to those parasitic on plants. He pointed out that they had "scarcely ever been observed.... One reason for this may be that the parts of plants have usually an acid reaction." In "Lectures on Bacteria" of the 146 pages were devoted to bacteria parasitic on plants.

The works of Wakker, Burrill, and Arthur were noted as well as those of Prillieux in France on I the bacterial decay of wheat grains and of Reinke and Berthold in Germany of on bacterial rot of potato tubers. That De Bary still regarded bacteria as of minor importance as pathogens of plants is shown in his statement that "saprophytic bacteria may also, under special conditions, attack the tissues of living plants as facultative parasites, produce disease in them and destroy them. But this only confirms what was above, that bacteria are not objects of great importance as contagia or diseases affecting plants."

The validity of Smith's work on bacterial diseases was challenged in 1897 by Alfred Fischer, professor of botany in the Koniglichen Gymnasium at Leipzig, who had studied with De Bary in 1881. In a book entitled "Lectures on Bacteria" he denied the reliability of the results of Smith and others who claimed to have seen bacteria in plant cells. Smith replied to Fischer in 1899, pointing out his complete lack of sound knowledge of the subject. Fischer replied the same year, and Smith replied in two rebuttal papers, which seemed to end the controversy. While this debate is now chiefly of historical interest, it is significant that more than 20 years after the work of Koch and of Burrill a rather prominent botanist still challenged the existence of bacterial plant pathogens.

EMPIRICAL STUDIES OF SELECTION IN PLANT PATHOLOGY

Though it is very difficult to measure fitness and estimate selection coefficients, plant pathologists have made many attempts because of the profound implications of these measurements for the utilization of resistance genes. If the fitness cost of a mutation from avirulence to virulence is very low, then pathogen populations will retain the virulence mutation at a significant frequency even after the resistance gene is removed from the agroecosystem. In these cases, it is likely that the resistance gene will never again be effective. But if the fitness cost of the virulence mutation is high, then resistance genes are likely to remain durable, and it is possible that resistance genes can be recycled regionally or through time even after they have been overcome.

Most studies of pathogen fitness cannot be generalized because of the difficulties of conducting these experiments. Fitness should be determined at the level of populations rather than for individuals because a pathogen population is often composed of many different genotypes. But plant pathologists almost always work with just a few individuals to evaluate fitness to simplify the experimental methods.

To measure fitness, sample sizes must be large when s (the selection coefficient) is small, but sample sizes are usually small (10-100 individuals) in field or greenhouse experiments due to practical limitations on the number of samples that can be handled. To further complicate the issue fitness can vary in different environments: the most fit isolate at 10° C may have the lowest fitness at 15° C.

Fungi with mixed reproductive systems present special challenges when measuring fitness because selection can operate on alleles (sexual populations), genotypes (asexual populations) or simultaneously on both alleles and genotypes (mixed reproduction systems). While selection always occurs on the individual through its phenotype, it can be difficult to distinguish between selection for specific alleles (e.g. a virulence allele) within an individual and selection for the entire genotype (e.g. a set of virulence alleles in combination with fungicide resistance alleles) within an individual.

Different alleles can occur in a single genetic background as a result of mutation within clonal lineages. And the same allele can occur in different genetic backgrounds as a result of the same mutation in different clonal lineages.

GREENHOUSE EXPERIMENTS TO MEASURE PATHOGEN FITNESS

Many greenhouse experiments have been conducted to compare fitness of pathogen strains carrying different virulence alleles. The following representative examples were chosen to show how the experiments have been conducted and how the results have not led to a clear interpretation of the fitness cost associated with unneeded virulence alleles.

Kolmer (1993) used a diverse collection of *Puccinia triticina* (*syn. Puccinia recondita* f. sp. *tritici* which causes wheat leaf rust) strains originating from a random-mating population to determine if unnecessary virulence genes were selected against on susceptible wheat plants. The diverse sexual progeny were cycled for 12 asexual generations through Roblin, Thatcher, and two isogenic Thatcher lines with different known resistance genes. Kolmer found no relationship between the number of unneeded virulence alleles and the fitness of the *P. recondita* population.

The most susceptible host, Thatcher, maintained the pathogen population with the greatest diversity of virulence alleles, indicating little or no selection against unnecessary virulence alleles. Kolmer suggested that differences in effective population size at the start of the experiment had a significant impact on the results, while selection in favour of necessary specific virulence alleles was more relevant than selection against unnecessary virulence alleles. It is also possible that at least some of these observations were due to selection for particular clones and clonal lineages carrying unnecessary virulence alleles in each host population.

Leonard (1969) used a diverse collection of *Puccinia graminis* f. sp. *avenae* (causing oat stem rust) originating from telia-bearing orchard grass stems adjacent to a barberry bush. The diverse pathogen population at the beginning of the experiment was derived from over 300 aecial infections. This rust population was passaged through oat cultivars Craig (susceptible) and Clintland A (with resistance gene A) for eight asexual generations and then tested for presence of virulence alleles corresponding to resistance alleles in a set of oat differentials.

Pustules indicating a resistance reaction (infection types 1 and 2) could be differentiated from pustules indicating a susceptible reaction (infection types 3 and 4). On every differential variety, Leonard found an increase in the frequency of resistant-type pustules with increasing numbers of pathogen generations, consistent with directional selection against pathotypes with unnecessary virulence alleles. He calculated that selection coefficients for unneeded virulence alleles ranged from 0.14 to 0.46 depending on the virulence allele under selection.

Prud'homme and Sackston (1990) used two pathotypes (Race 1 and Race 3) of the sunflower rust pathogen *Puccinia helianthi* inoculated at equal frequencies onto a susceptible sunflower line (CM 303) and cycled through eight asexual generations. The resulting populations were inoculated onto the susceptible line and two sunflower lines (CM29 and CM90RR) carrying known resistance genes.

The simple pathotype (Race 1 with no known virulence alleles) made up 95% of the pathogen population by the end of the 5th generation. In this case, it was not possible to distinguish between selection against unnecessary virulence and selection for a particular genotype.

Bronson and Ellingboe (1986) used two isolates of the wheat powdery mildew pathogen *Blumeria graminis* to inoculate an isogenic wheat line containing a known resistance gene. The virulence alleles present in the two pathogen strains were assayed on a set of isogenic wheat lines with known resistance genes. Over the course of 12 cycles of asexual reproduction, the frequency of the isolate with more virulence alleles decreased, consistent with a selection coefficient of 0.24 for the isolate with unnecessary virulence alleles.

The two pathogen strains were crossed and the F1 progeny were compared in 8-way mixtures for virulence and reproductive fitness. In the progeny, reduced fitness was found to segregate independently of virulence alleles, suggesting that the fitness of the progeny was not affected by unneeded virulence alleles. The authors concluded that there was no evidence for selection against unneeded virulence and that other genes were more important for determining fitness under these greenhouse conditions.

These examples demonstrate that there are cases where there is and cases where there is no fitness cost associated with maintenance of an avirulence allele in a pathogen population. These examples also highlight how difficult it is to investigate selection and related fitness costs in plant pathogen populations.

Field Experiments to Measure Pathogen Fitness

Field experiments to compare fitness of pathogen strains are much more difficult than greenhouse experiments because it is impossible to prevent immigration of non-inoculants into the experiment and it was difficult to differentiate among strains with certainty before the advent of molecular genetic marker systems.

Hence, these experiments remain rare in plant pathology. But these experiments are critical to obtain data on pathogen fitness in actual field environments.

Grant and Archer (1983) used published field survey data of virulence allele frequencies and a simple mathematical model to estimate selection coefficients for particular unneeded virulence alleles in *Puccinia graminis* f. sp. *tritici* (causes wheat stem rust) and *Blumeria graminis* f. sp. *hordei* (causes barley powdery mildew). For wheat stem rust, the estimate was based on changes in the frequency of virulence against the Sr6 resistance gene in Australia between 1948 and 1955.

For barley powdery mildew, the estimate was based on changes in the frequency of virulence against the Mla6 resistance gene in the United Kingdom between 1969 and 1975. In both plant pathosystems, values for s ranged from 0.04 to 0.06, suggesting that avirulent strains had greater fitness in the absence of corresponding resistance genes. Using the barley powdery mildew pathosystem (*Blumeria graminis* f. sp. *hordei*), Chin and Wolfe (1984) examined the frequencies of pathogen genotypes in naturally infected pure stands and mixtures of four barley cultivars with different resistance genes over a two-year period (1977-78).

The virulence complexity of each pathogen population sampled from the field plots was assayed in bulk by inoculating leaf segments of the four component barley cultivars. The authors concluded that in pathogen populations isolated from pure stands, selection favored strains with fewer virulence alleles, while in host mixtures selection favored pathogen strains with more virulence alleles. Huang et al. (1994) conducted a similar experiment with this pathosystem in 1989 and 1990 and obtained similar results.

Welz and Leonard (1993) sampled field populations of the maize leaf pathogen *Cochliobolus carbonum* and used changes in frequencies of three pathotypes over 32 or 51 days to estimate selection coefficients associated with each pathotype. In this pathosystem, pathotypes were differentiated based on lesion morphology and there was not evidence for genetic exchange between pathotypes, though there is some evidence for recombination within pathotypes. Pathotype 3 had lower fitness (s=0.17 on average) than Pathotype 2, and Pathotype 0 had the lowest fitness (s=0.58). The lower fitness of Pathotype 3 was surprising because it produces the largest lesions, but it also produces fewer spores in vitro. This shows that lesion size does not necessarily correlate with overall fitness.

Zhan et al. (2002) used DNA fingerprints to investigate the effect of host resistance and genetic diversity on the population genetic structure of *Mycosphaerella graminicola,* which causes *Septoria tritici* leaf blotch on wheat. They collected five strains of the pathogen from each of two wheat cultivars (Stephens and Madsen), and inoculated pure stands and a 1:1 mixture of Madsen and Stephens with an equiproportional mix of spores from all 10 strains at the start of the growing season.

They monitored changes in frequencies of the 10 pathogen isolates through the course of the growing season using RFLP fingerprints (Figure). The frequencies of the 10 strains changed significantly over the growing season, and different strains were selected on different host populations (Figure).

Changes in allele frequencies were used to calculate selection coefficients for each isolate in each host population. The selection coefficients differed on each host population, ranging from 0.01 to 0.55 (Table), with different strains showing higher adaptation on different host populations. The pathogen populations evolved more slowly on the moderately resistant cultivar Madsen than on the susceptible cultivar Stephens. There were no differences in the selection coefficients on the 1:1 host mixture, suggesting that disruptive selection occurred in the mixed host population (Table).

Table. Selection Coefficients for six *Mycosphaerella Graminicola* Strains used in a Replicated Field Experiment.

Strains	Madsen	Mixture	Stephens
M1	0.18 a	0.27 a	0.22b
M2	0.07bc	024 a	0.55 a
M3	0.12 ab	0.11 a	0.01 c
M5	0.00 c	0.16 a	0.09 c
S1	0.13 ab	0.13 a	0.26 b
S3	0.12 ab	0.10 a	0.04 c

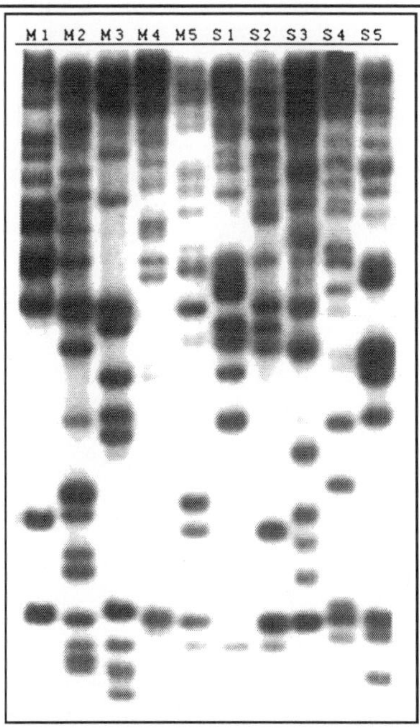

Fig. DNA Fingerprints of 10 *Mycosphaerella Graminicola* Strains used to Inoculate Field Plots Planted to Wheat Cultivars Madsen and Stephens.

M1-M5 were collected from the moderately resistant host Madsen. S1-S5 were collected from the susceptible host Stephens.

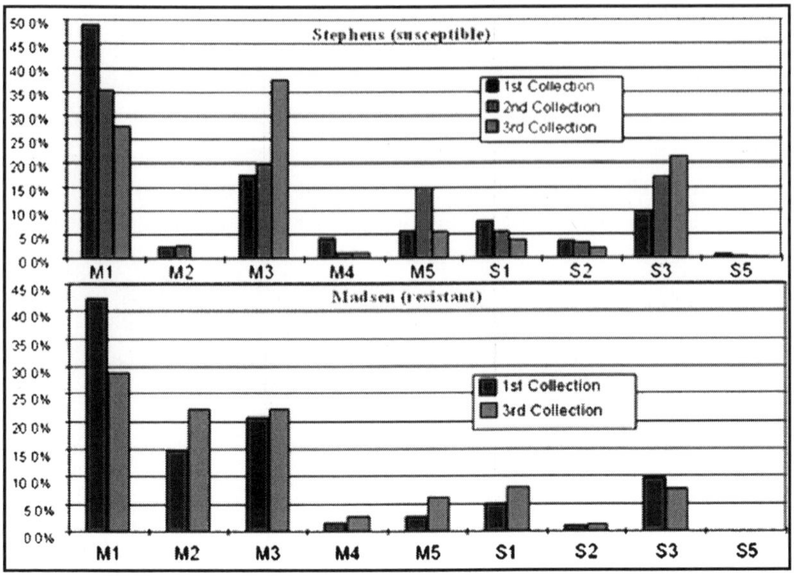

Fig. Frequency of 10 Inoculated *Mycosphaerella Graminicola* Isolates During the Growing Season on the Resistant Cultivar Madsen and the Susceptible Cultivar Stephens. Collections were made Approximately six Weeks Apart.

Mean values followed by different letters differ at $P < 0.05$. Strains with the lowest selection coefficient have the highest fitness to each host treatment. Selection coefficients of strains M4, S2, S4, and S5 were not estimated because they occurred at frequencies that were too low to be meaningful.

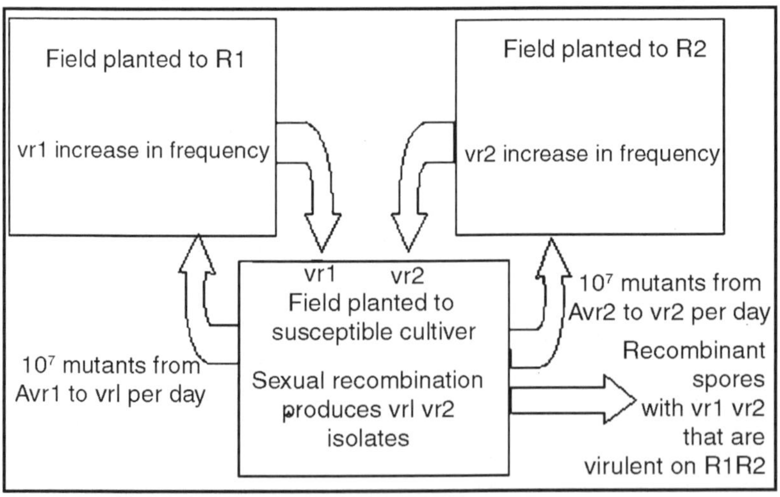

Fig. Virulence Alleles can Recombine Efficiently in Agroecosystems where Fields are Arranged as a Mosaic of Susceptible Cultivars and Resistant Cultivar with Different Resistance Genes.

This can lead to the breakdown of resistance gene pyramids. Assume that two barley fields are planted to two cultivars that have different resistance genes, R1 and R2. A third nearby field is planted to a barley cultivar with no resistance genes (or a previously defeated resistance gene). The third field (susceptible) is the source of spores that initiate the infections.

From the susceptible field, we expect to find 10^7 mutants from avirulence to virulence for vr1 each day that can infect the field planted to R1. Only the vr1 mutants can survive on R1, so they increase in frequency quickly by asexual reproduction. Similarly, we expect that 10^7 mutants from avirulence to virulence for vr2 each day can infect the field planted to R2. Only the vr2 mutants survive, and these increase in frequency on R2. After the vr1 and vr2 types cause epidemics on R1 and R2, their spores can move back into the field of susceptible barley, possibly at quite high numbers.

If the vr1 and vr2 mutants are in compatible mating types, they will have a chance to cross and recombine the vr1 and vr2 alleles during meiosis. While the number of successful matings between vr1 and vr2 mutants may represent only a small proportion of the matings in the population, if these genes are unlinked, then 25% of their progeny will contain both virulence alleles. These recombinants would be able to defeat a host genotype with both the R1 and R2 resistance genes.

A real-world experiment that illustrates the impact of sexual reproduction on the evolution of pathogen virulence is now underway. *Phytophthora infestans,* which was restricted to asexual reproduction as a single global lineage for more than 150 years, has recently become sexual. Will it become more difficult to control now that new combinations of genes can come together in agricultural ecosystems? Keep in touch with the plant pathology literature.

THE EVOLUTION OF PATHOGEN RACES OR PATHOTYPES

The effect of the cultivar mixtures on pathogen evolution can be analyzed based on two key questions:

- *Will a given resistance be more durable when deployed in a mixture than deployed in a monoculture?*
  The lower level of exposure to the pathogen population of the resistance gene in a mixture compared with monoculture will reduce selection pressure on the pathogen population and therefore increase the gene's durability.
- *Considering a given number of resistance genes, will they be more durable if deployed in a mixture than deployed sequentially in monoculture or combined into a single host genotype?*

Mixtures support more diverse pathogen populations than do pure stands, and that diversity is positively related to the degree of disease control provided by the mixture. The mixtures lead, generally, to the evolution of complex pathogen races potentially capable of overcoming many resistance genes and therefore reduce the efficacy of the crop mixture as a measure to reduce disease. However, the appearance of complex races does not mean that those races will be selected for within the mixtures. There are several factors that can prevent the prevalence of complex races. The relative efficacy and durability of the resistance genes deployed in crop mixtures compared with the other two options, depends on the rate of progress of those races in the mixture, which can be affected by several factors:

- *Diversity within pathotypes.* Increasing diversity reduces the rate of increase of complex pathotypes in crop mixtures.
- *Fitness cost associated with virulence.* The ability of complex races to attack multiple host genotypes is countered by a reduction in fitness associated with the lack of avirulence genes.
- *Differential adaptation.* Complex races can infect different host genotypes but have reduced infection efficiency compared with the specific simple races corresponding to each component of the mixture, reducing therefore their rate of progress.
- *Density dependence.* It is the decrease in pathogen multiplication rate associated with an increase in lesion density. It could affect simple and complex races differentially during an epidemic and reduce selection for complex races.

The progress towards complex pathogen races may be relatively slow. Single races or pathotypes capable of overcoming a single resistance gene should progress faster than complex races capable of overcoming multiple resistance genes, precluding the former of becoming dominant within the pathogen population.

CROPS AND DISEASES SUITED TO CULTIVAR MIXTURES

The effectiveness of cultivar mixtures has been demonstrated most commonly for foliar diseases of small grains in which host plants are small and for which there is frequent inoculum exchange among host genotypes. The effectiveness of cultivar mixtures in the control of foliar diseases of plants is related to the probability that pathogen propagules will fail to encounter susceptible tissues. Following Garrett and Mundt (1999) there are several inherent characteristics of each specific plant disease that affect this probability:

- Size of genotype unit area
- Steepness of dispersal gradient
- Ultimate lesion size
- Pathogen generation time
- Degree of host specialization.

Size of Genotype Unit Area

Genotype unit area (GUA) is defined as the area occupied by an independent unit of host tissue of the same genotype. Normally, GUA is the size of an individual plant, but not always. For example, if individual plants are intertwined in the field, then the effective GUA may be smaller. And in the case of clonally propagated plants, the GUA could be larger than the individual plant. On an individual plant basis, crop species with a small GUA include wheat, barley, oats and rice. Those with a moderate-sized GUA include beans, potatoes, and corn, whereas fruit trees have a large GUA. Within each dispersal gradient, propagule loads decrease with distance from the source.

Top) When the size of individual plants is large relative to the dispersal gradient or the gradient is steep, inoculum tends to land on and reinfect the host plant on which it originated. Bottom) When the size of individual plant is small relative to the dispersal gradient or the gradient is shallow, a greater share of inoculum falls on plants with a genotype different from the source host plant.

Mundt and Leonard (1986), working with bean rust, demonstrated that the least amount of disease was associated with the smallest genotype unit area. Mundt and Browning reported similar results (1985) with oat crown rust and maize common rust. Effect of spatial arrangement of susceptible and resistant snap beans on the increase of bean rust in mixtures of susceptible and resistant plants and in pure stands of susceptible plants in field plots.

Steepness of Dispersal Gradient

The shape of a pathogen's dispersal gradient and its interaction with genotype unit area influences the effectiveness of cultivar mixtures for disease suppression. Steeper dispersal gradients result in more inoculum loading on the plant (genotype) on which it was produced, with fewer spores lost as a result of the dilution or barrier effects. Consequently, splash-dispersed pathogens are less suited to control by cultivar mixtures than are wind-dispersed pathogens.

Computer simulations have illustrated that cultivar mixtures are most effective in reducing spread of diseases caused by pathogens with shallow spore dispersal gradients. This figure shows the predicted effect of the relative steepness of the spore dispersal gradient on the amount of disease in three mixtures of susceptible and resistant cultivars (ratio of susceptible to resistant cultivars = 1:1, 1:2, and 1:3) after six pathogen generations in a model simulating barley powdery mildew.

Ultimate Lesion Size

Continuous expansion of individual lesions, in the cases that it is possible, decreases the degree of disease suppression achieved by mixing host genotypes because it increases the rate of new infection of the susceptible genotype. Thus, analogous to genotype unit area, cultivar mixtures for disease suppression are most effective when the size of individual lesions remains small. Non-expanding lesions require a proportionally greater number of infections for the disease epidemic to progress. The dilution and barrier effects influence the number of new infections but not the rate of lesion expansion. Leaf rust of wheat (*Puccinia recondita*) is an example of a disease with a small determinant lesion size. Stripe rust of wheat (*Puccinia striiformis*) shows substantial lesion expansion and, thus, cultivar mixtures may be expected to have somewhat less impact on suppression of this disease.

Degree of Host Specialization

Most research on use of cultivar mixtures for disease suppression has focused on diseases caused by biotrophic, obligate pathogens that interact with their hosts on a gene-for-gene basis. The presence of differential, qualitative resistance to races of the pathogen in different host genotypes is commonly one of the most important criteria for selecting the cultivars that make up the mixture. With differential qualitative resistance, each host genotype potentially benefits from being in a mixed population; i.e., relative to a pure line stand of a genotype, the cultivar mixture will have a reduced proportion of tissue susceptible to the races that can infect it.

EFFECTS ON PLANT HEALTH INDEPENDENT OF DISEASE CONTROL

Growth enhancement. Several Q_0I fungicides are known to cause growth-promoting effects in certain plants. For example, kresoxim methyl has been shown to cause changes in the hormonal balance of wheat which results in increased grain yield, apparently from delayed leaf senescence and water-conserving effects. Growth-enhancing effects independent of disease control have been observed in Q_0I-treated plants of several species, although these effects are very much dependent on the crop, the fungicide used, and environmental conditions.

Phytotoxicity. While the Q_0I fungicides are very valuable for disease control, several are known to cause phytotoxicity in certain, limited circumstances; these are described in product labels. For example, apple cultivars with a genetic background which includes MacIntosh are extremely sensitive to azoxystrobin.

Indeed, these varieties are so sensitive that they can be injured when a sprayer is used to apply azoxystrobin to another crop (grapes, for example), rinsed, and then used to apply another fungicide to the apple crop! Another example: while trifloxystrobin may be used safely on most grapes, it can cause injury to Concord grapes. And kresoxim methyl is phytotoxic to certain sweet cherry varieties but not others. Producers should be aware of phytotoxicity concerns both for the treated crop and because of the possibility of injury via spray drift.

Another aspect of the phytotoxicity risk is the possibility that tank-mixes of Q_0I fungicides with materials that solubilize the cuticle-oils, surfactants, certain liquid formulations of insecticides-could increase their phytotoxicity potential. Although some of the active ingredient can be found within the host tissue, much of the dose of Q_0I fungicides remains on or within the plant cuticle. Application of a spray material that causes abnormally high levels of these fungicides to penetrate into the host tissue could potentially cause phytotoxicity on certain crops or varieties where none is normally expected. Obviously, when applying a previously unused tank-mix on a particular crop variety, a wise practice is to test-apply to small areas before treating large acreages.

RESISTANCE

All Q_0I fungicides share a common biochemical mode of action: they all interfere with energy production in the fungal cell. To be precise, they block electron transfer at the site of quinol oxidation (the Qo site) in the cytochrome bc_1 complex, thus preventing ATP formation. The preceding sentence may "seem like Greek" to even the most knowledgeable crop consultant, but it contains an important point-that the mode of action of the Q_0I fungicides is highly specific. Of the millions of biochemical reactions that occur in the fungal cell, these fungicides interfere with just one, very specific biochemical site.

It is a very important biochemical site for the fungus, to be sure, but it is just one site. Thus, these are called *site-specific fungicides*. This is important because, commonly, just one mutation at that biochemical site (the target site of the fungicide) can result in a fungicide-resistant strain. If such a fungicide-resistant strain occurs, repeated application of Q_0I fungicides can lead to buildup of a fungicide-resistant pathogen subpopulation.

Experience with the Q_0I fungicides worldwide indicates there is a high risk of development of resistant pathogen subpopulations. Worldwide, resistance has been reported in an increasing number of pathogens of field crops, fruit, vegetable, and nut crops, ornamentals and turfgrass.

There are two general types of fungicide resistance: quantitative and qualitative. With *quantitative resistance,* resistant strains are somewhat less sensitive to the fungicide as compared to the wild type, but they often can still be controlled with higher rates and/or more frequent applications (within labeled limits, of course). A good example of this type of resistance is that observed with strains resistant to the DMI (demethylation-inhibitor) fungicides, such as propiconazole or triadimefon.

With *qualitative resistance,* the resistant strain is vastly less sensitive to the active ingredient, and is no longer controlled at labeled field rates. The effect on disease control is the same as if one were spraying water on the crop instead of a fungicide. A good example of this type of resistance is that observed with the benzimidazole fungicides, such as benomyl or thiophanate methyl. Natural occurrences of resistance to the Q_0I fungicides indicate that most cases of control failure are due to resistance of the qualitative type, but that instances of quantitative resistance to certain Q_0I fungicides have also been recorded.

Fungicides that share a common biochemical mode of action for poisoning the fungus are thought to be in the same "fungicide family". When different fungicidal products share a common mode of action, the fungus does not distinguish between the fungicides, even if the chemical structure of the active ingredients is different and the fungicides are produced by different manufacturers.

Biochemically, the fungus sees them all as the same active ingredient. When a fungus is resistant to one fungicide in a chemical family, it is usually resistant to all fungicides in that family. This is called *cross resistance*. In many situations, fungal strains resistant to Q_0I fungicides exhibit cross-resistance to other Q_0I fungicides. In such cases, efficacy of all Q_0I fungicides may be compromised, even if some of them have never been used on that farm. Cross-resistance only applies within a given chemical family. Therefore, Q_0I-resistant subpopulations can be controlled with other fungicides not in the Q_0I family.

GUIDELINES FOR REDUCING RESISTANCE RISK

Start by understanding that there is nothing a grower can do that will *eliminate* the risk of fungicide resistance, except to never use the at-risk fungicide. One can *reduce* the risk of its development by following practices that delay development of a resistant subpopulation.

Guidelines for reducing the risk of resistance against fungicides are issued by the Fungicide Resistance Action Committee (FRAC). FRAC is an organization composed of scientists from manufacturers of the various at-risk fungicides. The guidelines issued by FRAC for each fungicide family are based on scientific principles and up-to-date research. Thus, FRAC guidelines provide a basis for understanding how to reduce the risk in the cropping situations where you work.

In addition to the key guidelines described below, growers should understand that reducing the risk of fungicide resistance begins by using non-fungicidal means for disease control: crop rotation, selection of varieties with reduced susceptibility, sanitation, pathogen-free seed, etc. These practices help reduce overall disease pressure. The occurrence of an adapted mutant with resistance to a fungicide is a matter of chance, like a "roll of the dice". The larger the pathogen population, the greater the chance that such a mutant will arise. Reducing disease pressure through non-chemical practices helps lower the chance that a fungicide-resistant mutant will occur; it does this by keeping the overall size of the pathogen population small.

Limit the number of applications of Q_0I fungicides in a given season. The basis of this guideline is this: the more often a Q_0I fungicide is used, the higher is the selection pressure towards the development of a resistant subpopulation. Limiting the number of applications reduces the opportunity for selection pressure, potentially extending the useful life of the Q_0I family of fungicides on a particular crop.

This guideline is indicated on product labels, and label restrictions on the seasonal total number of sprays apply to all Q_0I fungicides, not just to the product being used. For example, as of the date of publication, the label for

Flint 50WG™ instructed the apple grower not to apply more than four sprays of Flint 50WG™ or other strobilurin (=Q_oI fungicides) during the same season. If a grower makes two applications of Flint 50WG™ in a given season, only two more applications of any Q_oI fungicide could be made that season, whether it be Flint 50WG™ or Sovran 50WG™.

The seasonal limit is higher on crops that receive 8-12 sprays per year than those that receive less. Obviously, if the seasonal limit for Q_oI fungicides is reached and the crop still needs protection against disease for a longer period of time, alternative fungicides must be used. These alternatives must, of course, not be in the Q_oI group and must be from other fungicide families. Under the current FRAC guidelines, seed treatments and in-furrow treatments do not count towards the seasonal limit, because of the limited mobility of these fungicides within plants.

Limit the number of consecutive applications of a Q_oI fungicide. The product labels indicate the number of consecutive applications of Q_oI fungicides that are allowed on each crop, before the user must switch to an equal number of applications of non-Q_oI fungicides. For most crops, the number of consecutive applications will be limited to two before the grower must switch to a fungicide with a different mode of action. FRAC guidelines on certain crops are even more strict; for example, on cucurbits, it is advised never to apply Q_oI fungicides consecutively. Like the seasonal limit described above, this guideline is designed to reduce the opportunity for selection pressure towards resistance.

Mixing Q_oI fungicides with other fungicides can reduce selection pressure towards resistance. Mixtures do not prevent resistant mutants from arising on a farm. They can, however, can slow the rate of spread of these mutants. A proper mixing partner is one that provides satisfactory disease control when used alone on the target disease. Also, the mixing partner must be from some fungicide family other than the QoI group. Tank-mixing fungicides from the same chemical family does nothing to reduce the risk of fungicide resistance. The application rates of the components should not be reduced below the minimal labeled rate.

From the standpoint of fungicide resistance, wisest use of Q_oI fungicides may be to use them at the early stages of disease development. Some researchers believe that curative use of a fungicide increases the risk of resistance, because the producer is treating a much larger population of spores and mycelium (the body of the fungus) than would be treated preventively. Allowing a buildup of a large population of spores before treatment increases the chances that a resistant mutant will be present when the chemical is applied.

Additional guidelines. While translaminar movement is a wonderful feature of the Q_oI fungicides, potential problems arise when a Q_oI fungicide is tank-mixed with a contact (=protectant) fungicide for resistance-management purposes. Since contact fungicides remain on the treated leaf surface, poor coverage of the underside of crop foliage could result in biologically active levels of Q_0I fungicides there through translaminar movement without the presence of the mixing partner. On such leaf surfaces, a Q_0I-resistant strain could take hold and flourish, should it arise. Whenever tank-mixing a Q_0I fungicide with a contact fungicide, always strive for complete coverage of all plant surfaces.

As noted above, growth-enhancing effects independent of disease control have been observed in Q_0I-treated plants of several species. Where these cases occur, this could pose an incentive to use a product inappropriately or excessively. While optimizing plant health is always an important objective, inappropriate use or overuse use of a Q_0I fungicide for its growth-promoting qualities could be a violation of the product label, and it may also increase selection pressure towards fungicide resistance. Users should be very mindful not to overuse any at-risk fungicide. Once resistance to Q_0I fungicides develops on a farm, there is a very good chance that efficacy of these products against that disease will be compromised for quite some time.

MORPHOLOGY OF THE CAUSAL ORGANISM

Parasitic angiosperms and some fungal pathogens of plants are sufficiently large and distinct for diagnosis to be made on the basis of morphology seen with the naked eye. As mentioned above, the flowers of Rafflesia are hard to miss since they are the largest in the plant kingdom and the appalling smell of decaying flesh that it exudes is a further prompt .The sporophores of Ganoderma boninense, a fungal pathogen of African oil palm (Elaies guineensis), coconut (Cocos nucifera) and betelnut palm (Areca catechu) are also distinctive. Other pathogens will probably require the aid of the light microscope or the higher magnifications obtainable with the scanning or transmission electron microscope.

Several reference books for the identification of plant pathogens with keys are available. For parasitic angiosperms, the monograph of Kuijt (1969) still seems to be the standard text. Fungi often prove difficult to identify on the basis of morphology as seen under the light microscope since, when cultured on agar media, they frequently refuse to yield anything more interesting than wefts of mycelium. Various culture treatments can sometimes induce these species to sporulate but even then a skilled mycologist is often required to make a definite identification. Moreover, their morphology may be influenced by the medium.

For example, Adaskaveg and Hartin (1997) found that the morphology of conidia of isolates of Colletotrichum acutatum varied according to whether they were grown on potato dextrose agar, pea straw agar or almond fruit. For a guide to the literature on the identification of plant pathogenic fungi the reader is referred to Rossman, Palm and Spieman (1987). Nematodes are classified on the basis of their morphology as seen under the light microscope but for the other groups of plant pathogens, light microscopy is seldom sufficient. Even if additional morphological information from transmission and stereoscan electron microscopy is available, a positive identification usually requires data from the supplementary techniques which are described in the remainder of this chapter.

SELECTIVE MEDIA

Some plant pathogenic organisms may be cultured in vitro. For these, placing infected material on a suitable medium and inspecting the resulting colonies of the pathogen may be sufficient for a positive identification. However, the choice of medium is important. All media are selective and considerable effort has been made for some pathogens or groups of pathogens to adapt substrates so that only the organisms of interest will grow.

In early work, Kado and Heskett (1970) published details of five media which were selective for the five genera of plant pathogenic bacteria recognized at the time: Agrobacterium, Corynebacterium, Erwinia, Pseudomonas and Xanthomonas. Both their selectivity and plating efficiency have been improved by further work. For example, Fatmi and Schaad (1988) published a semi-selective medium for the isolation of Clavibacter (formerly Corynebacterium) michiganense subsp. michiganense from tomato seed which, besides nutrients, contained nalidixic acid, potassium tellurite and cycloheximide.

The medium gave recovery rates of 85–132 per cent when compared with nutrient-broth yeast-extract medium and allowed the detection of a single contaminated seed containing 50 colony-forming units in samples of 10 000 seeds. Similarly, Gitaitis and co-workers (1997) have published a diagnostic medium which was effective for the semi-selective isolation and enumeration of Pseudomonas viridiflava, the causal agent of bacterial streak and rot of onion. The medium contained tartrate as a carbon source and the antibiotics, bacitracin, vancomycin, cycloheximide, novobiocin and penicillin G. Selectivity was enhanced by incubation at 58C rather than higher temperatures.

Many media selective for particular fungi have been published. Here the problem is often to suppress fast-growing, saprophytic fungi and bacteria associated with diseased plant material, which would otherwise swamp the slowergrowing pathogens, while allowing these to grow out of the infected plant material.

For example, Manandhar, Hartman and Wang (1995) found that a semi-selective medium consisting of one-quarter strength potato dextrose agar and seven antibiotics, fenarimol, vinclozolin, choramphenicol, erythromycin, iprodione, neomycin and tetracycline was significantly less inhibitory to Colletotrichum capsicum and C. gloeosporioides than to Alternaria spp. and Fusarium spp. As a result C. gloeosporioides was detected more frequently in pepper seeds than when they were placed on moist filter paper, the normal means of detection. Similarly, Duffy and Weller (1994) supplemented diluted potato dextrose agar with rifampicin and tolclofosmethyl for isolation of Gaeumannomyces graminis var. tritici, the causal agent of take-all, an important disease of wheat.

Isolation of this organism is difficult since roots of the plant often harbour several other pathogenic fungi. Moreover, lesions caused by G. graminis var. tritici are rapidly invaded by many opportunistic soil fungi, which displace the pathogen. Another difficulty with the identification of G. graminis var. tritici is that it does not sporulate in culture, necessitating pathogenicity and molecular tests. However, the medium of Duffy and Weller (1994) was also of value in identification since the fungus altered its orange colour, imparted by rifampicin, to purple.

Often the most selective medium of all is the host plant itself and, in some instances, this has been exploited in a technique known as baiting. For example, Jeffers and Aldwinckle (1987) used apple cotyledons to detect Phytophthora cactorum in naturally infested soil. The soil to be tested was air-dried and remoistened before flooding several days later with water and adding the apple cotyledons. In soils that were infested with the Oomycete, the colonized cotyledons turned brown and abundant sporangia, characteristic of the fungus, could be seen with the aid of a microscope.

Baiting may also be used to detect viruses in soil samples. Fillhart, Bachand and Castello (1998) used Chenopodium amaranticolor to detect tobamoviruses in forest soils by growing the plant in the soils and assaying their roots for the viruses using DAS-ELISA.

BIOCHEMICAL MARKERS

All organisms have distinctive biochemical features and these can be used for diagnosis. Some characteristics are shared by large groups while, at the opposite end of the scale, others are unique to individual populations. The choice of character is therefore paramount in determining the taxonomic level to which an organism is defined.

Bacteria have long been identified by their metabolic functions, such as their ability to metabolize certain substrates and, more recently, by analysis of their fatty acid profiles. Additionally, soluble protein analysis by gel electrophoresis has been adopted both for bacteria and fungi. All three methods rely upon gene expression and, since this may be regulated by environmental factors, care has to be taken to standardize these.

Serological techniques were initially used for virus identification but are now applied to several of the other groups of plant pathogenic organisms. Finally, some of the methods currently used for analysing nucleic acids provide exquisite sensitivity in the detection and quantitative analysis of plant pathogens independent of environmental conditions. These molecular technologies have been reviewed by Martin, James and Levesque (2000). More specifically, Louws, Rademaker and de Bruijn (1999) have reviewed the application of the polymerase chain reaction to the analysis of the diversity of bacterial plant pathogens and their detection as well as the diagnosis of the plant diseases they cause.

Substrate Metabolism

Goor and co-workers (1984) were among the first to investigate the applicability of the 'galleries' of biochemical tests contained in the API (Appareils et Proce´de´s d'Identification) systems to the identification of phytopathogenic strains of Erwinia and Pseudomonas. Those useful for distinguishing strains of Erwinia were API 20E, API 50CHE and the oxidase, zym and aminopeptidase systems. Pseudomonas strains were differentiated using the API auxanographic systems 50CH, 50AO and 50AA. The authors concluded that, provided an adequate reference system was available, the tests provided useful alternatives to conventional procedures.

Biolog is an alternative to the API system and gave better differentiation of 204 bacterial pathogens associated with a sheath rot complex and grain discoloration of rice in the Philippines. Using this system, Cottyn and coworkers (1996) found that all the reported strains of Pseudomonas fuscovaginae were positive for the production of 2-ketogluconate but strains of Acidovorax avenae and Burkholderia glumae were negative. In contrast, B. glumae was positive for the production of acid from inositol but negative for the production of 2-ketogluconate and A. Avenae was negative for both these reactions. de Laat, Verhoeven and Janse (1994) identified the causal agent of bacterial leaf rot of a species of aloe (Aloe vera) as Erwinia chrysanthemi biovar 3 on the basis of its ability or failure to metabolize a number of substrates as well as its agglutination by an antiserum prepared against a defined strain of the organism. Similarly, Pernezny and co-workers (1995) were able to define the bacterial species causing a severe outbreak of bacterial spot in lettuce fields in Florida as Xanthomonas campestris on the basis of substrate utilization, the pathovar being defined as vitians by its fatty-acid profile.

In some instances organisms may be identified by their production of unusual metabolites. For example, strains of Aspergillus flavus that were aflatoxigenic were recognized by their production of volatile C15H24 compounds such as alpha-gurjunene, trans-caryophyllene and cadinene. These compounds were not produced by non-toxigenic strains.

Fatty Acid Profiles (FAME Analysis)

Identification of bacterial pathogens of plants by fatty acid methyl ester analysis is usually performed on pure cultures of the organism. About 40mg of wet cells are saponified and methylated. The fatty acid methyl esters (FAME) are extracted in an ether–hexane mixture and analysed by gas chromatography. Areas of the resulting peaks on the chromatograms are calculated and compared with profiles of known reference strains by computer programs. For example, the organism responsible for an outbreak of bacterial spot of lettuce was defined as Xanthomonas campestris pv. vitians as the fatty acid profiles of the strains collected from the field matched this pathovar most closely. In a more extensive study, Wells, Vanderzwet and Hale (1994a) and Wells, Vanderzwet and Butterfield (1994b) were able to differentiate the five species of Erwinia of the 'amylovora' group as well as the four species of the 'herbicola' group.

Norman and co-workers (1997) differentiated three species of Xanthomonas, X. albilineans, X. fragariae and some pathovars of X. campestris as well as Stenotrophomonas maltophilia using fatty acid analyses with 100 per cent accuracy. Other pathovars of X. campestris were more diverse and therefore FAME analysis was not suitable for them.

Protein Analysis

Electrophoresis of soluble proteins from plant pathogens often gives rise to complex patterns and these can be used for identification purposes. Instead of using a general protein stain, such as Coomassie Blue, particular proteins, which for example might have enzymic activity, may be revealed by appropriate staining methods.

MacNish and co-workers (1994), by staining for pectic enzymes, were able to place 4250 Australian isolates of Rhizoctonia solani in 10 groups which they termed zymograms. However, a major disadvantage of the technique is that, compared with immunological or PCR methods, relatively large quantities – 50–100mg – of the pathogen must be grown in order to obtain the enzymes either from culture filtrates or from extracts of the organism itself.

In contrast, peptide profiling may be performed on comparatively small amounts of pathogen and this technique has been used to define virus strains. Kittipakorn and co-workers (1993) investigated virus isolates from groundnut which caused a wide variety of symptoms varying from mild mottle to systemic necrosis and stunting. Virion proteins (150–190mg) were isolated, digested with trypsin and separated by high-performance liquid chromatography (HPLC). Despite the disparate symptoms all were found to have identical HPLC profiles and were identified as strains of peanut stripe potyvirus.

Serological Techniques

Serological techniques rely on the specificity of antigen–antibody binding and a sensitive method for detection of the resulting complex. They have been used for the detection of plant viruses for many years but more recently they have been applied to the diagnosis of fungi, Oomycetes, phytoplasmas and bacteria as well. In order to obtain reliable results, the antibody preparation used, whether polyclonal or monoclonal, must have the required specificity and high affinity for an appropriate epitope of the pathogen. Polyclonal antibody preparations are obtained by injecting an animal with antigen and, as their name implies, contain a population of different antibodies. As pointed out previously these are likely to be heterogeneous with regard to the antigen with which they bind, the affinity of the binding and their suitability for the attachment of reagents used to detect them. Purification of the antigen to a high level will eliminate some of this variation. In contrast monoclonal antibodies, although more difficult to obtain, may be screened for specificity to a single epitope and will provide a stable source of antibody since the hybridoma may be kept in perpetuity by culturing and storage in the frozen state.

In some instances the extreme specificity of a monoclonal antibody can be a disadvantage since it may not detect all members of a taxonomic group. Here a mixture of monoclonal antibodies may be used to overcome this problem. Advances in molecular immunology have allowed the development of specific recombinant monoclonal antibodies.

One such reagent has been used to detect Ralstonia solanacearum (formerly Pseudomonas solanacearum) race 3, a pathogen of potato, tomato and eggplant for which many countries have quarantine regulations. The antibody is specific to the lipopolysaccharide of the bacterium and holds promise for being sufficiently reliable to obviate the need for time-consuming bioassays, which are normally required to provide confirmation of positives detected by polyclonal antibodies. Antibody binding may be recorded and quantified by a variety of techniques but enzyme-linked immunosorbent assay (ELISA) in one of its variants is the one that is used predominantly. The most common of these are double-antibody sandwich ELISA (DAS-ELISA) and indirect ELISA (I-ELISA).

In DAS-ELISA, antibody is bound to a solid support, such as the wells of a microtitre test plate and the test sample, containing the antigen of interest, added. This is 'sandwiched' by adding another antibody which has been conjugated with an enzyme such as alkaline phosphatase. When substrate for the enzyme is added such as p-nitrophenyl phosphate a coloured reaction product is obtained. Normally the intensity of the colour is determined spectrophotometrically to give a measure of the amount of antigen, using a plate reader.

In I-ELISA, antibody is bound to a solid support, as in DAS-ELISA, and this binds the antigen. However, the bound antigen is recognized by an antibody raised in a second animal and this second antibody is recognized by a further antibody, which is conjugated to an appropriate enzyme. The advantage of I-ELISA is that the antibody used to recognize the antigen of interest is not constrained by conjugation with an enzyme and is likely to bind more completely.

For this reason it may be less strain specific than DAS-ELISA. Another variant of the technique involves trapping the antigen to the F(ab')2 fragments of specific IgG on a solid support and detecting it by intact specific immunoglobulin (IgG) to which enzyme has been conjugated in the Fc region.

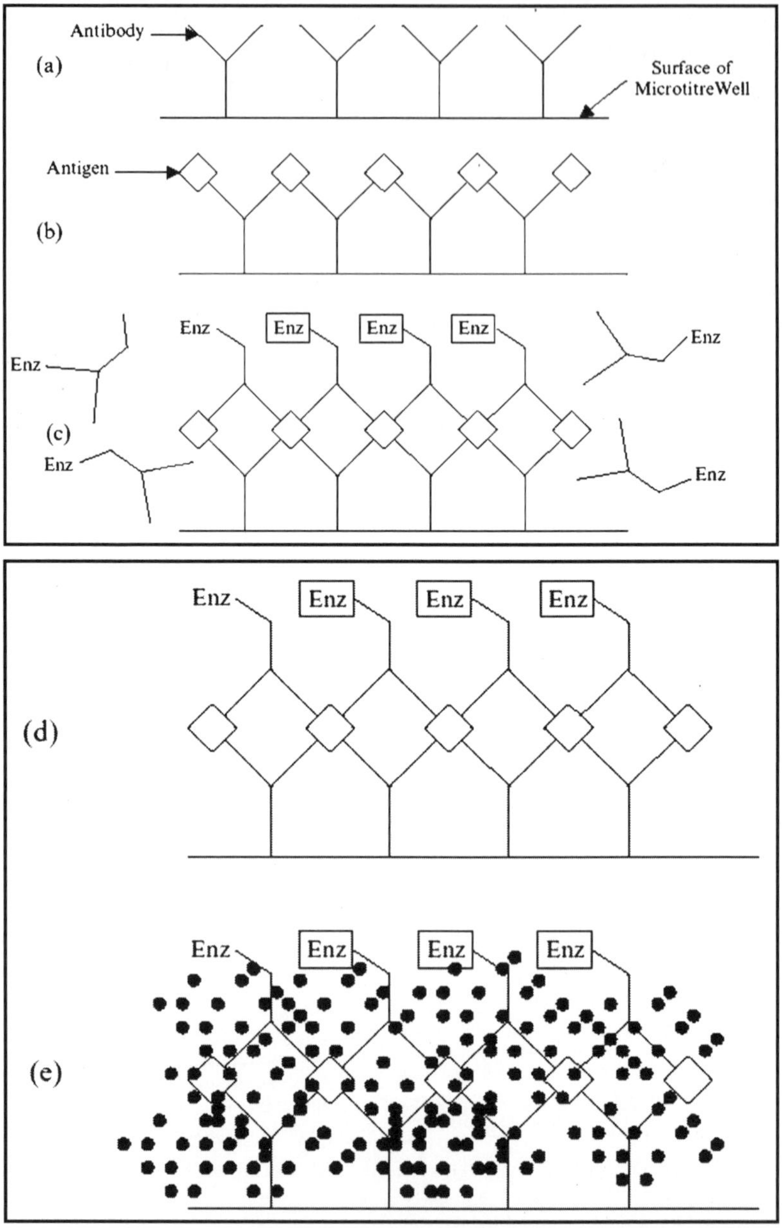

Fig. Double Antibody Sandwich-enzyme Linked Immunosorbent Assay (DAS-ElISA): (a)Antibody Bound to the Wells of a Microtitre Plate;(b)Sample Containing Antigen à of Interest Added and Wells Washed Well with Buffer;(e)Substrate for Enzyme Added and Coloured Product – Measured–the Intensity of the Colour Developed is Proportional to the Amount of Antigen.

Dot ELISA and dipsticks are two further developments of the ELISA technique. In both, antibodies are bound to a membrane and the membrane is incubated with a preparation suspected to contain the antigen. After washing, the membranes are treated with enzyme-conjugated antibody and incubated with the enzyme substrate. The membrane is then dried and the intensity of colour measured, for example, with a hand-held reflectometer.

Cahill and Hardham (1994a) describe a development of the dipstick technique which they combined with baiting to detect Phytophthora cinnamomi from soil. Soil samples (20g) were flooded with distilled water and baited with cotyledons from Eucalyptus sieberi. After incubation for 3 days the samples were cold shocked at 48C for 20–30 min to induce the release of zoospores and three dipsticks were floated, membrane side downwards, on the water for 1.5 h. Some of the released zoospores encysted on the dipsticks and were detected immunologically as follows.

The dipsticks were treated with a murine monoclonal antibody specific for an antigen of P. cinnamomi located on the cyst periphery; the antibody was recognized by an anti-mouse antibody conjugated to alkaline phosphatase and the conjugate was visualized by treatment of the dipsticks with a mixture of 4-chloro- 2-methylbenzene-diazonium salt and naphthol AS-MX phosphate to give a red insoluble dye, which was easily visible under a hand-held lens or dissecting microscope.

Other techniques for detecting bound antigens include gold-conjugated antibody, antibody conjugated with fluorescent dyes and, for biotinylated antibodies, avidin conjugates, since the glycoprotein avidin binds strongly to biotin. Milne and co-workers (1995) used immunogold labelling successfully to detect three unrelated phytoplasmas in plant tissue both before and after embedding and Jones, D.A.C. and co-workers (1994) used fluorescence labelling to detect Erwinia carotovora subsp. atroseptica in seed potato.

This organism is responsible for blackleg in cool temperate regions and the incidence of the disease is directly related to the incidence of contamination of the seed potato, the threshold for disease development being about 103 cells per tuber. Juice was extracted from the peel of potatoes and placed in the wells of 24-well tissue culture plates together with an agarized medium containing polygalacturonic acid. After incubation for 48 h the agar in the wells was dried to a thin film in a hot-air oven at 508C and stained with fluorescein isothiocyanate conjugated with antiserum specific to the bacteria. The films were washed before photographing under a fluorescence microscope and colonies of the bacterium were enumerated from the negatives using a commercial imaging system.

Immunofluorescence was also used by Chittaranjan and de Boer (1997) to detect Xanthomonas campestris pv. pelargonii in greenhouse nutrient solution. A murine monoclonal antibody, 2H5, was used and binding was detected by an antimouse antibody conjugated with indocarbocyanine.

Blotting is a very convenient technique and may be adapted for field applications. For example, Lin, Hsu and Hsu (1990) were able to detect several viruses and phytoplasmas in plant material by blotting freshly cut surfaces of the hosts onto nitrocellulose membranes and treating the membranes with biotinylated antibodies specific to the pathogens. Antibody binding was visualized in these instances by treatment of the membranes with avidin–enzyme conjugates and the appropriate enzyme substrates.

Nucleic Acid Techniques

The genome of every organism consists of a unique sequence of nucleotides but the degree of conservation of different domains is variable. Highly-conserved regions are found among many organisms which are distantly related whereas some regions such as satellite DNA sequences, which are repetitive and usually non-transcribed, are very variable. In consequence, a considerable degree of discrimination is possible. For example, Piotte and co-workers (1994, 1995) found two highly reiterated but different satellite DNA sequences with monomeric units of 295 bp and 169 bp in the nematodes, Meloidogyne incognita and M. hapla, respectively. The monomers were cloned and used as probes. That from M. incognita was non-specific since it hybridized with all populations belonging to M. incognita, M. arenaria and M. javanica.

In contrast, the monomer from M. hapla was specific and, because of its variability, could be used to differentiate three populations of this species as well as to distinguish it from sympatric populations of M. chitwoodi and M. incognita. Moreover, because of the high reiteration – about 15 000 monomers per haploid genome – identification of the organism was possible by direct hybridization with as little material as one squashed female nematode, obviating the need to extract DNA. For non-reiterative DNA, the selection of the sequence to be analysed is critical and, in practice, this means the choice of appropriate restriction enzymes or nucleic acid sequences for the probes or primers in the techniques described below.

A further advantage of nucleic acid analyses is that, unlike the techniques described so far, they are not dependent on gene expression and therefore are independent of environmental factors. In the past few years a large number of techniques has been published, many of which are extremely ingenious and which are revolutionizing the detection and assay of plant pathogens. In order to give some coherence to this rapidly developing field the techniques are described roughly in order of specificity.

Gel Fractionation

Nucleic acids fragments are separated according to size by gel electrophoresis. Agarose gels are normally used for separating fragments which are 300 nucleotides in length or more and polyacrylamide gels for shorter lengths. The resulting bands are visualized by staining with ethidium bromide, which is often included in the gel, and viewing under ultraviolet light. Usually the technique is used in conjunction with other procedures such as cutting extracted DNA with restriction enzymes or amplifying specific domains by the polymerase chain reaction. However, Hodgson, Wall and Randles (1998) were able to obtain preliminary identification of a viroid associated with the lethal Tinangaja disease of coconuts present on the island of Guam by two-dimensional polyacrylamide gel electrophoresis of RNA extracted from palms showing symptoms.

Restriction Fragment Length Polymorphism (RFLP)

RFLP analysis was one of the first techniques to be exploited in determining subtle differences in the DNA from two or more organisms. DNA preparations from the organisms of interest are digested with restriction enzymes and the resulting fragments run on a gel. Different organisms will give fragments of different size according to the distribution of their restriction sites.

In some cases the restriction enzymes used may delineate repetitive DNA, in which case banding patterns, visualized by staining with ethidium bromide and viewing under ultraviolet light, may be clear without any further treatment. More often, however, the differences are obliterated by a smear of thousands of differently sized DNA fragments. In these instances it is necessary to use a probe to pick out the fragments of interest.

Nucleic Acid Probes

Probes are selected on the basis of their complementarity to unique nucleotide sequences of the organism to be detected and will therefore bind wherever these sequences occur. Recognition that binding has occurred is achieved by labelling the probe. Frequently the label is 32P or 33P which enables binding to be recognized by autoradiography but non-radioactive probes are being developed which may have greater applicability in areas where elaborate laboratory facilities are not available. For example, biotin-labelled adenine may be introduced into the probe either by nick translation or chemically with photobiotin. The probe may then be recognized by avidin conjugated to an enzyme in an ELISA assay since avidin binds tightly to biotin.

Various methods for the preparation of probes have been used. For plant viruses, labelled complementary DNA (cDNA) can be prepared directly from the virus RNA using the enzyme reverse transcriptase. Alternatively, screening of a cDNA library of the organism of interest may allow the identification of a number of sequences which vary in their degree of conservation. The more highly conserved sequences will be common to related groups while strain variation may be detected among less conserved sequences. For example, Waterhouse, Gerlach and Miller (1986) found both general and luteovirus serotype specific probes in a cDNA library of barley yellow dwarf virus.

Three other examples will further illustrate the importance of choosing the appropriate fragment of DNA as a probe and the different techniques that have been used to produce them. Coronatine is a toxin produced by a number of pathovars of Pseudomonas syringae and is an important virulence factor. It would therefore be useful to be able to determine specifically the prevalence of toxigenic isolates. Cuppels and Elmhirst (1999), using the probe TPRI, derived from the Pseudomonas syringae pv. tomato gene cluster controlling production of the toxin, in conjunction with a semi-selective medium, were able to trace natural populations of the pathogen on tomato plants from just before planting to harvest.

Magnaporthe grisea, the cause of rice blast disease, was thought to be genetically unstable and this was deemed to be an explanation of the apparent instability of pathotype. Levy and co-workers (1991) were able to disprove this view by using the highly repetitious sequence (termed MGR for Magnaporthe grisea repeat), discovered by Hamer and co-workers (1989) as a probe. When the MGR probe was used with EcoR1 digests of strains of the fungus, which had been separated by agarose gel electrophoresis, the resulting MGR-DNA fingerprints distinguished and accurately identified the major pathotypes collected in the USA over 30 years. The fingerprints also defined the clonal lineages within and among pathotype groups.

Bacterial spot of tomatoes and peppers is caused by Xanthomonas axonopodis pv. vesicatoria and Xanthomonas vesicatoria, corresponding to the groups A and B of the former taxon, X. campestris pv. vesicatoria. Diagnosis of the diseases and identification of the pathogens is complicated by the occurrence of epiphytic xanthomonads which are non-pathogenic. Kuflu and Cuppels (1997) have developed a probe for the pathogens by genomic subtraction. DNA from a pathogenic strain was labelled with digoxigenin and denatured.

The labelled DNA was hybridized with denatured unlabelled DNA from three nonpathogenic strains. After one round of subtraction the preparation was found to hybridize only with Southern blots of restriction enzyme-digested DNA from pathogenic but not non-pathogenic strains, allowing the identification of a fragment which preferentially hybridized to DNA from only pathogenic strains.

Probes that are either radio, chemically or antigenically labelled may be used in a method which is analogous to dot ELISA. The nucleic acid to be tested is spotted and baked onto a solid matrix such as a nitrocellulose membrane and the remaining sites on the membrane are blocked with heterologous DNA, such as that from calf thymus, and protein (e.g. bovine serum albumin).

The probe is then hybridized with the test nucleic acid preparation under carefully controlled conditions. Variation of these conditions alters the stringency of the test (i.e. the amount of mismatching of bases) and hence the degree of binding of the probe. For example, Herdina-Roget (2000) has published an assay using a radioactive probe that detects DNA levels of Gaeumannomyces graminis var. tritici at less than 30 pg, 30–50 pg and more than 50 pg in 0.1 g soil organic matter corresponding to low, moderate and high levels of the disease, respectively. The assay allows farmers to have soil samples assessed before sowing and can be used to predict the potential yield loss.

Probes may also be used to recognize pathogens in tissue blots. For example, Duran-Vila, Romero-Durban and Hernandez (1996) pressed freshly cut stems or leaves of chrysanthemum on polyvinylidene membranes and were able to detect the presence of chrysanthemum stunt viroid with a digoxigenin labelled probe.

Polymerase Chain Reaction Techniques

The polymerase chain reaction (PCR) has facilitated advances in many areas of biology and the diagnosis of plant disease is no exception. Essentially the technique provides a method for amplifying in vitro specific nucleotide sequences from the organisms of interest, usually corresponding to 0.5–2.5 kDa, although sequences in excess of 10 kDa have been obtained. The basic procedure is now so well known that it hardly requires any introduction. Specificity of the region to be amplified is determined by the primers used to delineate the two ends of the sequence and the amplification itself is carried out in a thermocycler.

Normally target DNA for amplification is given a preliminary heating to ca. 948C in order to separate the strands. It is then mixed with the other constituents of the reaction – the primers, dNTPS (the deoxynucleotide triphosphates ATP, TTP, GTP and CTP), buffer, MgCl2 and a thermostable DNA polymerase. Usually this last is Taq polymerase from Thermus aquaticus but other heat stable polymerases are now available such as Vent which has the advantages of greater fidelity and products that are blunt ended (Taq DNA polymerase has a tendency to add an extra A residue to the 3′-end of each strand).

The reaction mixture is then cycled around 30 times through a temperature re´gime of 948C, 45–558C and 728C. At 948C the DNA strands separate and at 45–558C primer annealing to the separated strands occurs, the precise temperature depending on primer composition and the degree of stringency required by the experimenter. DNA complementary to the separated strands is synthesized at 728C.

The time at each temperature varies according to the experimental material but 30 s at the lowest and highest temperatures and 1 min at the intermediate temperature are commonly used. Allowing for ramping time among the three temperatures, 30 cycles of such a re´gime can be achieved in less than 3 h. After the reaction, the PCR products may be visualized by running them on a conventional agarose gel and staining with ethidium bromide.

Choice of primers The detection of variation within pathogen populations and the detection of specific pathogens are usually the applications of PCR of most interest to plant pathologists. For the former, Random Amplified Polymorphic DNA (RAPD) is often suitable since it does not presume any knowledge of DNA sequences of the organism. In this technique, rather than selecting primers of known complementarity to the pathogen, arbitrary sequences of about 10 nucleotides are used. Separation of the products by agarose gel electrophoresis often reveals polymorphisms among isolates.

For example, Serrano, Camargo and Teixeira (1999) were able to separate 48 isolates of Phytomonas species, of which 31 were obtained from plants and 17 from insects into five main clusters and show that they were grouped according to geographic origin. RAPDS are not so suitable for the diagnosis of pathogens for at least two reasons. First, the pattern of bands on the gels is contingent upon very precise conditions of amplification and it is difficult to reproduce these from one thermocycler to another. Even in the same instrument, the heating block may not be uniform. Second, in complex substrates such as plants or soil, DNA from these sources may interfere. In these instances, primers that are known to recognize only the organisms of interest are required.

One method for choosing primers is to design them on the basis of the sequence of specific amplicons obtained by RAPD analysis. For example, Parry and Nicholson (1996) found two RAPD amplicons that were common to all isolates of Fusarium poae tested. These were cloned and used to probe Southern blots of DNA from a range of isolates pathogenic to seeds and stem bases of wheat. One fragment was partially sequenced and the sequence data used to design primers. When tested, these were found to amplify DNA from all isolates of F. poae but not from a range of other fungal species associated with diseases of cereal ears or seed.

Similarly, Schilling, Moller and Geiger (1996) identified sequences which enabled them to design primers which gave rise to amplicons in a PCR reaction that allowed differentiation of Fusarium culmorum and Fusarium graminearum. In fungi, sequences from ribosomal genes are particularly favourable targets for detection and may also provide useful data concerning the phylogenetic relations of isolates. This is because ribosomal genes are present in high copy number (around 60–200 per haploid genome) and there is a wide disparity in the degree of conservation of the genes and their spacer regions.

Thus, although the rDNA genes themselves are conserved, there is considerable heterogeneity in the internal transcribed spacer (ITS) regions between the genes, more in the transcribed external spacer (ETS) regions and still more in the non-coding or intergenic spacer (IGS) units within rDNA repeat units. As a result, it is possible to use universal primers, based on sequences of the conserved genes, which give rise to amplicons that span regions of heterogeneity. One method of identifying differences among amplicons is to digest them with restriction enzymes and run gels to reveal differences in banding patterns. For example, Balesdent and co-workers (1998) investigated the Leptosphaeria maculans complex which is the cause of blackleg disease of oilseed rape. They were able to propose a scheme for the large-scale identification of species of the complex based on amplification of the ITS1–5.8S–ITS2 region of the fungi involved and digestion of the amplicons with four restriction enzymes.

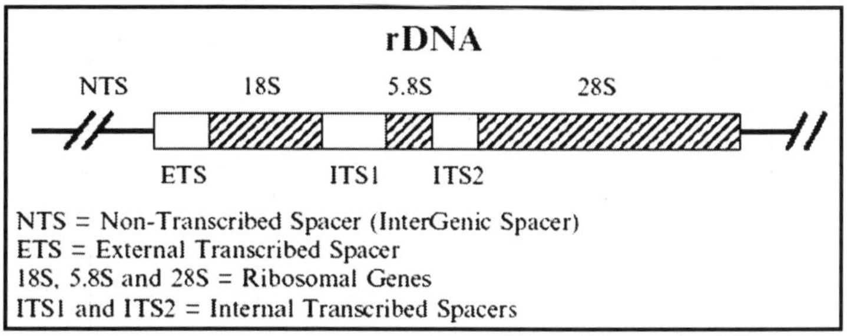

Fig. Diagram of Eukaryotic Ribosomal DNA: A Single Repeat Unit is Shown Which May Exist in Copy Numbers as High as 200 Making Conserved Regions Easy Targets for Polymerase Chain Reaction (PCR)Primers;When these Span Less Conserved Regions Such as the Internal Transcribed Spacers,ITSI and ITS2,these will also be Amplified,Allowing their Sequences to be used for Diagnosis Either by Restriction Enzyme Digestion and Gel Electrophoresis of the Product or Sequencing.

More specifically, sequence data from the ITS1–5.8S–ITS2 region may be used to design primers for specific amplification of sequences from the organism of interest. For example, Schilling, Moller and Geiger (1996) found that there was sufficient sequence variation, especially in ITS2, to construct primers for the specific amplification of F. avenaceum but that polymorphism between the ITS sequences of F. culmorum and F. graminearum was insufficient to design primers specific for these species.

In bacteria, three unrelated families of repetitive DNA sequences, known as BOX, ERIC (enterobacterial repetitive intergenic consensus) and REP (repetitive extragenic palindromic) are distributed throughout the genome. Thus, outwardly facing primers complementary to these sequences allows the amplification of the intervening regions giving rise to many amplicons of different sizes which may be separated on agarose gels. Using this technique, Louws and co-workers (1998) were able to separate Clavibacter michiganensis into five subspecies which corresponded to their host ranges.

Other workers have designed primers which recognize sequences required for virulence of pathogens. For example, bacteria such as Pseudomonas syringae pv. savastanoi, Agrobacterium tumefaciens and Erwinia herbicola form galls on their hosts as a result of the expression of genes which code for the plant-growth regulating compounds indole acetic acid (IAA) and cytokinins. Manulis and co-workers (1998) designed primers which recognized non-conserved sequences of the IAA and cytokinin biosynthesis genes that enabled them to detect specifically gall forming as opposed to saprophytic strains of Erwinia herbicola pv. gysophilae. Moreover, sensitivity of detection was boosted 100 fold by nested PCR in which, after one round of PCR with primers which recognized cytokinin biosynthesis genes and gave an amplicon of 607 bp, a second round was performed with primers which were complementary to sequences internal to this amplicon and gave a product of 522 bp.

Toxins are important pathogenicity or virulence factors . For example, Pseudomonas syringae pv. syringae produces small cyclic lipodepsipeptides which are phytotoxic and antifungal. Production of these compounds is dependent on the expression of the gene SyrB. Use of primers specific to sequences of the SyrB gene allowed the polymerization of a 752 bp fragment only from toxigenic isolates of the bacterium.

For RNA viruses the RNA is reverse-transcribed into cDNA before amplification (RT-PCR). For example, Haber and co-workers (1995) developed a technique for detecting double-stranded RNA associated with flame chlorosis of cereals in small amounts of tissue and candidate fungal vectors. Total RNA was extracted from milligram quantities of test tissue, reverse transcribed, and amplified by PCR. Two primer pairs were used which were predicted to yield 358- and 347-bp DNA fragments from a consensus 821-bp sequence covering an open reading frame.

Sequence data from PCR products Colletotrichum is a notoriously 'difficult' fungal genus with taxa that have been recorded on a wide range of plant species. In particular, C. gloeosporioides has been described as a 'group species' and the opinion expressed that 'no progress in the systematics and identification of isolates belonging to this complex is likely to be made based on morphology alone'.

Accordingly, Sherriff and co-workers (1994 and 1995) used rDNA sequence analysis to distinguish members of this genus. Their work revealed new species groupings and confirmed the distinction between C. graminicola, a pathogen of maize and C. sublinoleum, a pathogen of sorghum and Rottboellia. Also, Moses and co-workers (1996), using the same technique, were able to confirm that C. gloeosporioides was the cause of a dieback disease of cassava that is prevalent in West Africa.

Quantitative PCR PCR may also be used to quantitate target DNA sequences. One way of doing this is to include in the reaction mixture as an internal standard a known amount of a sequence that is heterologous to the target and differs in size while retaining the sequences that are recognized by the primers. On the assumption that both the standard and target sequences are amplified equally, the amount of the target sequence can be calculated from the amount of initial standard sequence and the ratio between the amounts of amplified products of both target and standard sequences. For example, Mahuku, Goodwin and Hall (1995) were able to quantify Leptosphaeria maculans during the development of blackleg symptoms in oilseed rape using a template derived from Leptosphaeria korrae as the internal standard. Alternatively, quantitative results may be obtained by the separate amplification of members of a dilution series. Amplification of target DNA will take place at the higher but not the lower concentrations and the amount present in the initial sample may be calculated by the Poisson distribution. However, the results should still be treated with caution since inhibitors of the PCR reaction may be present and interfere.

TISSUE CULTURE AND PATHOGEN FREE PLANTS

Another purpose for which plant tissue culture is uniquely suited is in the obtaining, maintaining, and mass propagating of specific pathogen-free plants. The concept behind indexing plants free of pests is closely allied to the concept of using tissue culture as a selection system. Plant tissues known to be free of the pathogen under consideration (viral, bacterial, or fungal) are physically selected as the explant for tissue culture. In most cases, the apical domes of rapidly elongating shoot tips are chosen.

These are allowed to enlarge and proliferate under the sterile conditions of *in vitro* culture with the resulting plantlets tested for presence of the pathogen (a procedure called indexing). Cultures which reveal the presence of the pathogen are destroyed, while those which are indexed free of pathogen are maintained as a stock of pathogen-free material. Procedures similar to these have been used successfully to obtain virus-free plants of a number of species and bacteria-free plants of species known to have certain leaf spot diseases. The impact of obtaining pathogen-free nursery stock can only be speculative, since little research documenting viral, bacterial, or fungal diseases transmitted through propagation of woody ornamentals is available.

SOMATIC HYBRIDIZATION

The ability to fuse plant cells from species which may be incompatible as sexual crosses and the ability of plant cells to take up and incorporate foreign genetic codes extend the realm of plant modifications through tissue culture to the limits of the imagination. Most such manipulations are carried out using plant "protoplasts". Protoplasts are single cells which have been stripped of their cell walls by enzymatic treatment. A single leaf treated under these conditions may yield tens of millions of single cells, each theoretically capable of eventually producing a whole plant. This concept has fueled speculation as diverse as the possibilities of obtaining nitrogen-fixing corn plants on the one extreme to discovering a yellow-flowered African violet on the other extreme.

The observation that has provided the impetus for most of this research is that when cells are stripped of their cell walls and brought into close contact, they tend to fuse with each other. This "somatic hybridization" is not subject to the same incompatibility problems that limit traditional plant breeding strategies. It is conceivable then that one could hybridize a Juneberry with a crabapple or a plum, but the fundamental research required to demonstrate such an event has yet to be conducted. The potential use of somatic hybridization to bring about novel combinations of genetic material has been demonstrated in the genera Petunia and Nicotiana. Research funded in part by the Horticultural Research Institute at the University of Wisconsin is investigating the feasibility of using such techniques with woody species. Brent McGown and co-workers have succeeded in obtaining naked cells from tissue cultures of Betula and Rhododendron, but as of yet, they have neither obtained plants from single cells not achieved cellular fusion.

However, further research in this area promises to have a tremendous impact on our concepts of woody plant diversity. Just as remarkable as the idea of fusing plant protoplasts is the idea of incorporating foreign genetic material into the genetic code of plant cells. Such transformations have been carried out in the so-called "gene-splicing" experiments where the information for making insulin was incorporated into bacteria. Not only is the desired information transmitted to succeeding generations of bacteria, but the bacterial cultures become synthesizers of insulin as well. Plant cells can be made to take up foreign genetic codes, but evidence that this can be transmitted into the daughter cells and serve the intended function is lacking.

Plant tissue culture research is multi-dimensional. While most nurserymen have been introduced to the techniques and advantages of micropropagation, few have ventured to use it as a propagation tool.

The applicability of micropropagation for woody trees has been demonstrated as feasible since all aspects of the technology have confirmed the fact that trees produced by this method look like and grow like their counterparts produced by traditional methods of cloning.Other dimensions of tissue culture research have been less well publicized. The potential for selecting pathogen free plants, for selecting stress-tolerant and pathogen-resistant clones of plants, and the novel genetic combinations to be achieved through somatic hybridization are all lines of research which can have a profound impact on the nursery industry.

UTILISING ANTHER CULTURE TO SELECT FOR COLD HARDINESS

Crosses between distantly related species can bring together novel gene combinations. However, the hybrid offspring can be few in number, genetically unstable and require years of further selection and screening before any advantageous characteristics can be brought near to commercial use.

Anther culture (androgenesis), to generate haploid plants from pollen microspores, is one way to shorten this process. It allows novel allele combinations, particularly ones involving recessive characters, to be assessed in intact plants. Useful individuals can then be developed into homozygous and fertile plants through chromosome doubling techniques, and brought into a breeding programme.We have recently been involved in a collaborative project with the Institute of Grassland and Environmental Research (IGER) to use this approach to improve cold-tolerance and fodder quality in grazing grasses.

Crosses between *Lolium multiflorum* (Italian ryegrass) and *Festuca arundinacea* (tall fescue) should offer valuable combinations of characteristics. The *Lolium* species should provide good growth characteristics, while the *Festuca* provides stress-tolerance. One hybrid individual (*Festulolium*) resulting such a cross had already shown drought-tolerance characteristics.

However, the out-breeding nature of these grass species, along with the hexaploid genome of *F. arundinacea* and autotetraploid *L. multiflorum* indicated that a lengthy breeding programme might be necessary.The research project therefore aimed to produce androgenic plants from the existing pentaploid *Festulolium* plant and assess them for cold tolerance.

This exploited the expertise in tissue culture at the University of Liverpool together with experience in breeding for stress-tolerance at IGER.Anthers containing immature pollen (microspores) are the starting material for androgenesis. Flowers have to be selected at the correct developmental stage, which varies from species to species. In addition, some individual genotypes may not be amenable to anther culture, or require specific pretreatments.

Careful microscopy and testing of successful pre-treatments of related species are therefore necessary when dealing with a new species. For the Graminae, microspores must be at the mononucleate stage and no pre-treatment is necessary. The cut flowers were surface sterilised and opened in sterile conditions under a binocular microscope. The anthers were dissected and transferred to a solid nutrient medium. Large numbers could be placed on each petri dish.

Callus developed, which was transferred to a different medium to initiate embryos and so generate haploid plants. Over 200 androgenic plants were produced at Liverpool, each originating from a different microspore. Each therefore represented a genetically different individual. Testing for phosphoglucosisomerase, where a different isozyme was contributed by each of the five chromosome groups within the *Festulolium* plant, indicated that the pollen-derived plants had a wide variety of chromosome combinations from each of the parents of the hybrid.

The freezing-tolerance of these plants varied considerably, with three individuals able to survive the extreme cold of -14 degrees Celsius. When the chromosome complement of two of these plants was examined using genomic *in situ* hybridisation (GISH), they carried virtually the whole genome of *F. pratensis*, a parent of *F. arundinacea* noted for its freezing-tolerance.

Unfortunately, the fertility of these two plants was not restored by chromosome doubling, so that they could not be used for further breeding. However, they demonstrated the potential of androgenesis for rapid assessment of the genetic potential available from a difficult breeding combination, indicating that this type of wide cross revealed characters of cold and drought tolerance that were worth pursuing.

Using Somaclonal Variation to Select for Disease Resistance

Plant tissue cultures isolated from even a single cell can show variation after repeated subculture. Distinct lines can be selected with their own particular morphology and physiology. It suggests that the tissue culture contains a population of genotypes whose proportion can be altered by imposing an appropriate selection pressure. This variation can be transmitted to plants regenerated from the tissue cultures, and is called somaclonal variation. It provides an additional source of novel variation for exploitation by plant breeders.

The carrot cultivar Fancy was used in our laboratory to generate a series of 197 regenerant progeny lines. These plants showed considerable morphological variation. They were tested for resistance to the leaf spot pathogen *Alternaria dauci,* which can cause total necrosis of mature leaves. They had a greater degree of variation in response than the parental cultivar.

DEMONSTRATION OF TISSUE CULTURE FOR TEACHING

The starting point for all tissue cultures is plant tissue, called an explant. It can be initiated from any part of a plant - root, stem, petiole, leaf or flower - although the success of any one of these varies between species. It is essential that the surface of the explant is sterilised to remove all microbial contamination.

Plant cell division is slow compared to the growth of bacteria and fungi, and even minor contaminants will easily over-grow the plant tissue culture. The explant is then incubated on a sterile nutrient medium to initiate the tissue culture. The composition of the growth medium is designed to both sustain the plant cells, encourage cell division, and control development of either an undifferentiated cell mass, or particular plant organs.

The concentration of the growth regulators in the medium, namely auxin and cytokinin, seems to be the critical factor for determining whether a tissue culture is initiated, and how it subsequently develops. The explant should initially form a callus, from which it is possible to generate multiple embryos and then shoots, forming the basis for plant regeneration and thus the technology of micropropagation.

The first stage of tissue culture initiation is vital for information on what combination of media components will give a friable, fast-growing callus, or a green chlorophyllous callus, or embryo, root or shoot formation. There is at present no way to predict the exact growth medium, and growth protocol, to generate a particular type of callus.

These characteristics have to be determined through a carefully designed and observed experiment for each new plant species, and frequently also for each new variety of the species which is taken into tissue culture. The basis of the experiment will be media and protocols that give the desired effect in other plant species, and experience.

FUNGI AS HUMAN PATHOGENS

Fungi that are pathogens are usually plant pathogenic Fungi. There are comparatively few species that are pathogenic to animals, especially mammals. According to Hawksworth (1992), there are approximate a little 1.5 million described species of fungi. A little more than 400 of these species are known to cause disease in animals, and far fewer of these species will specifically cause disease in people.

Many of the latter will only be superficial types of diseases that are more of a cosmetic than a health problem. Thus, there are not many species of fungi that are pathogenic to human that will be fatal. The study of Fungi as animal and human pathogens is medical mycology. There is also such a thing as veterinary mycology, but the types of diseases that are found in your pets often are the same as those that are found in people. Because of the rarity of human diseases caused by Fungi, most people have little, if any, knowledge of such diseases.

The diseases of warm-blooded animals caused by fungi are known as mycoses (sing.=mycosis). Although such diseases are relatively few, the fungi that cause them have a wide host as well as geographical range. Most of these diseases are not fatal, but once contracted, they may forever be a source of constant irritation and can lead to permanent scaring, which is why they are not such a pretty sight to view.

The successful treatment of fungal diseases is more difficult than those caused by bacteria. Because bacteria are prokaryotes, the makeup of their cells are very different than our own eukaryotic cells and pharmaceutical products, such as antibiotics, are able to successfully destroy bacteria without harming our cells, tissues and organs. However, because fungi are eukaryotes, finding a treatment that will kill the fungus and not harm our own cells is more difficult. Thus, most chemical treatments are also toxic us as well as the fungus. The most widely used drug for treating systemic mycosis and other fungal infections that do not respond to other drugs is Amphotericin B. Azole drugs are also widely used, but these only inhibited fungal growth and do not kill the fungus.

HISTORY ON THE THEORY OF HUMAN DISEASES

The history human diseases tells a story that parallels and overlaps that of plant diseases. Early man viewed disease as a work of demons, and in many tribal societies, the exorcism of demons was the job of the shaman, witch doctor, sorcerer, etc. With the development of agriculture and civilization, more elaborate stories came about. For example, the Greeks tell of the myth about Pandora opening a box given to her by Zeus and releasing a host of plagues on mankind. However, they also created Aesculapius, the god of healing and medicine who had two daughters, Hygeia and Panacea, who were goddesses of health.

During the 4th Century B.C., Aristotle introduced the theory that life could arise abiologically, i.e. without parents. This is the theory of spontaneous generation, a theory that survived well into the 1800s.

As early as 350 B.C., Hippocrates developed a theory that was not based on superstition and myth. His Humoral Theory suggested that diseases were the result of disproportionate relationship of body humors or fluids: blood, yellow bile, black bile and phlegm (saliva). In order to have good health, each body humor must maintain a certain proportion with the other body fluids. An individual's humors may become unbalanced when they becomes ill, with the changing of the season, over-eating, etc., and the balance in their humors must be adjusted to bring them back into balance. Adjustments were often made by inducing an individual to vomit or to draw out "bad blood" by cutting an artery or vein.

Thus, arose the practice of blood letting, which along with hair cuts, shaves and enemas, was carried out in barber shops. Even today, barbershops still utilize the red and white striped pole display as a means of letting the public know that it is a hair cutting establishment. However, meaning of this barbershop symbol has nothing to do with hair cutting. The red strip is to indicate the blood that is bled, the white corresponds to the tourniquet used to dilate the veins.

Roman scholar, Marcus Terentius Varro (116BC -27BC) suspected that disease was caused by little animals in the air. However, because of the dark age that followed, science progressed little until the 1500s. In 1546, Girolamo Fracastoro gave the first description of typhus and suggested that this disease could be contracted from one person to another through direct contact or individuals coming in contact with inanimate objects referred to as fomite, i.e., linen, eating utensils, clothing, etc.

In 1688, Francesco Redi, a Florentine scientist, was the first to challenge the idea of spontaneous generation. It was thought that maggots and flies arose spontaneously, from rotting animal flesh. Redi protected the rotted animal flesh from the air and other to sources of visible infection and observed that maggots and flies did not arise from the protected meat. His conclusion was that flies and maggots were present in the unprotected meat, but absent in the protected meat, because it served as a nesting area for the eggs of the flies as well as a food source once the eggs hatched.

In 1674, Anton Van Leeuwenhoek became the first person to see and describe various microorganisms. He would continue to describe observations of anything that he could place under the microscope until just before his death in 1723.

You should also recall that Anton de Bary's demonstration that the Potato Blight was caused by Phytophthora infestants came a year earlier than Pasteur's Germ Theory. However, the demonstration that fungi could cause disease in people preceded even de Bary. In 1835, Agostino Bassi discovered that a disease of silkworms known as muscardine could be transferred from silk worm to silk worm and was caused by a fungus, Beauvaria bassiana, a species that was later named for Bassi in honor of his discovery.

The fungus grows on the silk worm, covering it with white mycelium, eventually killing it. This is a significant problem in the silk industry since one infected silkworm can spread the fungus to all of the other silkworms. It was shortly after that, in 1841, David Gruby demonstrated for the first time that a fungus infection of the scalp, called favus, was caused by a fungus (in Rippon, 1988). The disease is characterized by thick yellow crusts over the hair follicles. At a time before the use of agar media, Gruby isolated the fungus causing favus, from infected individuals, grew the fungus on slices of potatoes and was able to reproduced the favus disease by carrying out inoculation experiments on healthy tissue. This experiment demonstrated, for the first time, that a microorganism was the cause of a human disease (remember Koch's Postulate?). However, Gruby's research in this areas has mostly been ignored, possibly due to strong anti-Semitic feelings, in medicine, at that time (Rippon, 1988).

SLOW KNOWLEDGE OF MEDICAL MYCOLOGY ADVANCEMENT

It would appear that with the knowledge gained from the research of Bassi and Gruby that in the study of diseases, the knowledge of fungal diseases should have progressed more rapidly than bacterial diseases. However, that is obviously not the case. Bacterial diseases are by far more well known. There are several reasons for this disparity in knowledge. Relative to bacterial diseases, fungal diseases are infrequent and while some mycoses can be severe to fatal, there have not been epidemics of such diseases as we have found in bacteria. For example, bubonic plague was responsible for approximately a third of the population of Europe, between 1346-1350 and 40% of the population of Constantinople, in 1542, and while other diseases such as tuberculosis, malaria, dysentery epidemics were not as severe, they, too, have caused numerous deaths.

The combination of the severity, frequency and the epidemics that have occurred in bacterial diseases have, undoubtedly, driven the progress in the study of bacterial diseases. Another problem that slowed the progress of fungal diseases, and one that should not be estimated, is the understanding of species concept in fungi. By 1890 Sabouraud began publishing large numbers of articles on fungus disorders of the skin which eventually culminated in an enormous contribution to the field of medical mycology. However, Sabouraud had difficulty understanding variations in forms fungi can often exhibit. Some species of fungi can take on several forms, a phenomenon known as pleomorphism. Thus, some of Sabouraud's published species as well as those of his contemporaries were merely different forms of existing species. This resulted in hundreds of species being described that would later be determined to represent already published species. Some of these species were based on careful observations where only slight variations in the form of the fungus was thought to represent new species, but some newly described species were also based on inaccurate and incomplete observations. The complication of nomenclature and classification of fungi also slowed progress. This was a problem, at that time, since doctors, were not trained in either mycology nor the systems of naming and classification of organisms.

It would not be until 1934 that species concepts of dermatophytes would be redefined by Chester Emmons, according to the Rules of Botanical Nomenclature, and current mycological standards of spore morphology and the structures on/in which they were borne. However, with the means of mass producing penicillin in the early 1940s and discovery of other antibiotics, many of the very serious bacterial diseases were being controlled. Although, to some extent, there have are chemical means of controlling some fungal diseases, they are by no means always successful ones.

The majority of most human pathogenic fungi appear to be soil inhabiting species where they live as saprobes, but given the appropriate conditions, i.e. if the person is not healthy, an open wound is present, direct injection of fungus into your system, a particular life-style, AIDS, etc., they will aggressively attack people. Thus, many fungal infections may be due to opportunistic fungi (=facultative parasites) rather than fungi that specifically cause human diseases.

For example Coprinus cinereus, a common mushroom has been recorded as causing endocarditis (Speller and MacIver, 1970), Ustilago maydis, the corn smut is known to cause skin lesions (Review by Lacaz, et al., 1996) and Schizophyllum commune, a wood decomposing fungus that has documented to cause several different types of infections, including meningitis and lung disorder (Reviewed in Kern and Uecker, 1986). These types of fungi can be more harmful than obligate parasites since a facultative parasite is not dependent on a live organism and have not evolved with any particular host, they aggressively attack their hosts, and there is a greater probability that they will kill their hosts than an obligate parasite. However, normally these types of fungi will not cause human mycoses unless their immune system have somehow become compromised.

We often come in contact with fungi during our everyday routines, some which are potentially pathogenic to human and others not. We may be exposed simply by walking by construction areas where the soil has been disturbed and scattered into the wind by the machinery, we are constantly exposed while we are hiking, jogging, hunting or fishing. During recreation when we injure ourselves, such as with puncture wound, abrasions, burns or even by inhalation of a large number of harmful spores. Fortunately, most of us have an immune system that will protect us from such infections by fungi, but some individuals will contract fungal diseases from such injuries.

So, what kind of fungi can be human pathogens? Probably all fungi can potentially be harmful, in this respect, if your immune system has been compromised. Fortunately, for most of us, this is not a problem and the probability of contracting a serious fungal disease is low relative to bacterial or viral diseases. However, in recent years, it has become more of a problem with the rising number of people with compromised immune systems such as people with AIDS, organ transplants, diabetes and treatment for various forms of cancer.

SOME FUNGAL HUMAN PATHOGENS

The most convenient way of classifying them is to categorize them according to the type of infection that has occurred: 1. Superficial infections, are caused by fungi that attack the skin or its appendages (nail, feathers and hair). Some examples of these infection include ringworms, jock-itch and athlete's foot. These fungi are known as dermatophytes. 2. Systemic infections, diseases that occur deep within the tissues, involving vital organs and/or the nervous system, and which may be fatal, but may also be chronic. Entry into the body is usually through inhalation of spores or open wounds. Blood circulation or respiratory system may then transmit fungus throughout body and additional infection of internal organ may occur. These fungi, are usually saprotrophic fungi, growing in the soil. A third, Intermediate infection, is sometimes also recognize and is intermediate between the two just discussed. The infection will occur below the skin, but will remain localized

Superficial Infections

The superficial mycoses are the most well known since they can be readily observed. They commonly occur on the hair, nails and skin of infected individuals. They have been recorded in various compilations of medical literature for well over a thousand years as ring worm, athlete's foot, jock itch and piedra. For each type of infection, there can also be a variety of species that may be causing the disease. Thus, we will only have a general discussion on this group of diseases.

Ringworm and Related Dermatophytes

Ringworm usually occurs on the exposed parts of the body, forming circular growths that may appear darker or lighter than the normal skin color, with symptoms that include skin lesion, rash and itching of the infected area. Ringworm infections are common where conditions are unsanitary and crowded with people and has been known since early historical time. There are indications that ringworm was more prevalent in the recent past than now because of improvements in sanitary conditions and health habits. The Greeks called it Herpes (=circular or ring form) and the Romans associated the disease with the larval stage of Tinea, the genus for clothes moth. The two names were eventually combined to "ringworm". Although the actual cause of ringworm was not known until the early 1800s, the practice of segregating infected individuals to prevent spread of the disease indicated that there was knowledge that this disease was contagious and prevalent, and could be passed from person to person was known prior to the cause of infection.

It was Gruby that isolated and described one of the ringworm fungi, Trichophyton, meaning "thread plant", and through inoculation on healthy parts of the scalp, was able to reproduce the disease. He also carried out the same experiment with several other human pathogenic fungi and inoculated himself with the pathogen, as well as others. Although this was a great accomplishment, Gruby also had a great deal of luck since, at the time, isolation of specific fungi was not common practice, and this was also 30 years prior to the development of techniques to grow fungi and bacteria in pure culture.

According to Ainsworth, more than 350 species of dermatophytes have been proposed and given approximately 1,000 names, which has caused some confusion in medical mycology. The proliferation in names have come about because different researchers have worked with the same species of a pathogen, in a different place and time, and each were familiar with that particular isolate of that particular species of fungus and probably not too familiar with that isolate. They grew it in culture briefly and published on it. Thus, each newly isolated dermatophytic fungus was given a name, sometimes according to the symptoms with which it was associated or according to the part of the body affected, i.e. top of the head, neck, face, hand, arm, leg and foot, or even the geographical region, or sometimes just for the sake of publication (due to the competitiveness of medical schools) . This led to a great deal of confusion in the understanding of mycoses.

Although the first species were described in the 1840s, they were little studied until the 1940s, when the United States military personnel, while fighting in the South Pacific, during WWII, contracted ringworm and other fungi in the humid tropics (an example of advancement of knowledge due to driven research). This led to an intensive study, by the government of such fungi with many species being reduced to synonyms. For example, 172 species were reduced to Candida albicans.

FUNGI PATHOGENIC SYSTEM

Most fungi can be easily seen. Some are large and conspicuous, like the fleshy mushrooms of field and forest. Some are stout and woody, like the bracket fungi on rotting timber. The pathogenic fungi, the ones that produce disease, more closely resemble the molds that overrun our shoes and stored leather goods in damp weather, damage foods, rot fabrics, and, in happier circumstances, produce valuable drugs and chemicals and give the desired flavour to our choicest cheeses. The most important detail to remember about fungi is that they are living organisms. They require, therefore, a source of energy.

Like animals, but unlike green plants, fungi cannot convert the energy of the sun and thus synthesize the food they need. They must therefore get food from some external source. In so doing, the slender, infinitely tiny, often colourless threads that make up the fungus itself live and grow at the expense of the environment in which they exist. If the environment is another plant, particularly if it is one of our cultivated agricultural plants, the resulting situation Is regarded as a plant disease.

Fig. Apple Scab Fungus

Thus the apple scab fungus, familiar to anyone who has tried to raise apples in practically any part of the United States, forms olive-green patches on the leaves and fruits of its host and Profits at its expense. The same is true of the grain rust fungi, the corn smut fungus, and a myriad of others. They grow thus because by their very nature they can grow successfully nowhere else.

A pathogenic fungus is but one of countless individuals in the organic world. To maintain its position it must live and grow, reproduce, be scattered about, and, under certain circumstances, survive periods during which the environment is relatively unfavourable to it. When one knows the essentials of how fungi accomplish this assignment, one knows the essentials of how fungi cause disease in plants.

Structurally, fungi are mostly very tiny but very complex. Most of them are composed of microscopic filaments, which may be colourless or variously coloured but which never contain chlorophyll, the green pigment responsible for binding the sun's energy in the formation of sugar.

Considered individually, the filaments are called hyphae (singular: hypha). In the aggregate they are spoken of as a mycelium. The filaments are essentially like the cells of which other plants are composed, each being surrounded by a thin membranous wall and composed of a jellylike living protoplasm.

Although the filamentous growth form is common to practically all plant pathogenic fungi, the actual appearance of the fungi is enormously varied.

Some, like the rhizopus soft rot fungi of sweetpotatoes and strawberries or the scab fungi of wheat, have no particularly characteristic form and appear as a downy growth over the surface of the affected host.

Others develop aggregates of filaments with a recognizable size and shape, of distinctive colour, and often decidedly firm to the touch. Such aggregates are commonly associated with the actively reproducing phase of development, and the "fruiting body" so formed serves as one of the most useful means of distinguishing one plant pathogenic fungus from another.

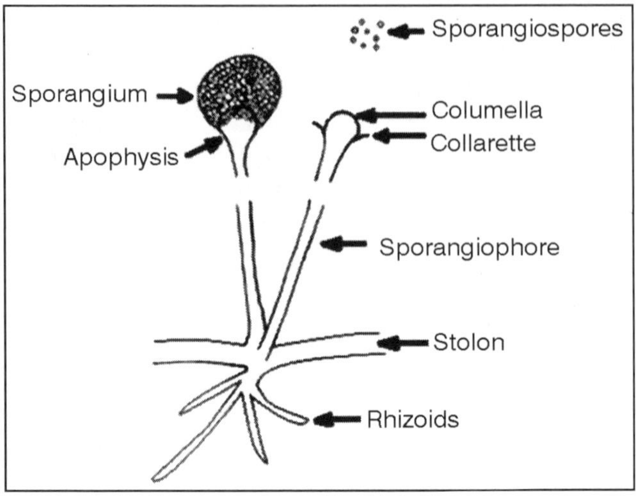

Fig. Rhizopus

The physiology of fungi (like the physiology of all plants and animals) is by no means completely understood. We have enough information to suggest that the physical and chemical processes that take place in the protoplasm of the fungi are essentially the same as those of any other living organism. Indeed, those processes, summed up in the term metabolism, are strikingly similar throughout the organic world. Because the fungi lack chlorophyll, their nutrition is most like that to be found in the colourless, or rather nongreen, parts of higher plants, except that many fungi secrete digestive enzymes as do the cells of the digestive tract of man and other animals.

Because fungi depend for their food on outside sources, and because some establish themselves on other living plants and some on nonliving food sources, a great deal has been made of the terms parasite and saprophyte, supposedly to describe these two alternative conditions. Attempts to apply the terms soon demonstrate their inadequacy, at least when used in unmodified form. It is much more useful to speak of obligate parasites, which can grow only on living host plants, and facultative parasites, which can grow on nonliving food supplies but which at times do establish themselves on a living host. True saprophytes, which live exclusively on nonliving substrata, can hardly be of importance in a discussion of plant diseases.

Fig. Saprophytes

Besides an ability to utilize food and grow, a fungus must reproduce. Because it is the most intriguing phase of the life history and because it furnishes most of the information necessary for identification, the reproductive aspects of fungi have received much attention.

By reproduction we ought always to mean an increase in the total number of individuals of a given species, a process which usually represents successive generations and thus assures the perpetuation of the species in time as well as its maintenance in space. It is an error in logic, however, to conclude that fungi reproduce in order to survive; one can say simply that they survive because they reproduce.

5

Plant Metabolism

INTRODUCTION

All biological organisms have defence mechanisms to protect them from the negative effects of small quantities of foreign compounds (xenobiotics). These foreign compounds in plants include pesticides. Herbicides are a type of pesticide toxic to plants specifically, because they inhibit metabolic pathways unique to plants (*e.g.* photosynthesis). Herbicides are used successfully in weed management systems because they selectively harm sensitive weeds while leaving crops undamaged.

Many factors contribute to successful herbicide action (*e.g.* successful placement or absorption and movement to target sites); however, the major reason herbicides are selective against weeds in crops is because crop plants are able to metabolize the herbicide to a non-toxic form. The basis for herbicide selectivity relies on enzymatic systems used in the plant's normal metabolic processes. The relative rates of herbicide absorption, translocation, and metabolism usually determine whether or not a herbicide will elicit a phytotoxic response.

To better understand how these processes may each influence herbicidal action, they should be considered separately. Absorption has primary control over translocation, metabolism and phytotoxic action because the total amount of herbicide available for these processes is determined by the amount of herbicide absorbed by the plant. Metabolism influences both herbicide absorption and phytotoxic action. Metabolism generally converts the herbicide to a form with reduced phytotoxicity, thus, increasing the concentration gradient of the parent herbicide, so more is absorbed. Herbicide metabolism also influences phytotoxic action by either rendering the herbicide less or more active.

EVOLUTION OF SECONDARY METABOLISM

Structure of Azadirachtin, a terpenoid produced by the Neem plant, which helps ward off microbes and insects. Many secondary metabolites have complex structures Plant secondary metabolites are low molecular weight compounds, sometimes with complex structures that have no essential role in primary metabolism.

They function in processes such as anti-herbivory, pollinator attraction, communication between plants, allelopathy, maintenance of symbiotic associations with soil flora and enhancing the rate of fertilization. Secondary metabolites have great structural and functional diversity and many thousands of enzymes may be involved in their synthesis, coded for by as much as 15–25 per cent of the genome.

Many plant secondary metabolites such as the colour and flavour components of saffron and the chemotherapeutic drug taxol are of culinary and medical significance to humans and are therefore of commercial importance.

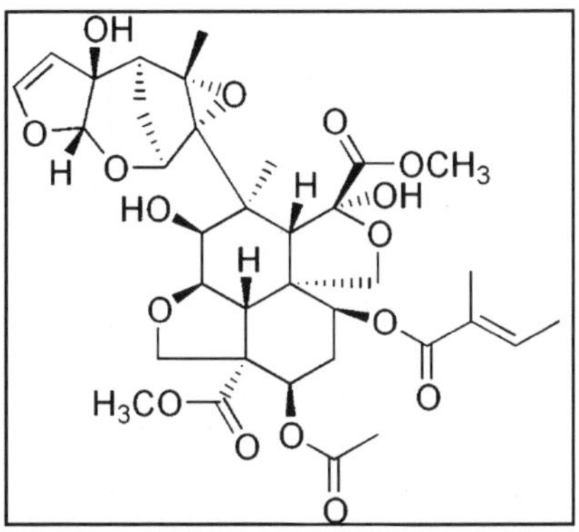

Since bacteria possess the ability to make secondary metabolites such as the antibiotic penicillin, their production began quite early during evolution. But they assume more significant roles in eukaryotic and multicellular organisms. In plants they seem to have diversified using mechanisms such as gene duplications, evolution of novel genes and the development of novel biosynthetic pathways. Studies have shown that diversity in some of these compounds may be positively selected for.

Cyanogenic glycosides may have been proposed to have evolved multiple times in different plant lineages, and there are several other instances of convergent evolution. For example, the enzymes for synthesis of limonene–a terpene–are more similar between angiosperms and gymnosperms than to their own terpene synthesis enzymes. This suggests independent evolution of the limonene biosynthetic pathway in these two lineages.

MECHANISMS AND PLAYERS IN EVOLUTION

While environmental factors are significantly responsible for evolutionary change, they act merely as agents for natural selection. Change is inherently brought about via phenomena at the genetic level–mutations, chromosomal rearrangements and epigenetic changes. While the general types of mutations hold true across the living world, in plants, some other mechanisms have been implicated as highly significant.

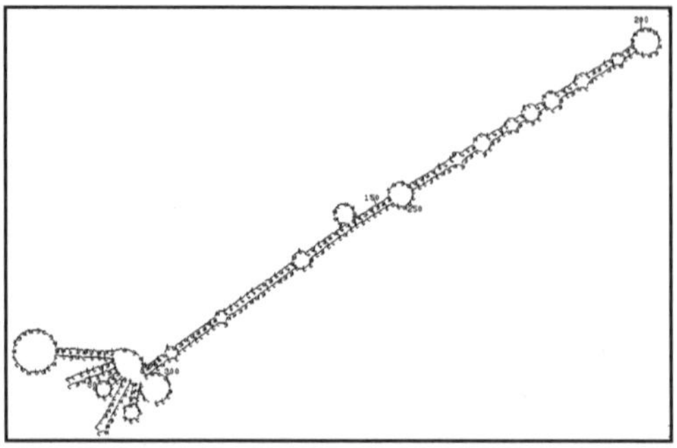

Fig. The Stem-loop Secondary Structure of a Pre-micro RNA from *Brassica Oleracea.*

Fig. Extreme Left: Teosinte; Extreme Right: Maize; Middle: Maize-teosinte Hybrid.

Polyploidy is a very common feature in plants. It is believed that at least half (*and probably all*) plants are or have been polyploids. Polyploidy leads to genome doubling, thus generating functional redundancy in most genes. The duplicated genes may attain new function, either by changes in expression pattern or changes in activity. Polyploidy and gene duplication are believed to be among the most powerful forces in evolution of plant form. It is not known though, why genome doubling is such a frequent process in plants. One probable reason is the production of large amounts of secondary metabolites in plant cells. Some of them might interfere in the normal process of chromosomal segregation, leading to polypoidy.

In recent times, plants have been shown to possess significant microRNA families, which are conserved across many plant lineages. In comparison to animals, while the number of plant miRNA families is less, the size of each family is much larger. The miRNA genes are also much more spread out in the genome than those in animals, where they are found clustered. It has been proposed that these miRNA families have expanded by duplications of chromosomal regions. Many miRNA genes involved in regulation of plant development have been found to be quite conserved between plants studied.

Domestication of plants such as maize, rice, barley, wheat etc. has also been a significant driving force in their evolution. Some studies have looked at the origins of the maize plant and found that maize is a domesticated derivative of a wild plant from Mexico called teosinte. Teosinte belongs to the genus *Zea*, just as maize, but bears very small inflorescence, 5–10 hard cobs, and a highly branched and spread-out stem.

Fig. Cauliflower: *Brassica Oleracea var Botrytis.*

Crosses between a particular teosinte variety and maize yield fertile offspring that are intermediate in phenotype between maize and teosinte. QTL analysis has also revealed some loci that when mutated in maize yield a teosinte-like stem or teosinte-like cobs. Molecular clock analysis of these genes estimates their origins to some 9000 years ago, well in accordance with other records of maize domestication. It is believed that a small group of farmers must have selected some maize-like natural mutant of teosinte some 9000 years ago in Mexico, and subjected it to continuous selection to yield the maize plant as known today. Another case is that of cauliflower. The edible cauliflower is a domesticated version of the wild plant *Brassica oleracea*, which does not possess the dense undifferentiated inflorescence, called the curd, that cauliflower possesses.

Cauliflower possesses a single mutation in a gene called *CAL*, controlling meristem differentiation into inflorescence. This causes the cells at the floral meristem to gain an undifferentiated identity, and instead of growing into a flower, they grow into a lump of undifferentiated cells. This mutation has been selected through domestication at least since the Greek empire.

ALTERNATION OF GENERATIONS

Alternation of generations (also known as alternation of phases or metagenesis) is a term primarily used to describe the life cycle of plants (taken here to mean the Archaeplastida). A multicellular gametophyte, which is haploid with *n* chromosomes, alternates with a multicellular sporophyte, which is diploid with $2n$ chromosomes, made up of n pairs. A mature sporophyte produces spores by meiosis, a process which reduces the number of chromosomes to half, from $2n$ to n.

Because meiosis is a key step in the alternation of generations, it is likely that meiosis has a fundamental adaptive function. The nature of this function is still unresolved (see Meiosis), but the two main ideas are that meiosis is adaptive because it facilitates repair of DNA damages and/or that it generates genetic variation.

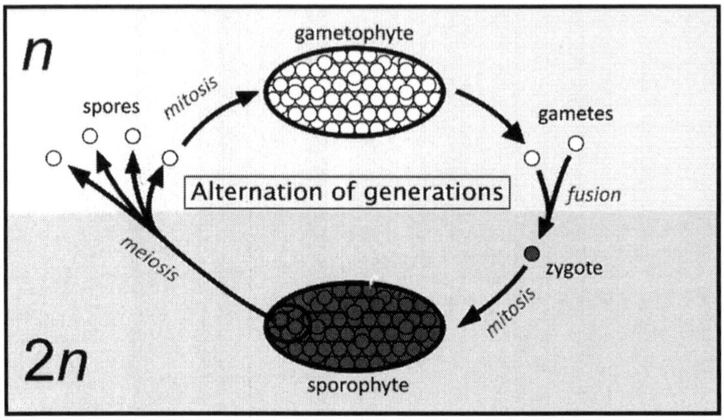

Fig. Diagram Showing the Alternation of Generations between a Diploid Sporophyte (Bottom) and a Haploid Gametophyte (Top).

The haploid spores germinate and grow into a haploid gametophyte. At maturity, the gametophyte produces gametes by mitosis, which does not alter the number of chromosomes. Two gametes (originating from different organisms of the same species or from the same organism) fuse to produce a zygote, which develops into a diploid sporophyte.

This cycle, from gametophyte to gametophyte (or equally from sporophyte to sporophyte), is the way in which all land plants and many algae undergo sexual reproduction.

The relationship between the sporophyte and gametophyte varies among different groups of plants. In those algae which have alternation of generations, the sporophyte and gametophyte are separate independent organisms, which may or may not have a similar appearance. In liverworts, mosses and hornworts, the sporophyte is less well developed than the gametophyte and is largely dependent on it. Although moss and hornwort sporophytes can photosynthesise, they require additional photosynthate from the gametophyte to sustain growth and spore development and depend on it for supply of water, mineral nutrients and nitrogen.

By contrast, in all modern vascular plants the gametophyte is less well developed than the sporophyte, although their Devonian ancestors had gametophytes and sporophytes of approximately equivalent complexity. In ferns the gametophyte is a small flattened autotrophic prothallus on which the young sporophyte is briefly dependent for its nutrition. In flowering plants, the reduction of the gametophyte is much more extreme; it consists of just a few cells which grow entirely inside the sporophyte.

All animals develop differently. A mature animal is diploid and so is, in one sense, equivalent to a sporophyte. However, an animal *directly* produces haploid gametes by meiosis. No haploid spores capable of dividing are produced, so neither is a haploid gametophyte. There is no alternation between diploid and haploid forms.

Other organisms, such as fungi, can have life cycles in which different kinds of organism alternate. The term 'alternation of generations' has also been applied to these cases. Life cycles of plants and algae with alternating haploid and diploid phases are referred to as diplohaplontic (the equivalent terms haplodiplontic, diplobiontic or dibiontic are also in use). Life cycles, such as those of animals, in which there is only a diploid multicellular phase are referred to as diplontic. (Life cycles in which there is only a haploid multicellular phase are referred to as haplontic.)

DEFINITION

The discussion of 'alternation of generations' above treats the alternation of a multicellular diploid form with a multicellular haploid form as the defining characteristic, regardless of whether these forms are free-living or not. In some species, such as the alga *Ulva lactuca,* the diploid and haploid forms are indeed both free-living independent organisms, essentially identical in appearance and therefore said to be isomorphic. The free-swimming, haploid gametes form a diploid zygote which germinates into a multicellular diploid sporophyte. The sporophyte produces free-swimming haploid spores by meiosis that germinate into haploid gametophytes.

However, in some other groups, either the sporophyte or the gametophyte is very much reduced and is incapable of free living. For example, in all bryophytes the gametophyte generation is dominant and the sporophyte is dependent on it. By contrast, in all modern vascular land plants the gametophytes are strongly reduced, although the fossil evidence indicates that they were derived from isomorphic ancestors In seed plants, the female gametophyte develops totally within the sporophyte which protects and nurtures it and the embryo sporophyte that it produces. The pollen grains, which are the male gametophytes are reduced reduced to only three cells. Here the notion of two generations is less obvious; as Bateman and Dimichele say "[s]porophyte and gametophyte effectively function as a single organism". The alternative term 'alternation of phases' may then be more appropriate.

ALTERNATION OF GENERATIONS IN PLANTS

Fundamental Elements

The diagram below shows the fundamental elements of the alternation of generations in plants. The many variations found in different groups of plants are described by use of these concepts later in the article.

Starting from the right of the diagram, the processes involved are as follows:

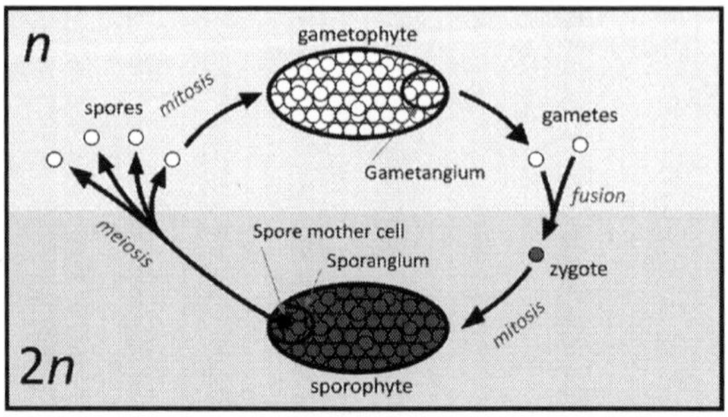

- Two single-celled haploid gametes, each containing *n* unpaired chromosomes, fuse to form a single-celled diploid zygote, which now contains *n* pairs of chromosomes, *i.e.* 2*n* chromosomes in total.
- The single-celled diploid zygote germinates, dividing by the normal process (mitosis), which maintains the number of chromosomes at 2*n*. The result is a multi-cellular diploid organism, called the *sporophyte* (because at maturity it produces spores).
- When it reaches maturity, the sporophyte produces one or more sporangia (singular: sporangium) which are the organs that produce diploid spore mother cells (sporocytes). These divide by a special process (meiosis) that reduces the number of chromosomes by a half. This results in four single-celled haploid spores, each containing *n* unpaired chromosomes.
- The single-celled haploid spore germinates, dividing by the normal process (mitosis), which maintains the number of chromosomes at N. The result is a multi-cellular haploid organism, called the *gametophyte* (because it produces gametes at maturity).
- When it reaches maturity, the gametophyte produces one or more gametangia (singular: gametangium) which are the organs that produce haploid gametes. At least one kind of gamete possesses some mechanism for reaching another gamete in order to fuse with it.

The 'alternation of generations' in the life cycle is thus between a diploid (2*n*) generation of sporophytes and a haploid (*n*) generation of gametophytes. Gametophyte of the fern *Onoclea sensibilis* (the flat thallus at the bottom of the picture) with a descendant sporophyte beginning to grow from it (the small frond at the top of the picture).

The situation is quite different from that in all animals, where the fundamental process is that a diploid (2*n*) individual *directly* produces haploid (*n*) gametes by meiosis. Spores (*i.e.* haploid cells which are able to undergo mitosis) are not produced, so neither is a haploid multi-cellular organism. The single-celled gametes are the only entities which are haploid.

Variations

The diagram shown above is a good representation of the life cycle of some multi-cellular algae (*e.g.* the genus *Cladophora*) which have sporophytes and gametophytes of almost identical appearance and which do not have different kinds of spores or gametes. However, there are many possible variations on the fundamental elements of a life cycle which has alternation of generations. Each variation may occur separately or in combination, resulting in a bewildering variety of life cycles. The terms used by botanists in describing these life cycles can be equally bewildering. As Bateman and Dimichele say "[...] the alternation of generations has become a terminological morass; often, one term represents several concepts or one concept is represented by several terms."

Possible Variations are

- *Relative importance of the sporophyte and the gametophyte.*
  - *Equal* (homomorphy or isomorphy). Filamentous algae of the genus *Cladophora,* which are predominantly found in fresh water, have diploid sporophytes and haploid gametophytes which are externally indistinguishable. No living land plant has equally dominant sporophytes and gametophytes, although some theories of the evolution of alternation of generations suggest that ancestral land plants did.
  - *Unequal* (heteromorphy or anisomorphy).

Gametophyte of *Mnium hornum*, a moss.

i. *Dominant gametophyte* (gametophytic). In liverworts, mosses and hornworts, the dominant form is the haploid gametophyte. The diploid sporophyte is not capable of an independent existence, gaining most of its nutrition from the parent gametophyte, and having no chlorophyll when mature.

Fig. Sporophyte of *Blechnum discolour*, a fern.

ii. *Dominant sporophyte* (sporophytic). In ferns, both the sporophyte and the gametophyte are capable of living independently, but the dominant form is the diploid sporophyte. The haploid gametophyte is much smaller and simpler in structure. In seed plants, the gametophyte is even more reduced (at the minimum to only three cells), gaining all its nutrition from the sporophyte. The extreme reduction in the size of the gametophyte and its retention within the sporophyte means that when applied to seed plants the term 'alternation of generations' is somewhat misleading: "[s]porophyte and gametophyte effectively function as a single organism". Some authors have preferred the term 'alternation of phases'.

- *Differentiation of the gametes.*
  - *Both gametes the same* (isogamy). Like other species of *Cladophora, C. callicoma* has flagellated gametes which are identical in appearance and ability to move.
  - *Gametes of two distinct sizes* (anisogamy).

i. *Both of similar motility*. Species of *Ulva*, the sea lettuce, have gametes which all have two flagella and so are motile. However they are of two sizes: larger 'female' gametes and smaller 'male' gametes.

ii. *One large and sessile, one small and motile* (oogamy). The larger sessile megagametes are eggs (ova), and smaller motile microgametes are sperm (spermatazoa, spermatozoids). The degree of motility of the sperm may be very limited (as in the case of flowering plants) but all are able to move towards the sessile eggs. When (as is almost always the case) the sperm and eggs are produced in different kinds of gametangia, the sperm-producing ones are called antheridia (singular antheridium) and the egg-producing ones archegonia (singular archegonium).

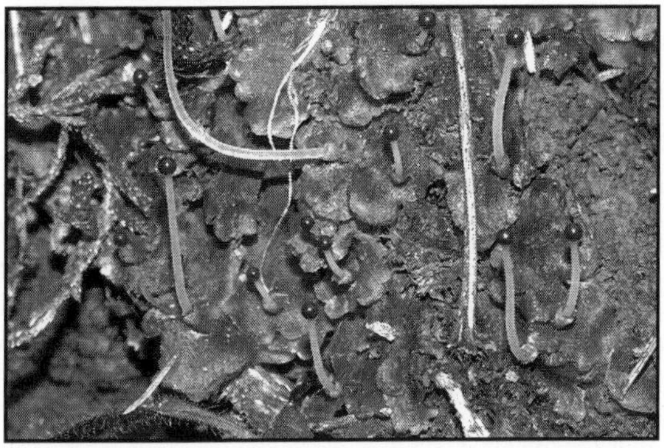

Fig. Gametophyte of *Pellia epiphylla* with Sporophytes Growing from the Remains of Archegonia.

iii. *Antheridia and archegonia occur on the same gametophyte,* which is then called monoicous. (Many sources, including those concerned with bryophytes, use the term 'monoecious' for this situation and 'dioecious' for the opposite. Here 'monoecious' and 'dioecious' are used only for sporophytes.) The liverwort *Pellia epiphylla* has the gametophyte as the dominant generation. It is monoicous: the small reddish sperm-producing antheridia are scattered along the midrib while the egg-producing archegonia grow nearer the tips of divisions of the plant.

iv. *Antheridia and archegonia occur on different gametophytes,* which are then called dioicous. The moss *Mnium hornum* has the gametophyte as the dominant generation. It is dioicous: male plants produce only antheridia in terminal rosettes, female plants produce only archegonia in the form of stalked capsules. Seed plants are also dioicous. However, the gametophytes are extremely reduced in size; the archegonium consists only of a small number of cells, and the entire male gametophyte may be represented by only two cells.

- *Differentiation of the spores.*
  - *All spores the same size* (homospory or isospory). Horsetails (species of *Equisetum*) have spores which are all of the same size.
  - *Spores of two distinct sizes* (heterospory or anisospory): larger megaspores and smaller microspores. When the two kinds of spore are produced in different kinds of sporangia, these are

called megasporangia and microsporangia. A megaspore often (but not always) develops at the expense of the other three cells resulting from meiosis, which abort.

i. *Megasporangia and microsporangia occur on the same sporophyte,* which is then called monoecious. Most flowering plants fall into this category. Thus the flower of a lily contains six stamens (the microsporangia) which produce microspores which develop into pollen grains (the microgametophytes), and three fused carpels (the megasporangia) which produce megaspores which develop into ovules (the megagametophytes). In other plants, such as hazel, some flowers have only stamens, others only carpels, but the same plant (*i.e.* sporophyte) has both kinds of flower and so is monoecious.

Fig. Flowers of European Holly, a Dioecious Species: Male Above, Female below (Leaves Cut to Show Flowers more Clearly).

ii. *Megasporangia and microsporangia occur on different sporophytes,* which are then called dioecious. An individual tree of the European holly (*Ilex aquifolium*) produces either 'male' flowers which have only functional stamens (microsporangia) producing microspores which develop into pollen grains (microgametophytes) or 'female' flowers which have only functional carpels (megasporangia) producing megaspores which develop into ovules (megagametophytes).

There are some correlations between these variations, but they are just that, correlations, and not absolute. For example, in flowering plants, microspores ultimately produce microgametes (sperm) and megaspores ultimately produce megagametes (eggs). However, in ferns and their allies there are groups with undifferentiated spores but differentiated gametophytes. For example, the fern *Ceratopteris thalictrioides* has spores of only one kind, which vary continuously in size. Smaller spores tend to germinate into gametophytes which produce only sperm-producing antheridia.

A COMPLEX LIFE CYCLE

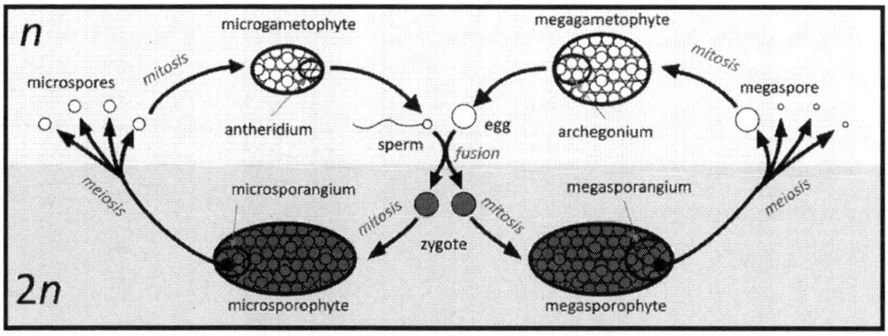

Fig. Graphic Referred in Text.

The diagram shows the alternation of generations in a species which is heteromorphic, sporophytic, oogametic, dioicous, heterosporic and dioecious. A seed plant example might be a willow tree (most species of the genus *Salix* are dioecious). Starting in the centre of the diagram, the processes involved are:

- An immobile egg, typically remaining in the archegonium, fuses with a mobile sperm, released from an antheridium. The resulting zygote is either 'male' or 'female'.
  - A 'male' zygote develops by mitosis into a microsporophyte, which at maturity produces one or more microsporangia. Microspores develop within the microsporangium by meiosis. In a willow (like all seed plants) the zygote first develops into a seed within the ovule (megasporangium). Later the seed is shed and grows into a mature tree. A 'male' willow tree (a microsporophyte) produces flowers with only stamens, the anthers of which are the microsporangia.
  - Microspores germinate producing microgametophytes; at maturity one or more antheridia are produced. Sperm develop within the antheridia. In a willow, microspores are not liberated from the anther (the microsporangium), but develop into pollen grains (microgametophytes) within it. The whole pollen grain is moved (*e.g.* by an insect or by the wind) to an ovule (megagametophyte), where a sperm is produced which moves down a pollen tube to reach the egg.
  - A 'female' zygote develops by mitosis into a megasporophyte, which at maturity produces one or more megasporangia. Megaspores develop within the megasporangium; typically one of the four spores produced by meiosis gains bulk at the expense of the remaining three, which disappear. 'Female' willow trees (megasporophytes) produce flowers with only carpels (the megasporangia).

- Megaspores germinate producing megagametophytes; at maturity one or more archegonia are produced. Eggs develop within the archegonia. In a willow, megaspores develop into ovules (megagametophytes) within the carpels (megasporangia). An archegonium develops within the ovule and produces an egg. All of this happens within the carpel (the megasporangium). The whole of the gametophytic 'generation' remains within the protection of the sporophyte except for pollen grains (which have been reduced to just three cells).

LIFE CYCLES OF DIFFERENT PLANT GROUPS

The term 'plants' is taken here to mean the Archaeplastida, *i.e.* the glaucophytes, red and green algae and land plants. Alternation of generations occurs in almost all multicellular red and green algae, both freshwater forms (such as *Cladophora*) and seaweeds (such as *Ulva*). In most, the generations are homomorphic (isomorphic) and free-living. Some species of red algae have a complex triphasic

Land plants all have heteromorphic (anisomorphic) alternation of generations, in which the sporophyte and gametophyte are distinctly different. All bryophytes, *i.e.* liverworts, mosses and hornworts, have the gametophyte generation as the most conspicuous. As an illustration, consider a monoicous moss. Antheridia and archegonia develop on the mature plant (the gametophyte). In the presence of water, the biflagellate sperm from the antheridia swim to the archegonia and fertilisation occurs, leading to the production of a diploid sporophyte. The sporophyte grows up from the archegonium. Its body comprises a long stalk topped by a capsule within which spore-producing cells undergo meiosis to form haploid spores. Most mosses rely on the wind to disperse these spores, although *Splachnum sphaericum* is entomophilous, recruiting insects to disperse its spores.

In ferns and their allies, including clubmosses and horsetails, the conspicuous plant observed in the field is the diploid sporophyte. The haploid spores develop in sori on the underside of the fronds and are dispersed by the wind (or in some cases, by floating on water). If conditions are right, a spore will germinate and grow into a rather inconspicuous plant body called a prothallus. The haploid prothallus does not resemble the sporophyte, and as such ferns and their allies have a heteromorphic alternation of generations. The prothallus is short-lived, but carries out sexual reproduction, producing the diploid zygote that then grows out of the prothallus as the sporophyte.

In the spermatophytes, the seed plants, the sporophyte is the dominant multicellular phase; the gametophytes are strongly reduced in size and very different in morphology. The entire gametophyte generation, with the sole exception of pollen grains (microgametophytes), is contained within the sporophyte. The life cycle of a dioecious flowering plant (angiosperm), the willow, has been outlined in some detail in an earlier section (A complex life cycle).

The life cycle of a gymnosperm is similar. However, flowering plants have in addition a phenomenon called 'double fertilization'. Two sperm nuclei from a pollen grain (the microgametophyte), rather than a single sperm, enter the archegonium of the megagametophyte; one fuses with the egg nucleus to form the zygote, the other fuses with two other nuclei of the gametophyte to form 'endosperm', which nourishes the developing embryo.

EVOLUTIONARY EMERGENCE OF THE DOMINANT DIPLOID PHASE

It has been proposed that the basis for the emergence of the diploid phase of the life cycle (sporophyte) as the dominant phase (*e.g.* as in vascular plants) is that diploidy allows masking of the expression of deleterious mutations through genetic complementation. Thus if one of the parental genomes in the diploid cells contained mutations leading to defects in one or more gene products, these deficiencies could be compensated for by the other parental genome (which nevertheless may have its own defects in other genes). As the diploid phase was becoming predominant, the masking effect likely allowed genome size, and hence information content, to increase without the constraint of having to improve accuracy of DNA replication. The opportunity to increase information content at low cost was advantageous because it permitted new adaptations to be encoded. This view has been challenged, with evidence showing that selection is no more effective in the haploid than in the diploid phases of the lifecycle of mosses and angiosperms.

PATHOGEN RECOGNITION

Plants use a vast array of signals originating from micro-organisms and the environment to recognise pathogens and elicit plant defence responses. Non-specific elicitors of biotic and abiotic origin induce host defences in a broad range of host species. Abiotic elicitors such as heavy metal ions or UV light can induce stress responses in exposed tissues, which may provide an additional barrier to invading pathogens or alternatively, increase the plant's susceptibility to infection. Biotic elicitors include cell wall fragments released from fungi and bacteria, hydrolytic enzymes of plant or pathogen origin, some peptides, glycoproteins and polyunsaturated fatty acids. These elicitors induce defence responses in a range of host species. Often, non-specific elicitors act as a general indication that the cell has been damaged in some way (for example, the release of fragments of the host's own cell wall can elicit defence responses).

Specific elicitors enable defence against a very specific pathogen, and are conditioned by *avirulence genes in that pathogen.* Avirulence genes determine the pathogen's host range, but are only able to function in the presence of another set of genes, the 'hypersensitive response and pathogenicity' (Hrp) gene cluster. Some Hrp gene products are involved in disguising the pathogen from host recognition, thus playing a role in both virulence and avirulence. For a biotroph to form a successful infection, it must establish a basic compatibility with its host.

The pathogen may also produce compatibility factors that delay, avoid or negate recognition by a normally resistant host plant. Virulent strains appear to be able to suppress the resistance mechanisms of the host, but are not able to halt resistance responses once they are activated. Incompatibility between a host and a pathogen results in the recognition of the pathogen and activation of defence mechanisms, while compatibility results in infection. Specific elicitors are encoded by avirulence genes, and these peptides are believed to bind to receptor peptides, encoded by host resistance genes. Recognition of the avirulence gene products by the host triggers signal transduction pathways that cause a massive shift in gene transcription and plant cell metabolism, and local and systemic signals are released that prime the rest of the plant against further infection. The presence of non-specific elicitors, such as the release of host and pathogen wall fragments, during this process may amplify the defence response.

Host-parasite specific resistance is determined by the interaction of between products of pathogen avirulence genes, Specific elicitors and products of host resistance genes. The defence responses of plants can be very rapid. Host gene expression begins within minutes, or even seconds, of exposure to elicitors or pathogens. A diverse range of elicitors can induce a common set of responses in the host, suggesting that second messengers are involved in the signalling pathway between pathogen attack and host response.

RAPID ACTIVE DEFENCES

Almost every host-parasite interaction is unique in the details of the activation, localisation, timing and magnitude of the defence responses.

At the Membrane

The host membrane appears to be involved in the earliest stages of pathogen recognition and signal transduction. A change in membrane permeability after exposure to a pathogen causes fluxes in ions, such as K+, H+ and Ca2+, and results in changes to gene activation and the triggering of the defence responses. Also at the membrane, the 'oxidative burst', which involves the generation of reactive oxygen species, such as hydrogen peroxide, triggers signals that affect gene expression, cross-linking in the host cell wall and initiation of later defence responses. The reactive oxygen species at the site of infection are also produced in quantities capable of killing micro-organisms directly.

At the Cell Wall

Preparations for the reinforcement of the cell wall, which can improve host resistance, begin very quickly after a pathogen attempts to penetrate a host cell. This is characterised by an intensification of cytoplasmic streaming and the accumulation of host cytoplasm around the site of attempted penetration.

The cytoplasmic aggregates are thought to contain cellular apparatus for the synthesis of cell wall fortifications. If the host cell can repair and reinforce its cell walls quickly enough, it might reduce the penetration efficiency of the pathogen. Several types of reinforcement are produced by host cells. A papilla is a deposit of callose, silicon, lignin and proteins between the cell wall and cell membrane, directly below the point of attempted penetration, while lignitubers are lignified callose reinforcements that ensheath invading hyphal tips.

Hydroxyproline-rich glycoproteins are structural cell wall proteins involved in secondary cell wall thickening. The expression of genes governing their production is activated ahead of invading hyphae, reinforcing walls. Cross-linking of hydroxyproline-rich glycoproteins caused by the release of hydrogen peroxide in the oxidative burst also reinforces cell wall compartments.

Rapid deposition of lignin and suberin following infection also increases resistance to pathogens in many plants. Lignin can also bind to hyphal tips and bacteria, physically restraining them and restricting the diffusion of their enzymes and toxins into, and the extraction of water and nutrients out of, the host cell. Cell wall reinforcements tend to be larger and more quickly formed in resistant hosts than in susceptible hosts, and inhibition of the production of callose or lignin synthesis by the pathogen enhances its penetration efficiency.

The Hypersensitive Response

Hypersensitive cell death is another widespread mechanism used by hosts to prevent the spread of a pathogen. Infected cells and those surrounding them "suicide", preventing further spread, and in some cases, killing the pathogen. It is often associated with the initiation of other responses, such as lignification and the synthesis of anti-microbial compounds. The success of hypersensitive cell death as a resistance mechanism depends on the nutritional requirements of the specific pathogen and the timing, magnitude and location of the host response.

Antibiotic Compounds

Phytoalexins are low molecular weight antibiotics produced by many (but not all) plants in response to infection. There are many biotic elicitors of phytoalexin production, such as cell wall components, as well as abiotic elicitors, such as heavy metals and ultraviolet light. Phytoalexins inhibit the growth of bacteria and fungi *in vivo* and *in vitro*, and production of these antibiotics during an infection can induce resistance to subsequent infections by that pathogen. Over 350 phytoalexins are known in over 100 plant species. They include pterocarpans, sesquiterpenes, cryptophenols, isocoumarins, isoflavenoids, and others. Phytoalexins may be produced by any part of the plant, although different phytoalexins can accumulate in different organs.

Generally, related plant species produce structurally-related phytoalexins, and many produce more than one, enabling the plant to present a toxic cocktail to invading pathogens. Phytoalexins are produced in cells surrounding an infection site and delivered to the infected cell packaged in lipid vesicles, creating a toxic micro-environment in the infected cell and, hopefully, preventing disease establishment. Phytoalexin accumulation is often associated with hypersensitive cell death, although only living cells can synthesise phytoalexins. Some plants can also sequester phytoalexins into vacuoles as stores of inactive sugar-conjugates, which can be cleaved and released quickly if initial defence responses are unsuccessful.

DELAYED ACTIVE DEFENCES

Delayed active defences include containment of the pathogen, wound repair, expression of pathogenesis-related proteins and the acquisition of systemic resistance. These mechanisms restrict the spread of the pathogen after infection is established and contain the damage to host tissues.

Physical Responses

The ability to repair wounds can help protect the plant from further infection by other, opportunistic pathogens. A secondary meristem in fleshy tissues, fruits, roots and bark, the cork cambium, can produce cork cells, which have thick, suberised walls. These cells can create a barrier to further colonisation by the pathogen and, in some cases, develop an abscission layer around the site of infection, causing the infected tissue separate from the healthy tissue. Wounded tree trunks often secrete protective gums that seal the wound from further infection. The formation of tyloses, ingrowths if the protoplasm in xylem parenchyma can also restrict the spread of pathogenic propagules in the xylem, although they also tend to reduce the movement of water through the vessels, causing water stress in the plant.

Pathogenesis-related Proteins

There is a range of novel proteins synthesised in response to infection, many of which have b-glucanase, chitinase or lysozyme activity. Some pathogenesis-related proteins disrupt pathogen nutrition. The presence of low levels of these proteins in healthy plants suggests that they might have other roles in plant growth and development aside from disease resistance. Chitinase and glucanase accumulate in the vacuoles, and glucanase is also sometimes secreted into the intercellular space. They dissolve the fungal cell wall, fragments of which then elicit hypersensitive cell death. The breakdown of the vacuole during decompartmentalisation of the cytoplasm results in a flood of hydrolytic enzymes, which have antiviral, antifungal and antibacterial activity. The accumulation of pathogenesis-related proteins peaks around 7-10 days after initial infection. The presence of these proteins *before* infection increases the plant's resistance to pathogens, as in the case of systemic acquired resistance.

Systemic Acquired Resistance

Systemic acquired resistance, or induced resistance, is characterised by the increased resistance of a plant to a wide range of pathogens following infection by one pathogen. It is therefore fundamentally different from the specific antigen-antibody mechanism of resistance seen in the immune response of mammals. Rather than providing immunity *per se*, systemic acquired resistance reduces the severity of later diseases.

The development of systemic acquired resistance usually requires the development of a slowly expanding necrotic lesion and other localised responses to infection, the release of a phloem-translocated signal originating from the infection site, and the subsequent priming of the plant against further attacks, allowing a more rapid response in the case of future infections. The nature of the signal that triggers systemic acquired resistance is as yet unknown, and is likely to be a complex signal transduction pathway mediated by a number of stress signals. Salicylic acid plays a key role, via interaction with salicylic acid-binding proteins that can cause build-up of reactive oxygen species or activate gene expression. The levels of salicylic acid increase around necrotic lesions and remain high in plants that have acquired resistance. It is not, however, salicylic acid itself that acts as the signal that is translocated systemically throughout the plant.

Co-ordination of Defence Responses

The success of defence responses is increased if activated in combination. Passive mechanisms, coupled with rapid active responses and slower follow-up defences provide a broad defence front to the plant. The specific interaction between host and pathogen is of course crucial to the success of the plant's resistance or the pathogen's invasion, and is mediated by the many pathways involved in producing or detecting elicitors, enhancers, suppressors and secondary signals.

PHOTOSYNTHESIS

Photosynthesis is the process by which light energy is captured, converted and stored in simple sugar molecule. This process occurs in chloroplasts and other parts of green organisms. It is a backbone process, in the sense that all life on earth depends on it's functioning.

The following equation sums up the process:

$$6CO_2 + 12\ H_2O + \text{light energy} \rightarrow C_6H_{12}O_6 + 6O_2 + 6H_2O$$

Carbon water Glucose Oxygen Water

As you see from the equation, this process is vital to us as humans, because it transforms carbon dioxide into oxygen—which we enjoy with every breath!

CARBON DIOXIDE (CO_2)

The earth's atmosphere contains approximately 79% nitrogen, 20% oxygen and the remaining 1% is a mixture of less common gases—including 0.039% carbon dioxide. Carbon dioxide in the atmosphere reaches plant mesophyll via the stomata. The carbon dioxide dissolves on the thin film of water that covers the outside of cells. The carbon dioxide then diffuses through the cell wall into the cytoplasm in order to reach the chloroplasts. The oceans hold a large reservoir of carbon dioxide, which keeps the atmospheric levels essentially constant. Although there are some indicators that the atmospheric levels of CO_2 are rising and adding to the global warming issue. That is a whole other topic though.

WATER

Water is plentiful on earth, however, it may or may not be plentiful at the location of each individual plant. Therefore, plants will close their stomata, if need be, which reduces the CO_2 supply to the mesophyll. Not even 1% of the water that is absorbed by plants is used in photosynthesis, the remainder is either transpired or incorporated into protoplasm, vacuoles or other cell materials. The water utilized in photosynthesis is the source of oxygen released as a photosynthetic byproduct.

LIGHT

Light has a dual nature, in that it exhibits properties of both waves and particles. The energy from the sun comes to earth in various wavelengths, the longest being radio waves and the shortest are gamma rays. Approximately 40% of the radiant energy the earth receives from the sun is visible light. Visible light ranges from red, 780 nanometers to violet, 390 nanometers.

The violet to blue and red to orange ranges are the most often used in photosynthesis. Most light in the green range is reflected. Of the visible light that reaches a leaf, approximately 80% is absorbed. Light intensity varies widely. Time of day, temperature, season of year, altitude, latitude and other atmospheric conditions all play roles in the intensity of the radiant energy that will reach the earth and it's organisms. High intensity light isn't necessarily a beneficial thing for plants. In high intensity light, photorespiration may occur, which is a type of respiration that uses oxygen and releases carbon dioxide but differs from standard aerobic respiration in the pathways that it utilizes.

CHLOROPHYLL

A few things to know about chlorophyll before we get into the nitty gritty of photosynthesis and respiration. There are more than one type of chlorophyll, however, they all have one atom of magnesium in the center. In some ways the chlorophyll is quite analogous to the heme structure in hemoglobin (the iron containing pigment that carries oxygen in blood). Chlorophyll has a long lipid tail that anchors the molecule in the lipid layers of the thylakoid membranes—recall that thylakoids are coin-like discs in stacks within the stroma of the chloroplasts. The chloroplasts of most plants contain two types of chlorophyll imbedded in the thylakoid membranes.

The formula for bluish-green chlorophyll a is $C_{55}H_{72}MgN_4O_5$ and the formula for yellow-green chlorophyll b is $C_{55}H_{70}MgN_4O_6$. In general, most a chloroplast has about three times as much chlorophyll a than b. The main role of chlorophyll b is to broaden the spectrum of light available for photosynthesis: chlorophyll b absorbs light energy and transfers the energy to a chlorophyll a molecule.

Other pigments are contained in chlorophyll c, d, and e and take the place for chlorophyll b in some cases. Note that all the chlorophyll molecules are related to each other and differ only slightly in molecular structure. Light-harvesting complexes contain 250 to 400 pigment molecules and are referred to as a photosynthetic unit. There are countless numbers of these units spread throughout the grana of a chloroplast. In the chloroplasts of green plants, two types of these harvesting units operate together in order to bring about the first phase of photosynthesis. The photosynthetic process occurs in two successive processes: the light reactions and the carbon-fixing reactions.

The Light Reactions

The light reactions involve light striking the chlorophyll molecules embedded in the thylakoids of chloroplasts. The subsequent reaction results in the conversion of some light energy to chemical energy. In the light reactions, water molecules are split apart into hydrogen ions and electrons and oxygen gas is released.

In addition, ATP (adenosine triphosphate) molecules are created and the hydrogen ions derived from the water molecules are involved in "loading" NADP which carries the hydrogen as NADPH. NADPH is integral in providing the hydrogen ions used in the second series of major photosynthetic reactions: the carbon-fixing reactions.

The Carbon-fixing Reactions

The carbon-fixing reactions used to be called the dark reactions because light does not play a direct role in their functioning. The reactions take place in series outside of the grana in the stroma of the chloroplast. These reactions only occur if the end products of the light reactions are available for use. Depending on the plant involved, the carbon-fixing reactions may develop or progress in different ways. The most common type of carbon-fixing reactions in plants is the process called the Calvin cycle. In the Calvin cycle, carbon dioxide from the atmosphere is combined with a 5-carbon sugar—RuBP, or ribulose bisphosphate.

The combined molecules are converted via several steps into a 6-carbon sugar, such as glucose. The ATP and NADPH molecules from the light reactions provide the energy and resources for the reactions. Some of the sugars produced are further combined into polysaccharides (strings of simple sugars) or are stored as starch within the plant. There are other variations, including the 4-carbon pathway which is usually found in desert plants (C_4 plants).

Before getting into respiration let's take a closer look at what happens in both the light reactions and the carbon-fixing reactions.

NITTY-GRITTY OF LIGHT REACTIONS

Einstein called the discrete particles of light photons. Particles (photons) and waves are both currently accepted aspects of light. The quantum (energy) of photons is different depending on what kind of light they are in. Longer wavelength light has lower photon energies, while light with shorter wavelengths have higher photon energies. As mentioned earlier, every pigment colour has a different distinctive pattern of light absorption—called the pigment's absorption spectrum. The energy levels of some of the pigment's electrons are raised when the pigment absorbs light. If energy is emitted immediately upon absorption, the effect is called fluorescence. The red part of light does this characteristically, as demonstrated when chlorophyll is placed in light it will appear red. If the absorbed energy is emitted as light after a delay, then the effect is called phosphorescence. The energy may be converted to heat or stored, as in photosynthesis within chemical bonds.

OXIDATION-REDUCTION REACTIONS

OIL RIG, a cute little mnemonic device to remember that oxidation is loss and reduction is gain. Perhaps better put, oxidation results in the net loss of an electron or electrons, while reduction results in a net gain of an electron or electrons. The electrons come from compounds within the process or donated in from previous processes. These types of chemical reactions are found scattered throughout the processes within photosynthesis and respiration.

PHOTOSYSTEMS

The two types of photosynthetic units in most chloroplasts are what constitute photosystem I and photosystem II.

Photosystem

contains photosynthetic units with 200 or more molecules of chlorophyll a, small amounts of chlorophyll b, protein saddled carotenoid pigment, and a pair of specialized reaction-center molecules of chlorophyll called P_{700}. All pigments in a photosystem are capable of absorbing photons, however, only the reaction-center molecules can really utilize the light energy. The other pigments aren't worthless in the system, as they act sort of like an antenna in gathering and passing light energy along to the reaction-center. Iron-sulphur complexed proteins initially receive electrons from P_{700} and serve as primary electron acceptors for the unit.

Photosystem

contains chlorophyll a, protein saddled beta-carotene, a small amount of chlorophyll b and special pair of reaction-center molecules of chlorophyll a otherwise called, P_{680}. The photosystem has a primary electron acceptor called pheophytin or Pheo.

For the record, the 680 and 700 in the names of the reaction-center molecules stands for the peaks in the absorption spectra of light waves of 680 nm and 700 nm.

PHOTOLYSIS

A photon of light strikes the photosystem II reaction-center, the P_{680} molecule to be exact near the inner surface of a thylakoid membrane. The received light energy excites an electron (boosts it to a higher energy level) which is an unstable reaction and thus most of the energy is lost to heat. Up to four photons at a time can strike the P680 molecule, however, it can only accept one electron at a time. The molecule of pheophytin picks up the excited electron, which then crosses the thylakoid membrane and is passed along to another acceptor called plastoquinone or Pq near the outside surface of the thylakoid membrane. Protein Z extracts electrons from water and replaces the ones lost by the P680 molecule. Protein Z contains manganese which is required in order to split water molecules. Simultaneously, as two water molecules are split and molecule of oxygen and four protons are produced. This enzyme-mediated water splitting process is called photolysis.

PHOTOPHOSPHORYLATION

Pq, the acceptor molecule, releases the excited electron into the care of an electron transport system that is sort of like a downhill bucket brigade. The transport system moves electrons extracted from water temporarily to a high-energy storage molecule called nicotinamide adenine dinucleotide phosphate ($NADP^+$). $NADP^+$ is an electron acceptor for photosystem. The transport chain is essentially iron-containing pigments, cytochromes, a copper containing protein called plastocyanin and other electron transferring molecules. As electrons are passed through the chain and protons are being shuffled through a coupling factor, ATP molecules are assembled from ADP and phosphate in a process called photophosphorylation.

A similar series of events occurs in photosystem I. After a photon of light strikes a P_{700} molecule, the resulting excited electron is passed along to an iron-sulphur molecule Fe-S which in turn passes it to another acceptor molecule ferrodoxin, (Fd). The ferrodoxin molecule releases the electron to a carrier molecule called flavin adenine dinucleotide (FAD) and then eventually on to $NADP^+$. A reduction occurs and NADP+ becomes NADPH. Electrons from photosystem II and the activities of the electron transport system replace any electrons removed from the P_{700}molecule. Because the electrons move in one direction, the movement of electrons from water to photosystem II to photosystem I to $NADP^+$ are said to be part of noncyclic electron flow. Any ATP that is produced are designated noncyclic phosphorylation.

It should be noted that photosystem I can operate independently of photosystem II. When this occurs, the electrons boosted from P_{700} reaction-center molecules (photosystem I) are passed through an intermediary acceptor molecule called P_{430} and then on to the electron transport chain. This is rather then to the ferrodoxin and $NADP^+$. After being passed through the electron transport chain, the electron is dumped back into the reaction-center of photosystem I. This process demonstrates cyclic electron flow and any ATP generated by cyclic electron flow is termed cyclic phosphorylation. Note, that no water molecules are split and no NADPH or oxygen is produced.

CHEMIOSMOSIS

Earlier we mentioned in passing a coupling factor. The enzyme necessary for mediation of the splitting of water molecules is on the inside of the thylakoid membrane. As a result of this, a proton gradient forms across the membrane and the movement of these protons is thought to be a source of energy for generating ATP. The motion is thought to be similar to molecular movement during osmosis and has hence been termed chemiosmosis. As the protons move across the membrane, they are assisted in crossing by protein channels called ATPase or coupling factor. Because of the proton movement, ADP and phosphate combine which produces ATP.

NITTY-GRITTY OF CARBON-FIXING REACTIONS

Both ATP and NADPH are important products of the light reactions and both of them play roles in the synthesis of carbohydrates from atmospheric carbon dioxide. Although the carbon-fixing reactions do not require daylight, they generally are conducted during daylight hours as there is some indication that some of the enzymes required for the processes in carbon-fixing may require some level of light. These reactions take place in the stroma of the chloroplast. Three known mechanisms of converting carbon dioxide to sugar.

The Calvin Cycle or the 3-carbon pathway

The Calvin cycle is the most common of the three mechanisms and has four main results:

1. With the assistance of the enzyme rubisco (RuBP carboxylase), six molecules of atmospheric carbon dioxide combine with six molecules of ribulose 1, 5-bisphosphate (RuBP)
2. The result of the first step is six unstable 6-carbon complexes, which immediately split into two 3-carbon molecules of 3-phosphoglyceric acid or 3PGA. This is the first stable compound in photosynthesis.
3. NADPH and ATP from the light-reactions, supply the energy required to convert the 3PGA to 12 molecules of glyceraldehydes 3-phosphate (GA3P), which is a 3-carbon sugar phosphate complex.
4. Finally, of the 12 molecules formed; 10 are restructured into six 5-carbon molecules of RuBP—the sugar that the process started with.

The sugars produced can either add to an increase in the sugar content (carbohydrate content) of the plant or they can be used in pathways that lead to the production of lipids and amino acids.

4-Carbon Pathway—C_4 plants

These plants use a 4-carbon molecule called oxaloacetic acid in place of the 3-carbon 3-phosphoglyceric acid used in step two of the Calvin cycle. Oxaloacetic acid is produced from a 3-carbon compound PEP—phosphoenolpyruvate and carbon dioxide. This process is enzyme mediated and occurs in the mesophyll cells of the leaf. Some species will convert the resulting oxaloacetic acid to aspartic, malic or other acids.

Note that the acids do not substitute for 3PGA. The 4-carbon acids migrate to the bundle sheaths surrounding the vascular bundles, where they are further converted to pyruvic acid and carbon dioxide. In returning to the mesophyll cells and interacting with ATP molecules, the pyruvic acids molecules are able to produce additional PEP. In the bundle sheath cells, the carbon dioxide formed converts into 3PGA and other molecules, by combining with RuBP. The other molecules formed are similar to the other ones formed in the Calvin cycle.

The C_4 cycle furnishes carbon dioxide to the Calvin cycle in a more roundabout way than the C_3 pathway, but there is an advantage to this extra pathway. The extra pathway greatly reduces photorespiration in C_4 plants, and this is a good thing because photorespiration is in direct competition with the Calvin cycle and even takes place in the light while the Calvin cycle is functioning. During photorespiration, RuBP reacts with oxygen to create carbon dioxide; in contrast, during photosynthesis RuBP and carbon dioxide are used to form carbohydrates.

C_4 plants are able to pick up carbon dioxide in very low concentrations via the mesophyll cells. The Calvin cycle occurs in the bundle sheath where carbon dioxide is readily available. Because of the location, the enzyme rubisco will be in a prime spot to catalyze the reaction between RuBP and carbon dioxide, rather than oxygen. As a result C_4 plants have photosynthetic rates that are two to three times higher than C_3 plants.

There are a few other characteristic features of C_4 plants worth noting:

1. C_4 plants have two types of chloroplasts and an alternate pathway for using carbon dioxide. C_3 plants only have one type of chloroplast and one pathway. Chloroplasts with starch grains and are large with very little grana, and sometimes none, in the bundle sheath cells. In the mesophyll, the small, but numerous chloroplasts have no starch grains and contain highly developed grana.
2. PEP carboxylase is found in high concentration in the mesophyll cells which permits ready conversion of carbon dioxide to carbohydrate at lower concentrations than does rubisco (in bundle sheath cells) of the Calvin cycle.
3. The temperature ranges for C_4 plants are much higher than C_3 plants which enables C_4 plants to live well in conditions that would likely kill a C_3 plant.

CAM Photosynthesis

Crassulacean acid metabolism is a modified photosynthetic system that is somewhat similar to C_4 photosynthesis in that 4-carbon compounds are produced during the carbon-fixing reactions. CAM plants accumulate malic acid in their chlorenchyma tissues at night, which is converted back to carbon dioxide during the day. In the daytime, malic acid diffuses out of the vacuoles and is converted to carbon dioxide for use in the Calvin cycle. PEP carboxylase is responsible for converting the carbon dioxide plus PEP to malic acid at night.

This modification allows for a greater amount of carbon dioxide to be converted to carbohydrate during the day than would be otherwise converted given the conditions CAM plants generally grow in. CAM plants generally close their stomata during the day in order to reduce water loss.

There are more than 20 families that contain CAM plants, including cacti, stonecrops, orchids, bromeliads and many succulents growing in regions of high light intensity. There are some succulents that do not have CAM photosynthetic capabilities, as well as non-succulents that do have the ability.

There are great resources available that go into even greater detail on these reactions. If you are interested in these titles, please don't be afraid to ask on the forum for direction.

PLANT RESPIRATION

Respiration is the group of processes that utilizes the energy that is stored through the photosynthetic processes. The steps in respiration are small enzyme-mediated steps tha release tiny amounts of immediately available energy, the energy released is usually stored in ATP molecules which allow for even more efficient use of an organism's energy. Respiration occurs in the mitochondria and cytoplasm of cells. There are several forms of respiration: aerobic—which requires oxygen, anaerobic—which occurs in the absence of oxygen, and fermentation—which also occurs in the absence of oxygen.

Aerobic respiration is the most common form of respiration and cannot be completed without oxygen gas. The controlled release of energy is the main event in aerobic respiration.

Certain types of bacteria and other organisms without oxygen gas carry on anaerobic respiration and fermentation. Compared to aerobic respiration the amount of energy released is quite small.

The main difference between aerobic respiration and fermentation is in the way hydrogen is released and combined with other substances. Two very common forms of fermentation are summed up by the following equations:

$$C_6H_{12}O_6 \rightarrow \text{(with enzymes)} \rightarrow 2C_2H_5OH + 2CO_2 + \text{energy (ATP)}$$

glucose ethyl alcohol carbon dioxide

$$C_6H_{12}O_6 \rightarrow \text{(with enzymes)} \rightarrow 2C_3H_6O_3 + \text{energy (ATP)}$$

glucose lactic acid

Note the first equation is particularly valuable to the brewing industry.

MAJOR STEPS IN RESPIRATION

Glycolysis

The first step does not require oxygen gas (O_2) and takes place in the cytoplasm. The glycolytic phase is subdivided into three main steps and several smaller ones. Each step is mediated by an enzyme. A small amount of energy is released and hydrogen atoms are removed from compounds derived from glucose.

The main gist of the steps are:

A. The glucose molecules goes through several steps and becomes a double phosphorylated fructose molecule.

B. The 6-carbon fructose molecule is split into two 3-carbon fragments, each with a phosphate, GA3P

C. Hydorgen, energy and water are removed from the GA3P molecules leaving pyruvic acid.

Glycolysis requires two molecules of ATP to get the process started. In the processes, four ATP molecules are created, with a net gain of 2 ATP molecules at the end of glycolysis. The hydrogen ions and electrons that are released are held by an acceptor molecule called NAD—nicotinamide adenine dinucleotide. The overall end products of gylcolysis is: 2-ATP molecules, 2-NADH molecules, and pyruvic acid.

The next step depends on the kind of respiration involved—aerobic, true anaerobic or fermentation.

AEROBIC RESPIRATION (WITH OXYGEN PRESENT)

The Krebs Cycle (or Citric Acid Cycle)

The Krebs cycle takes place in the fluid matrix of the cristae compartments of the mitochondria. It is called the citric acid cycle because of all the intermediate acids in the cycle. The pyruvic acid product of glycolysis is restructured, some of the CO_2 is lost and becomes acetyl CoA which then dumps into the Krebs cycle. During the restructuring of pyruvic acid, a molecule of NADH is produced. The Krebs cycle removes energy, CO_2 and hydrogen from acetyl CoA via enzyme mediated reactions of organic acids.

The hydrogen removed during the Krebs cycle is picked up by FAD and NAD acceptor molecules. The end result of the metabolizing of two acetyl CoA molecules in the Krebs cycle is: 2-ATP molecule, oxaloacetic acid (to further drive the cycle), 6-$NADH_2$ molecules, 2-$FADH_2$ molecules and $2CO_2$ molecules.

The NAD and FAD molecules and the hydrogens that they carry will be dumped into the next step in respiration in order to extract the energy from the molecules.

FACTORS REGULATING RATE OF RESPIRATION

Temperature

To a point, the higher the temperature the faster respiration occurs. At some temperature, enzymes will become inactivated, although there are thermophilic (heat-loving) organisms that do quite well in high-temperature environments. Energy from sugar is released faster as the rate of respiration increases which results in a net weight loss. Plants offset the weight loss by increasing photosynthetic production of sugar. Note that during respiration, some of the energy is lost as heat, which results in an overall increase in organism temperature—not necessarily detectible by human hands.

Water

Enzymes generally operate in the presence of water, and reduced water in a plant will reduce the rate of respiration. Seeds usually have a water content of less than 10%, while mature living cells usually are in excess of 90% water. Seeds keep better if they are kept dry as the respiration rate remains quite low. However, if a seed comes into contact with water and via imbibition swells, the respiration rate will skyrocket. The temperature could increase to the point of killing the seeds. Spontaneous combustion can occur from the respiration generated heat when a fungi or bacterium is permitted to grow on wet seeds. Kind of a neat little trivia fact to tuck away.

Oxygen—Oxygen is an important regulator of respiration. If oxygen is drastically reduced, respiration may drop off to the point of retarding growth or death. Low oxygen concentrations can lead to the onset of fermentation processes.

ASSIMILATION AND DIGESTION

Assimilation is the conversion of the sugar produced by photosynthesis to fats, proteins, complex carbohydrates and other substances. While digestion is the breakdown of large insoluble molecules by hydrolysis to smaller soluble forms that can be transported to various parts of the plant.

Summary of key differences between photosynthesis and respiration:

Photosynthesis

- Energy stored in sugar molecules
- Carbon dioxide and water used
- Increases weight
- Requires light
- Occurs in chlorophyll
- In green organisms, produces oxygen
- With light energy, produces ATP

Respiration

- Energy released from sugar molecules
- Carbon dioxide and water released
- Decreases weight
- Can occur in light or darkness
- Occurs in all living cells
- Uses oxygen (aerobic respiration)
- With energy released from sugar, produces ATP.

PRINCIPAL OF PLANT MORPHOLOGY

Despite its history of over two centuries, German plant morphology has had only four principal figures who were significant in its development: Johann Wolfgang von Goethe, Wilhelm Hofmeister, Karl von Goebel, and Wilhelm Troll. I now look at these gentlemen individually and collectively to have some sense of the path of development of plant morphology and the reasons it took the directions it did.

Troll felt that the pinnacle of morphology would be knowing the diversity of forms so well that one could predict morphologies that had not yet been discovered or described. The best analogy for Troll's goal of plant morphology is the development of the equivalent of the Periodic Table in chemistry. According to such a perspective, the range of variants could be derived from the type by quantitative variations in growth distribution, a point of view represented by Goethe's principle of variable proportions In fact, Troll's idealistic morphological accounts of the variations in plant form in his treatises are very reminiscent of D'Arcy Thompson's theoretical derivations of differences in animal morphology. Many of Troll's constructs were purely hypothetical and not tested by actual developmental studies. In a few cases, they turned out to be wrong because they were based on the false assumption that all developmental changes were strictly quantitative in nature, when in fact qualitative changes in development can also be significant. Thus, while Troll's idealistic morphological theorizing can be useful pedagogically, in other instances it can be misleading. Nevertheless, because he used variations in development as an underpinning of his typological deductions, Troll's treatises and papers are a useful resource for information on comparative plant development.

Troll's typological approach became especially murky with reference to questions of phylogeny. From von Goebel he inherited a skepticism towards the more simplistic phylogenetic deductions of the time. He had also developed his own skepticism about relying on fossil plants for a definitive picture of plant evolution; he saw the fossil record as being too fragmentary and lacunate to be able draw any valid phylogenetic conclusions from it. Such views put Troll at loggerheads with the noted German paleobotanist/ phylogenist Walter without resolution because of its partisan nature.

In reality, Troll's views on Darwinian evolution were complex. Looking at his typological philosophy superficially, with its quasi-religious overtones, it would be easy to paint him as anti-evolutionary, and there are some who have done so. However, in reality, Troll conceded that evolution was the best explanation we have of the succession of forms found in the earth's history. Darwin's recognition of the basic unity of type in different groups served as a pointof harmony between

Troll and Darwin. They differed, however, in their views about what this unity of type represented. Darwin felt it reflected the commonality of descent, whereas Troll considered it to be a form principle that was more fundamental than geneology. Because he saw his typologies as fundamentalnatural expressions, Troll felt that classification systems could be built upon them. Furthermore, since they involved a consideration of entire organisms and the integration of their form relationships, Troll believed the typologies were more significant than the individual, isolated characteristics upon which systematists tended to base their conclusions.

While Troll conceded that natural selection did play a role in the origin of some characteristics of organisms, he did not believe that it could explain all of the variations in form. Certainly, modern views of structural evolution would incorporate the idea that the organizational theme or "Bauplan" of the organismal group in question have to be taken into account in any consideration of plant evolution. Thus, Troll inveighed against a strictly random conception of the evolutionary process. Interestingly, Troll postulated that major jumps in the evolution of forms could occur, as opposed to the gradualism represented in Darwin's views. Eldredge and Gould's theory of "punctutated equilibrium" would have found some resonance with Troll, even thoughtheir philosophies were very different. Despite starting from altogether different first principles and being colored by an element of skepticism about the simplistic features of phylogenetic thinking and Darwinian dogmatism, Troll's views of plant phylogeny are not as extreme as they might seem.

Troll's comparative morphology treatise, an amalgamation of his own research and the work of others, was the most comprehensive to date. The formal work was intended to be a programmatic presentation of the complete range of vascular plant morphology, both vegetative and reproductive. Only the first three volumes dealing with vegetative morphology were published. Volume I dealt with vegetative shoot morphology, Volume II dealt with vegetative leaf morphology, and Volume III with the morphology of roots and root systems. The original intention was to follow with volumes on reproductive shoots ending with seed and seedling morphology. Unfortunately, World War II interrupted the work.

Troll's principal activity in his postwar position in Mainz was the continuation of this programmatic work, beginning with inflorescence morphology. Unfortunately, Troll's contributions in this area became so excessive in detail and quantity that he never completed the remainder of the programme. Not even his most active disciple, Focko Weberling, was able to complete the inflorescence programme, so massive and extensive was this undertaking.

In the final analysis, despite having a number of notable disciples, Troll's idiosyncratic philosophy was not practiced or promoted actively by his students. His students tended to represent much more conventional perspectives and made syntheses with other areas of interest. The only long-term benefit of Troll's typological orientation was his belief that only way to deduce types was through the rigorous and accurate characterization of nature. Thus, Troll's lasting contributions were his exceptionally detailed, lucidly presented comparative studies of plant morphology and his effective organization of the subject. While it was good to have such an extensive subject represented by a single, uniform perspective, the negative side has been that such an exhaustive treatment gives the impression that everything has been studied and therefore the subject of plant morphology is closed. Nothing could be further from the truth. No matter how extensively a given discipline may be studied by an individual, the work of that person is restricted to a particular time in history.

With the advent of new technologies and accompanying new perspectives, every discipline, regardless of its age, is subject to revision. No matter how empirically based they may have been, Troll's contributions contain uncertainties that need to be revisited if for no other reason because they were colored by his distinct philosophical biases.

MORPHOLOGY IN THE PTERIDOPHYTES

Given that the prevailing interpretation of the relationship of leaf morphology between pteridophytes and seed plants seemsquestionable, I want to look at other basic features of shoot morphology in contemporary pteridophytes to see if they exhibit any primitive features that differ fundamentally from those of contemporary seed plants. In order to provide the criteria for comparison, I first review the principles of seed-plant shoot organization that come from the long history of plantmorphology we have just traced. The fundamentality of these principles is further reinforced by demonstrating their expression in all of these pteridophyte groups.

PRINCIPLE I: RELATIONSHIP OF LEAF TO STEM

The shoots of higher plants are typically characterized as being differentiated into nodes and internodes. The nodes, by definition, are the sites of leaf insertion, whereas the internodes are considered to be the stem units that typically elongate between the points of leaf insertion. The model for such a clear delineation between node and internode is illustrated in. However, a more accurate model of leaf–stem relationships is illustrated in. Here, leaf insertion is not localized at the node.

Each internode is not just stem, but a compound structure consisting of decurrent leaf bases, that run along the length of the internode below it. These decurrent leaf bases corticate the shoot axis, and the photosynthetic tissue one sees at the periphery of a stem transection is actually leaf tissue that is adnatewith the shoot axis.

Because leaf and axis components are inseparable, one uses the term "shoot" to refer to this compound, axial unit. The decurrent leaf bases evident along the lengths of the internodes in *Buxus sempervirens* (Buxaceae), an angiosperm shrub, and *Taxus baccata* (Taxaceae), agymnospermous shrub. Even if a marginal seam is not conspicuously delimiting the decurrent leaf base along the length of the internode, it is doubtless true that all internodes are corticated along their length by leaf base tissue. When a leaf abscises, only the lamina/petiole region of the leaf (upper leaf zone) falls leaving its decurrent leaf base (lower leaf zone) clothing the axis. According to such a conception of shoot organization, the shoot is not clearly delineated into "nodes" and "internodes" but between the locus of leaf projection and its subjacent basal extension.

EVOLUTION OF PLANT MORPHOLOGY

OVERVIEW OF PLANT EVOLUTION

Evidence suggests that an algal scum formed on the land 1,200 million years ago, but it was not until the Ordovician period, around 500 million years ago, that land plants appeared.

These begun to diversify in the late Silurian period, around 420 million years ago, and the fruits of their diversification are displayed in remarkable detail in an early Devonian fossil assemblage known as the Rhynie chert.

This chert preserved early plants in cellular detail, petrified in volcanic springs. By the middle of the Devonian period most of the features recognised in plants today are present, including roots, leaves and seeds. By the late Devonian, plants had reached a degree of sophistication that allowed them to form forests of tall trees. Evolutionary innovation continued after the Devonian period.

Most plant groups were relatively unscathed by the Permo-Triassic extinction event, although the structures of communities changed. This may have set the scene for the evolution of flowering plants in the Triassic (~200 million years ago), which exploded the Cretaceous and Tertiary. The latest major group of plants to evolve were the grasses, which became important in the mid Tertiary, from around 40 million years ago. The grasses, as well as many other groups, evolved new mechanisms of metabolism to survive the low CO_2 and warm, dry conditions of the tropics over the last 10 million years.

MERISTEMS

The meristem architectures differ between angiosperms, gymnosperms and pteridophytes. The gymnosperm vegetative meristem lacks organization into distinct tunica and corpus layers. They possess large cells called Central Mother Cells in the meristem. In angiosperms, the outermost layer of cells divides anticlinally to generate the new cells, while in gymnosperms, the plane of division in the meristem differs for different cells. However, the apical cells do contain organelles like large vacuoles and starch grains, like the angiosperm meristematic cells.

Pteridophytes, like fern, on the other hand, do not possess a multicellular apical meristem. They possess a tetrahedral apical cell, which goes on to form the plant body. Any somatic mutation in this cell can lead to hereditary transmission of that mutation.

The earliest meristem-like organization is seen in an algal organism from group *Charales* that has a single dividing cell at the tip, much like the pteridophytes, yet simpler. One can thus see a clear pattern in evolution of the meristematic tissue, from pteridophytes to angiosperms. Pteridophytes, with a single meristematic cell; gymnosperms with a multicellular, but less defined organization and finally, angiosperms, with the highest degree of organization.

ORIGINS OF THE LEAF

Leaves are the primary photosynthetic organs of a plant. Based on their structure, they are classified into two types-microphylls, that lack complex venation patterns and megaphylls, that are large and with a complex venation. It has been proposed that these structures arose independently. Megaphylls, according to the Telome hypothesis, have evolved from plants that showed a three dimensional branching architecture, through three transformations—planation, which involved formation of a planar architecture, webbing, or formation of the outgrowths between the planar branches and fusion, where these webbed outgrowths fused to form a proper leaf lamina. Studies have revealed that these three steps happened multiple times in the evolution of today's leaves.

It has been proposed that the before the evolution of leaves, plants had the photosynthetic apparatus on the stems. Today's megaphyll leaves probably became commonplace some 360mya, about 40my after the simple leafless plants had colonized the land in the early Devonian period. This spread has been linked to the fall in the atmospheric carbon dioxide concentrations in the Late Paleozoic era associated with a rise in density of stomata on leaf surface.

This must have allowed for better transpiration rates and gas exchange. Large leaves with less stomata would have gotten heated up in the sun's heat, but an increased stomatal density allowed for a better-cooled leaf, thus making its spread feasible.

FACTORS INFLUENCING LEAF ARCHITECTURES

Various physical and physiological forces like light intensity, humidity, temperature, wind speeds etc. are thought to have influenced evolution of leaf shape and size. It is observed that high trees rarely have large leaves, owing to the obstruction they generate for winds. This obstruction can eventually lead to the tearing of leaves, if they are large. Similarly, trees that grow in temperate or taiga regions have pointed leaves, presumably to prevent nucleation of ice onto the leaf surface and reduce water loss due to transpiration.

Herbivory, not only by large mammals, but also small insects has been implicated as a driving force in leaf evolution, an example being plants of the genus *Aciphylla*, that are commonly found in New Zealand. The now extinct Moas fed upon these plants, and its seen that the leaves have spines on their bodies, which probably functioned to discourage the moas from feeding on them. Other members of *Aciphylla* that did not co-exist with the moas, do not have these spines.

GENETIC EVIDENCES FOR LEAF EVOLUTION

At the genetic level, developmental studies have shown that repression of the KNOX genes is required for initiation of the leaf primordium. This is brought about by *ARP* genes, which encode transcription factors. Genes of this type have been found in many plants studied till now, and the mechanism *i.e.* repression of KNOX genes in leaf primordia, seems to be quite conserved. Expression of KNOX genes in leaves produces complex leaves. It is speculated that the *ARP* function arose quite early in vascular plant evolution, because members of the primitive group Lycophytes also have a functionally similar gene Other players that have a conserved role in defining leaf primordia are the phytohormone auxin, gibberelin and cytokinin. One feature of a plant is its phyllotaxy. The arrangement of leaves on the plant body is such that the plant can maximally harvest light under the given constraints, and hence, one might expect the trait to be genetically robust. However, it may not be so. In maize, a mutation in only one gene called *abphyl (ABNORMAL PHYLLOTAXY)* was enough to change the phyllotaxy of the leaves. It implies that sometimes, mutational tweaking of a single locus on the genome is enough to generate diversity. The *abphyl* gene was later on shown to encode a cytokinin response regulator protein.

Once the leaf primordial cells are established from the SAM cells, the new axes for leaf growth are defined, one important (and more studied) among them being the abaxial-adaxial (lower-upper surface) axes. The genes involved in defining this, and the other axes seem to be more or less conserved among higher plants. Proteins of the *HD-ZIPIII* family have been implicated in defining the adaxial identity. These proteins deviate some cells in the leaf primordium from the default abaxial state, and make them adaxial. It is believed that in early plants with leaves, the leaves just had one type of surface-the abaxial one.

This is the underside of today's leaves. The definition of the adaxial identity occurred some 200 million years after the abaxial identity was established. One can thus imagine the early leaves as an intermediate stage in evolution of today's leaves, having just arisen from spiny stem-like outgrowths of their leafless ancestors, covered with stomata all over, and not optimized as much for light harvesting.

How the infinite variety of plant leaves is generated is a subject of intense research. Some common themes have emerged. One of the most significant is the involvement of KNOX genes in generating compound leaves, as in tomato. But this again is not universal.

For example, pea uses a different mechanism for doing the same thing. Mutations in genes affecting leaf curvature can also change leaf form, by changing the leaf from flat, to a crinky shape, like the shape of cabbage leaves.

There also exist different morphogen gradients in a developing leaf which define the leaf's axis. Changes in these morphogen gradients may also affect the leaf form. Another very important class of regulators of leaf development are the microRNAs, whose role in this process has just begun to be documented. The coming years should see a rapid development in comparative studies on leaf development, with many EST sequences involved in the process coming online.

Flower-like structures first appear in the fossil records some ~130 mya, in the Cretaceous era. The flowering plants have long been assumed to have evolved from within the gymnosperms; according to the traditional morphological view, they are closely allied to the gnetales. However, recent molecular evidence is at odds to this hypothesis, and further suggests that gnetales are more closely related to some gymnosperm groups than angiosperms, and that gymnosperms form a distinct clade to the angiosperms.

Molecular clock analysis predicts the divergence of flowering plants (anthophytes) and gymnosperms to ~300 mya

EVOLUTION OF FLOWERS

The main function of a flower is reproduction, which, before the evolution of the flower and angiosperms, was the job of microsporophylls and megasporophylls. A flower can be considered a powerful evolutionary innovation, because its presence allowed the plant world to access new means and mechanisms for reproduction.

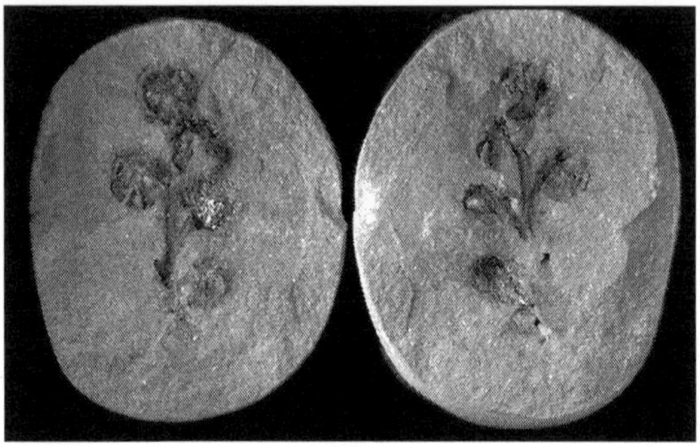

Fig. The Pollen Bearing Organs of the Early Flower *Crossotheca.*

ORIGINS OF THE FLOWER

The family Amborellaceae is regarded as the sister family of all living flowering plants. That means members of this family were most likely the first flowering plants. It seems that on the level of the organ, the leaf may be the ancestor of the flower, or at least some floral organs. When we mutate some crucial genes involved in flower development, we end up with a cluster of leaf-like structures. Thus, sometime in history, the developmental programme leading to formation of a leaf must have been altered to generate a flower. There probably also exists an overall robust framework within which the floral diversity has been generated.

An example of that is a gene called *LEAFY (LFY)*, which is involved in flower development in *Arabidopsis thaliana*. The homologs of this gene are found in angiosperms as diverse as tomato, snapdragon, pea, maize and even gymnosperms.

Expression of *Arabidopsis thaliana* LFY in distant plants like poplar and citrus also results in flower-production in these plants. The *LFY* gene regulates the expression of some gene belonging to the MADS-box family. These genes, in turn, act as direct controllers of flower development.

EVOLUTION OF THE MADS-BOX FAMILY

The members of the MADS-box family of transcription factors play a very important and evolutionarily conserved role in flower development.

According to the ABC Model of flower development, three zones-A,B and C- are generated within the developing flower primordium, by the action of some transcription factors, that are members of the MADS-box family. Among these, the functions of the B and C domain genes have been evolutionarily more conserved than the A domain gene. Many of these genes have arisen through gene duplications of ancestral members of this family. Quite a few of them show redundant functions.

The evolution of the MADS-box family has been extensively studied. These genes are present even in pteridophytes, but the spread and diversity is many times higher in angiosperms. There appears to be quite a bit of pattern into how this family has evolved. Consider the evolution of the C-region gene *AGAMOUS (AG)*. It is expressed in today's flowers in the stamens, and the carpel, which are reproductive organs. It's ancestor in gymnosperms also has the same expression pattern.

Here, it is expressed in the strobili, an organ that produces pollens or ovules. Similarly, the B-genes' *(AP3 and PI)* ancestors are expressed only in the male organs in gymnosperms. Their descendants in the modern angiosperms also are expressed only in the stamens, the male reproductive organ. Thus, the same, then-existing components were used by the plants in a novel manner to generate the first flower. This is a recurring pattern in evolution.

FACTORS INFLUENCING FLORAL DIVERSITY

How is the enormous diversity in the shape, colour and sizes of flowers established? There is enormous variation in the developmental programme in different plants. For example, monocots possess structures like lodicules and palea, that were believed to be analogous to the dicot petals and carpels respectively.It turns out that this is true, and the variation is due to slight changes in the MADS-box genes and their expression pattern in the monocots. Another example is that of a plant called *Linaria vulgaris*, which has two kinds of flower symmetries-radial and bilateral. These symmetries are due to epigenetic changes in just one gene called *CYCLOIDEA*.

Arabidopsis thaliana has a gene called *AGAMOUS* that plays an important role in defining how many petals and sepals and other organs are generated. Mutations in this gene give rise to the floral meristem obtaining an indeterminate fate, and many floral organs keep on getting produced. We have flowers like roses, carnations and morning glory, for example, that have very dense floral organs.

These flowers have been selected by horticulturists since long for increased number of petals. Researchers have found that the morphology of these flowers is because of strong mutations in the *AGAMOUS* homolog in these plants, which leads to them making a large number of petals and sepals. Several studies on diverse plants like petunia, tomato, Impatiens, maize etc. have suggested that the enormous diversity of flowers is a result of small changes in genes controlling their development. Some of these changes also cause changes in expression patterns of the developmental genes, resulting in different phenotypes. The Floral Genome Project looked at the EST data from various tissues of many flowering plants.

The researchers confirmed that the ABC Model of flower development is not conserved across all angiosperms. Sometimes expression domains change, as in the case of many monocots, and also in some basal angiosperms like *Amborella.*

Different models of flower development like the *The fading boundaries model,* or the *Overlapping-boundaries model* which propose non-rigid domains of expression, may explain these architectures. There is a possibility that from the basal to the modern angiosperms, the domains of floral architecture have gotten more and more fixed through evolution.

FLOWERING TIME

Another floral feature that has been a subject of natural selection is flowering time. Some plants flower early in their life cycle, others require a period of vernalization before flowering. This decision is based on factors like temperature, light intensity, presence of pollinators and other environmental signals. We know that genes like *CONSTANS (CO), Flowering Locus C* (*FLC*) and *FRIGIDA* regulate integration of environmental signals into the pathway for flower development.

Variations in these loci have been associated with flowering time variations between plants. For example, *Arabidopsis thaliana* ecotypes that grow in the cold, temperate regions require prolonged vernalization before they flower, while the tropical varieties, and the most common lab strains, don't. We now know that this variation is due to mutations in the *FLC* and *FRIGIDA* genes, rendering them non-functional.

Quite a few players in this process are conserved across all the plants studied. Sometimes though, despite genetic conservation, the mechanism of action turns out to be different. For example, rice is a short-day plant, while *Arabidopsis thaliana* is a long-day plant. Now, in both plants, the proteins *CO* and *FLOWERING LOCUS T (FT)* are present. But in *Arabidopsis thaliana, CO* enhances *FT* production, while in rice, the *CO* homolog represses *FT* production, resulting in completely opposite downstream effects.

THEORIES OF FLOWER EVOLUTION

There are many theories that propose how flowers evolved. Some of them are described below. The *Anthophyte Theory* was based on the observation that a gymnospermic family Gnetaceae has a flower-like ovule. It has partially developed vessels as found in the angiosperms, and the megasporangium is covered by three envelopes, like the ovary structure of angiosperm flowers. However, many other lines of evidence show that Gnetophytes are not related to angiosperms.

The *Mostly Male Theory* has a more genetic basis. Proponents of this theory point out that the gymnosperms have two very similar copies of the gene *LFY* while angiosperms only have one. Molecular clock analysis has shown that the other *LFY* paralog was lost in angiosperms around the same time as flower fossils become abundant, suggesting that this event might have led to floral evolution. According to this theory, loss of one of the *LFY* paralog led to flowers that were more male, with the ovules being expressed ectopically. These ovules initially performed the function of attracting pollinators, but sometime later, may have been integrated into the core flower.

6

Biotechnological Approaches to Weed Management

WEED MANAGEMENT

Weed management is essential for good quality and quantity of food production. The presence of weeds in general may reduce crop yield by about 30% or more. Being potential pests, weeds have to be removed employing mechanical, chemical cultural and biological means. Weed management through application of chemicals (herbicides) is, through effective, not eco-friendly. The herbicides may pollute the environment and many alter the natural equilibrium. They also endanger and alter plant and microbial biodiversity. Effective non-chemical weed management may include various biotechnological approaches using plants and/or their products. These are useful both for the natural equilibrium and the environment. in addition they also condition the soil and improve crop production. Some of the recently evolved biocontrol and biotechnological methods of weed management developed and tried at the National Research Centre for Weed Science, Indian Council of Agricultural Research, Jabalpur are outlined in the following text for ready use for trials and for working our their cost effectiveness.

Management of Phalaris Minor in wheat by Using Fungi

At present Phalaris minor is one of the most predominant weeds in wheat fields. The Weed spreads through seeds. The weed has developed resistance against popular herbicides like isoproturon. Biotechnological approach using fungi offers an alternative possibility of management of the weed through biocontrol.

Control of Phararis Minor using Trichoderma Viride

Trichoderma viride is used as a seed dressing fungi in various crops including in wheat for management of seed borne/soil borne pathogens. Interestingly, the fungus has been found to inhibit germination of seeds of Phalaris minor. Application of Trichoderma, viride grown in saw dust and neem oil cake as a base consistently inhibits germination of phalaris minor seeds and reduces vigour of the seedlings of the weed. At the same time it improves yield of wheat by about 5%. The fungus has potential of controlling phalaris minor. Soil application of the fungus, wheat seed treatment and spray of the fungus were all effective in controlling the weed.

Management of Parthenium (Parthenium HysterophorusL) by Marigold (Tagetes PatulaL)

Parthenium is an obnoxious weed of worldwide occurrence, harmful to human and animal health, agriculture, environment and the natural biodiversity. The weed reduces yields of crops and grasses. When shown in parthenium infested field or area, marigold inhibits seed germination and growth of the weed. It also reduces population of the weed. In the next generation, reduction of parthenium population is as hight as 85-100%. This has been found primarily due to competition and through allelopathy (affecting parthenium plants by releasing chemical substances including thicophene into soil through roots.) Built up of the phytotoxic substances of marigold origin prevent parthenium seed germination, growth of plants, and flower and seed production completely in 2-3 years. This technology appreas to be economically rewarding for marigold flowers have good market. This can be achieved simply by spraying marigold seeds over parthenium infested area. The technology is simple, effective, economically rewarding, and ensures self perpetuating control of parthenium in wastelands. Marigold to parthenium plant ratio of 0.5 to 4 aoppear effective for near complet control of parthenium. Adequate moisture availability in soil facilitates establishment of marigold in parthenium infected area.

MANAGEMENT OF PARTHENIUM BY FUNGI

Two fungal species namely Gliocladium virens and Trichoderma viride which have been widely used as seed dressing to present pathogenic damage have also been found effective for parthenium control.

Control of Parthenium by Gliocladium Virus

Glioladium vieus and 10% neem oil separately as well as in combinations spray control pathenium and other broad leaf weeds like chenopondium album, Melilotus alba and Medicago sp.

In wheat crop. The fungus could be explored as a self perp control measure for non-cropped area also.

Control of parthenium by Trichoderma Viride and Need Oil

Thrichoderma viride and neem oil separately as well as in combinations spray control parthenium under field conditions. The fungi has potential of becoming a self perpetuating means for control of parthenium under non-cropped area. Trichoderma viride also acts as an antidote for pathogens of parthenium that can possibly be used as biocontrol agents. If any of the pathogens attack crops, the Trichoderma virid can be used to overcome the problem.

Management of Parthenium by Utilising as a Source of Nutrients

Parthenium is able to extract nutrients even from nutrient deficient soils in which it grows. It has very hight levels of nitrogen (3%), phosphorus (0.2%), potassium (4.5.%) and other macro and micro-nutrients. Uprooted, dried and cut to small pieces or powdered parthenium plants at preflowering stage can be applied directly to crop fields. Apart from increasing crop growth and yields, the parthenium residue conditions soil with the organic supplement. Keeping safety considerations, it may serve as an incentive for participation of public in its eco-friendly control.

Weed Management through mid-summer Irrigation and Covering the beds with black Polythene

Weed problem in most cases is due to seeds of weeds present in soil seed bank which germinate before, with or after planting the crops. Mid summer irrigation and covering the beds or fields with black polythene throughout the summer kill most of the readily geminable seeds in soil seed bank. Polythene covering can be removed after summer before planting the crops. The treatment may provide near complete weed control. This technique is suitable for areas where summer temperature exceeds 40 °C..

Weed Management through Black Polythene Stripe Mulching

Where crop is grown in rows, inter row space could be covered with black polythene mulching. This kills weeds appearing in the area due to interception of light necessary for survival of the plants.

The are many other ways of biotechnological management of weeds such as boiling water jet treatment of soil, by intensive cropping, by increasing plant density and choosing allelopathic crop species etc.

BIOTECHNOLOGYCAL APPLICATION

In this field biotechnological applications are extremely relevant if we are to made rapid progress in raising the productivity of these small ruminants.

Techniques such as embryo transfer, dilution of semen and its cryo preservation and artificial insemination (A.I.) are well developed and expertise in these fields is available. This technology is extensively used in the case of cows and to some extent Buffaloes.

The main lacuna in the case of sheep and goats is the non availability of improved breeds which could be utilised. As these animals are owned by the poor strata of society and there is a mistaken prejudice that they are responsible for environmental degradation no sustained efforts have been made to improve indigenous breeds and thus make available genetic stock for upgrading. In spite of this the numbers of these animals have been increasing rapidly (*2.5% per year in the case of goats) due to an increasing demand for meat. Our breeds have developed over the centuries to survive and adapts themselves to harsh environmental conditions but do not universally possess ability to respond to better management and nutrition.

Outstanding work in the case of the Sirohi goat was carried out from 1981 to 1993 by the Indo-Swiss Goat Development and Fodder Production Project in Rajasthan where they showed that outstanding animals are available within this breed. However, before the impact of this project could be felt the project was discontinued.

Goats

Out institutes is a very small NGO with only three technically qualified staff, one of who is in an honorary capacity and one on deputation from the Government of Maharashtra. We decided to initially concentrate on crossbreeding using the outstanding South African Boer Goat the worlds only meat goat as an improve breed. With assistance from the Government of Australia we imported 20 embryos of this breed from Australia implanted them in local goats and now have a small flock of Boer goats.

This goat unlike many temperate breeds of Dairy goats have proved to be extremely hardy, producing excellent crossbreed when crossed with local goats. In order to utilise this germplasm effectively we have developed and modified the techniques of semen cryopreservation so that it can be done with low cost equipment producing pellets instead of straws. We have also improvised. A.I,. techniques using cervical insemination to get better conception rates. We have so far trained over 50 trainees among them farmers. We have supplied pure Boer goat frozen semen to Governmental agencies in Orissa, Tamil Nadu as well as to may private goat rearers in Maharashtra.

Our veterinarian is capable of carrying out multiple ovulation and embyo transfer if there is need for rapid expansion of our flock. However, this technology is expensive and only to be used if costs justify its use.

A I technology in goats has been spread to near by villages where fresh diluted semen as well as frozen semen have been used to produce 800 crossbreeds over the past five years. This technology is viable and easily transferable but unfortunately dependent on the sincerity of the inseminator.

Sheep

We were unable to identify a suitable improver breed to be imported which had the most desirable but greatly neglected attribute of prolificacy. We did locate a prolific and unusual breed in the hot humid sundarbands of West Bengal called the Garole. Up to 1993 this breed had remained unnoticed by the community of sheep breeders. We did however import a small flock of Awassi Dairy Sheep from Israel. This breed has produced an outstanding crossbred animal when crossed with the Malpura of Rajasthan. This crossbred has been developed by the Central Sheep and Wool Research Institute (CSWRI) using Awassi rams obtained from us.

The Australian Center for International Agricultural Research has sponsored a collaborative

project between the University of New England, Australia, The national Chemical Laboratory, Pune and the Nimbkar Agricultural Research Institute (A SISTER NGO of Maharashtra Goat and sheep Research and Development Institute), Phaltan entitled, ''Prolific Worm Resistant Meat Sheep for Maharashtra, India.

This project is making full use of Biotechnology. We are comparing three breeds of Indian sheep the Grole, Deccan and the bannur. We hope to evaluate lamb production including prolificacy and resistance to parasites and try and determine the genetic basis of these differences by DNA analysis. We also hope to account for the interaction between measurement of resistance to parasitic nematodes and environmental effects. The project has completed one year and the results so far are encouraging.

Our ultimate objective is to provide shepherds with rams incorporating the prolificacy of the Garole, the growth potential of the Awassi and the meat quality and hardiness of the Bannur.

A very important but at present neglected application of Biotechnology which requires considerable investment is in the field of disease diagnosis and vaccine preparation. Once productive animals that respond to inputs are available and the readers can afford to pay for the above services I am confident that the private sector will enter this field to provide this input.

It is imperative to review the cost benefit ratios of Government institutions and their contribution to the Goat and Sheep industries I think you will find that funds given to NGO's are more productive. There is much talk of helping NGO's but real assistance only seems to come from foreign aid agencies which is unfortunate.

BACTERIAL BIOTECHNOLOGY

The Production of high quality agricultural products in quantities sufficient to feed the world's people is clearly a matter of vital importance. Any general improvement in agricultural practices can have a tremendous economic impact, and the methods of biotechnology suggest at least two important strategies for bringing about widespread improvements.

IMPROVEMENT THROUGH THE USE OF SYMBIOTIC OR POTENTIALLY PATHOGENIC MICROORGANISMS

Plants live in intimate relationship with many microorganisms. Leaf surfaces are covered by layers of symbiotic and sometimes pathogenic bacterial and yeasts, and the soil close to the root area is populated by bacteria species very different from the species found in soil away from plants. There have been a number of attempts to modify symbiotic microorganisms so as to make them even more beneficial to plants.

IMPROVEMENT THROUGH THE PRODUCTION OF TRANSGENIC PLANTS

Plants now being cultivated represent the end result of many years of pain staking efforts at improvement. Traditional plant breeding programmes rely on two methods; the selection of advantageous spontaneous mutations and the introduction of desirable traits from closely related species by crossbreeding. Recombinant DNA Technology is bringing revolutionary change to these endeavors.

It enables scientists to select precisely defined genes with well-characterized properties, clone them, and transfer them to a given plant. These genes can be taken from organisms unrelated to the target plant -even from bacteria or animals.

Transferring cloned genes from prokaryotes (bacteria) into higher eukaryotes, such as plants, is difficult. However, there are two species of bacteria, Agro-bacterium tumefaciens and A rhizogenes, that transfer a small piece of DNA into plants as part of their normal life cycle.

Use of Symbionts and Pathogenes

There are many Symbiotic bacterial associated with specific organs of various plants, a simple plant might be to modify such bacterial and then use such interactions with the plants to introduce the modified traits at the appropriate locations. This alternative approach is technically easier, because the engineering of bacterial DNA through recombinant DNA methods is now routine, whereas the manipulation of plant DNA is yet to be perfected.

Protection of plants from frost Damage via Engineered Symbiotic Bacteria

One of the earliest examples of successful modification of a symbiotic bacterium was performed in Pseudomonas syringe, which is found at high concentrations on the leaves of many plants. Many strains of this bacterium produce an ice nucleation protein that is apparently located on the surface of the bacteria cell, and the presence of this protein cause the formation of ice at temperature only little below 0 °C, inflicting significant frost damage on important crop plants and so facilitating invasion of plant tissues by the bacteria.

Use of Nitrogen Fixing Bacteria to Improve Crop Yields

All animals and plants and most bacteria depend on the availability in their environment of some form of "combined nitrogen" nitrate (NO_3), or such nitrogen containing organic compounds as amino acids. The huge amount of N_2 that exist in the atmosphere are unavailable to the biological world except through the process of nitrogen fixation.

The biological process of nitrogen fixation does not require the consumption of fossil fuels or electricity does not produce environmental pollution. Thus increasing the extent of biological nitrogen fixation has been an important goal for biotechnology. According to one estimate, the biological process fixes times more nitrogen (24 x 10^{17} tons/year) than is converted into chemical fertilizers by industrial processes. Thus even a small improvement in exploiting the biological process would have been a global impact.

The biological process of nitrogen fixation is complex and consumes a large number of ATP, molecules, because the enzymes involved in it must overcome the same huge activation energy barriers that faces the chemical synthesis of ammonia. Two enzymes are required:component I (nitrogenase) and component II (nitrogenase reductase). After component Ii is reduced by a strong biological reductant, 16 molecules of ATP are hydrolysed to accomplish the reduction of component I. Reduced component II alone is not capable of overcoming the activation energy barrier. Reduced component I finally reduced N_2 to two molecules of NH_3. Nitrogen fixation is strongly reductive reaction, and the enzymes involved are usually irreversibly inactivated when they are exposed to oxygen.

The species of closiridium and klebsilla fix nitrogen only under anaerobic conditions. Other free living baueria, however, can fix nitrogen even under aerobic conditions. The cyanobaueria, which carry out oxygen - evolving photosynthesis, perform nitrogen fixation only in speciallized cells called heterocyst's, which do not produce oxygen. Azotobactor consumes oxygen at an extremely high rate, and in doing so apparently protects its nitrogen fixing machinery. Another group of bacteria fix nitrogen only when they are in a symbiotic nitrogen fixer is Rhizobium, which invades the root tissues of teguminous plants, such as alfalfa, pea, clover and soyabean and lives in intracellular vacuoles. The vacuoles become filled with an oxygen-binding protein, leghomoglobin produced by the plants. This creates an environment low in free oxygen where Rhizobium is able to carry out nitrogen fixation.

Production of Transgenic Plants

To produce transgenic plants, the crucial steps are the introduction of the cloned genetic material into the plant cell nuclean and the facilitated integration of the cloned gene into the plant chromosome. It is interesting that the best method for doing so uses a system that already exists in nature, the system by which the plant pathogenic bacterium Agro-bacterium tumefaciens injects a portion of its plasmid DNA into plants and inserts it into the plant become.

The Agro-bacterium system is an excellent system for introducing foreign genes into the chromosomes of plants. The transfer into intact plant cells occur at high frequency, the T-DNA is usually integrated into the plant chromosomes at high frequencies without undergoing structural alterations, and the cells that have received the T-DNA can be selected easily by using antibiotic resistance such as the neomycin resistance maricer. Finally the transgenic plants produced in this manner are quiet stable for at least several generations. Consequently almost all existing transgenic plants with potentially desirable traits have been obtained by this method.

Herbicide Resistance Plants -

The advantages of making crop plants resistant to herbicides are obvious. Although there is a fear that such resistant may eventually increase the use of herbiude chemicals, there are also reasons to except that these transgenic plants will promote the use of sager, more biodegradable herbicides, perhaps in smaller amounts one example involves the herbicide glyphosate which inhibits 5-enol-pyruvylshikimate 3-phophate synthesis - an enzyme involved in the biosynthesis of aromatic amino acids. This enzyme has been purified from crop plants and sequenced and DNA Probes corresponding to its amino acid sequence have been synthesized. These probes were used to isolate CDNA for the enzyme from the CDNA library of a plant cell line known to overproduce 5-enol-pyruvylshikimate 3-phophate synthase. The CDNA was then cloned behind the strong CaMv promoter, and the promoter gene complex was introduced into plant cells via a disarmed Ti plasmid vector. The transgenic plants produce a much higher level of the target enzyme and thus are significantly more resistant to glyphosate. These results are encouraging because glyphosate has very low toxicity to animals is rapidly degraded in soil.

Insect Resistant Plants

Bacillus thuringiensis is used in the biological control of caterpillars because its speculating cells contain toxic proteins. The gene for one toxic protein was cloned behind promoters that are effectively expressed in plants and was introduced into plant cells via a Ti plasmid vector, thus producing plants toxic to caterpillar. The major problem with this approach has been the low level of expression of the toxin protein in plants, which is presumably related to the fact that the gene come from a bacterium,. Nevertheless, the method has produced tomato and tobacco plants that proved quiet resistant to caterpillars in field tests.

Cotton however, is often attacked by insects that are more resistant to the B. thuringiensis toxin and the low level of expression of this toxin in transgenic cotton plants provided the plants with little if any protection. Recently, the coding sequence of the toxin was altered extensively to replace codons that are rarely used in plants as well as to preclude the formation of a strong secondary structure in MRNA. When cotton plants were provided with this modified gene, their production of the bacterial toxin increased 100-fold and they showed impressive resistant to common bepidopteran insects than damage unmodified plants. Some seeds contain high concentrations of protease inhibitors, which are thought to interfere with the digestive process in insects. The cloning of cowpea trypsin inhibitor genes, for example and their transfer to tobacco plants have resulted in good resistance to a wide variety of leaf-eating insects.

Viral resistant plants

The method most frequency used for producing virus-resistant plants sprang from the observation that, oftentimes, plants infected by nearly avirulent virus are thereafter resistant to super infection by a related, highly virulent one. Thus, deliberate infection with avirulent virus strain has been used to produce protection in crop plants. The method is not totally safe, however, because the avirulent strains many mutate to produce strains that are significantly pathogenic. Most plant viruses are positive strand RNA viruses covered by coat protein sub units. When the virus enters an injured plant cell the replication process begins, starting with the progressive uncoating of the virus from the 5-end of the RNA. Transgenic plants, whose genomes contain introduced tobacco mosaic virus (TMV) coat protein genes and which continuously synthesize the coat proteins, show resistance to virus infection. The phenotype of these plants is consistent with the idea that resistance is a result of interference with the uncoating of the virus particle. Although the plant cells are resistant to infect TMV particles, they remain sensitive to the TMV RNA or to partially uncoated TMV particles.

The coat protein gene is usually put behind a strong promoter such as the 35S promoter of CaMV and is introduced into a plant's genome via the Agro-bacterium Ti system. Transgenic plants showing significant resistance to TMV alfalfa mosaic virus, cucumber mosaic virus, tobacco streak virus and tobacco rattle virus have already been produced in this manner. In a recent experiment two virus and potato virus Y, were introduced simultaneously into a commercially important potato cultivar, and one of the resulting transgenic plants provide quiet resistant to both viruses under field test conditions.

WEED CONTROL

Weed control is the botanical component of pest control, using physical and chemical methods to stop weeds from reaching a mature stage of growth when they could be harmful to domesticated plants and livestock. In order to reduce weed growth, many "weed control" strategies have been developed in order to contain the growth and spread of weeds. The most basic is ploughing which cuts the roots of annual weeds. Today, chemical weed killers known as herbicides are widely used.

A plant is often termed weed when it has one or more of the following characteristics:

- Little or no value (as in medicinal, nutritional, or energy)
- Very high growth rate and/or ease of germination
- Exhibits competition to crops, for space, light, water and nutrients

EFFECTS ON OTHER PLANTS

Weeds can compete with productive crops or pasture, or convert productive land into unusable scrub. Weeds are also often poisonous, distasteful, produce burrs, thorns or other damaging body parts or otherwise interfere with the use and management of desirable plants by contaminating harvests or excluding livestock.

Weeds tend to thrive at the expense of the more refined edible or ornamental crops. They provide competition for space, nutrients, water and light, although how seriously they will affect a crop depends on a number of factors. Some crops have greater resistance than others- smaller, slower growing seedlings are more likely to be overwhelmed than those that are larger and more vigourous.

Onions are one of the crops most susceptible to competition, for they are slow to germinate and produce slender, upright stems. Quick growing, broad leafed weeds therefore have a distinct advantage, and if not removed, the crop is likely to be lost. Broad beans however produce large seedlings, and will suffer far less profound effects of weed competition other than during periods of water shortage at the crucial time when the pods are filling out. Transplanted crops raised in sterile seed or potting compost will have a head start over germinating weed seeds.

Weeds also differ in their competitive abilities, and can vary according to conditions and the time of year. Tall growing vigourous weeds such as fat hen (*Chenopodium album*) can have the most pronounced effects on adjacent crops, although seedlings of fat hen that appear in late summer will only produce small plants. Chickweed (*Stellaria media*), a low growing plant, can happily co-exist with a tall crop during the summer, but plants that have overwintered will grow rapidly in early spring and may swamp crops such as onions or spring greens.

The presence of weeds does not necessarily mean that they are competing with a crop, especially during the early stages of growth when each plant can find the resources it requires without interfering with the others. However, as the seedlings' size increases, their root systems will spread as they each begin to require greater amounts of water and nutrients. Estimates suggest that weed and crop can co-exist harmoniously for around three weeks, therefore it is important that weeds be removed early on in order to prevent competition occurring.

Weed competition can have quite dramatic effects on crop growth. Harold A Roberts cites research carried out with onions wherein *"Weeds were carefully removed from separate plots at different times during the growth of the crop and the plots were then kept clean. It was found that after competition had started, the final yield of bulbs was being reduced at a rate equivalent to almost 4 per cent per day. So that by delaying weeding for another fortnight, the yield was cut to less than half that produced on ground kept clean all the time."* (*The Complete Know And Grow*

Vegetables, Bleasdale, Salter and others, OUP 1991). He goes on to record that *"by early June, the weight of weeds per unit area was twenty times that of the crop, and the weeds had already taken from the soil about half of the nitrogen and a third of the potash which had been applied"*.

Perennial weeds with bulbils, such as lesser celandine and oxalis, or with persistent underground stems such as couch grass (*Agropyron repens*) or creeping buttercup (*Ranunculus repens*) are able to store reserves of food, and are thus able to grow faster and with more vigour than their annual counterparts. There is also evidence that the roots of some perennials such as couch grass exude allelopathic chemicals which inhibit the growth of other nearby plants.

Weeds can also host pests and diseases that can spread to cultivated crops. Charlock and Shepherd's purse may carry clubroot, eelworm can be harboured by chickweed, fat hen and shepherd's purse, while the cucumber mosaic virus, which can devastate the cucurbit family, is carried by a range of different weeds including chickweed and groundsel.

However, at times the role of weeds in this respect can be overrated. As far as insect pests are concerned, often the species that live on weeds are not the same as those that attack vegetable crops; *"Tests with the common cruciferous weeds such as shepherds purse have shown that they do not act as hosts for the larvae of the cabbage root fly. One exception was found to be the wild radish, but this is not usually a weed of established vegetable gardens"* (Roberts, *The Complete Know And Grow Vegetables*). However pests such as cutworms may first attack weeds then move on to cultivated crops.

While charlock, a common weed in southeastern USA, may be considered a weed by row crop growers, it is highly valued by beekeepers, who seek out places where it blooms all winter, thus providing pollen for honeybees and other pollinators. Its bloom is resistant to all but a very hard freeze, and even that will only kill it back briefly. By feeding an array of pollinators during a seasonal dearth, it can redound to the farmer's advantage. Many weeds are likewise highly beneficial to pollinators.

METHODS

In domestic gardens, methods of weed control include covering an area of ground with several layers of wet newspaper or one black plastic sheet for several weeks. In the case of using wet newspaper, the multiple layers prevent light from reaching all plants beneath, which kills them. Saturating the newspaper with water daily speeds the decomposition of the dead plants. Any weed seeds that start to sprout because of the water will also be deprived of sunlight, be killed, and decompose. After several weeks, all germinating weed seeds present in the ground should be dead. Then the newspaper can be removed and the ground can be planted. The decomposed plants will help fertilise the plants or seeds planted later. In the case of using the black plastic sheet, the greenhouse effect is used to kill the plants beneath the sheet. A 5–10 cm layer of wood chip mulch on the ground will also prevent most weeds from sprouting. Also, gravel can be spread over the ground as an inorganic mulch. In agriculture, irrigation is sometimes used as a weed control measure such as in the case of paddy fields. Many people find that although the black plastic sheeting is extremely effective at preventing the weeds in areas where it covers, but in actual use it is difficult to achieve full coverage.

Fig. Weeds are Removed Manually in Large Parts of India.

Knowing how weeds reproduce, spread and survive adverse conditions can help in developing effective control and management strategies. Weeds have a range of techniques that enable them to thrive; Annual and biennial weeds such as chickweed, annual meadow grass, shepherd's purse, groundsel, fat hen, cleaver, speedwell and hairy bittercress propagate themselves by seeding. Many produce huge numbers of seed several times a season, some all year round. Groundsel can produce 1000 seed, and can continue right through a mild winter, whilst Scentless Mayweed produces over 30,000 seeds per plant. Not all of these will germinate at once, but over several seasons, lying dormant in the soil sometimes for years until exposed to light. Poppy seed can survive 80–100 years, dock 50 or more. There can be many thousands of seeds in a square foot or square metre of ground, thus and soil disturbance will produce a flush of fresh weed seedlings.

"STALE SEED BED" TECHNIQUE

One technique employed by growers is the 'stale seed bed', which involves cultivating the soil, then leaving it for a week or so. When the initial flush of weeds has germinated, the grower will lightly hoe off before the desired crop is planted. However, even a freshly cleared bed will be susceptible to airborne seed from elsewhere, as well as seed brought in by passing animals which can carry them on their fur, or from freshly imported manure. The organic solution to the problem of spreading annual weeds lies in regular, properly timed weeding, preferably just before flowering (fortuitously, this is also the time at which they will be of the most value in composting). This technique is also quite often used by farmers who let weeds germinate then return the soil before crop sowing.

Perennial weeds also propagate by seeding; the airborne seed of the dandelion and the rose-bay willow herb are parachuted far and wide. But they also have an additional range of vegetative means of spreading that gives them their pernicious reputation. Dandelion and dock put down deep tap roots, which, although they do not spread underground, are able to regrow from any remaining piece left in the ground. Removal of the complete tap root is the only sure remedy.

The most persistent of the perennials are those that spread by underground creeping rhizomes that can regrow from the tiniest fragment. These include couch grass, bindweed, ground elder, nettles, rosebay willow herb, Japanese knotweed, horsetail and bracken, as well as creeping thistle, whose tap roots can put out lateral roots. Other perennials put out runners that spread along the soil surface. As they creep along they set down roots, enabling them to colonise bare ground with great rapidity. These include creeping buttercup and ground ivy. Yet another group of perennials propagate by stolons- stems that arch back into the ground to reroot. Most familiar of these is the bramble.

All of the above weeds can be very difficult to eradicate- thick black plastic mulches can be effective to a degree, although will probably need to be left in place for at least two seasons. In addition, hoeing off weed leaves and stems as soon as they appear can eventually weaken and kill the plants, although this will require persistence in the case of plants such as bindweed. Nettle infestations can be tackled by cutting back at least three times a year, repeated over a three-year period. Bramble can be dealt with in a similar way. Some plants are said to produce root exudates that suppress herbaceous weeds. *Tagetes minuata* is claimed to be effective against couch and ground elder, whilst a border of comfrey is also said to act as a barrier against the invasion of some weeds including couch.

USE OF HERBICIDES

The above described methods of weed control avoid using chemicals. They are often used by farmers. However, these methods may damage a fragile soil by restructuring it, hence are not always used. They are those preferred by the organic gardener or organic farmer. However weed control can also be achieved by the use of herbicides. Selective herbicides kill certain targets while leaving the desired crop relatively unharmed. Some of these act by interfering with the growth of the weed and are often based on plant hormones.

Herbicides are generally classified as follows:

- Contact herbicides destroy only that plant tissue in contact with the chemical spray. Generally, these are the fastest acting herbicides. They are ineffective on perennial plants that are able to re-grow from roots or tubers.

- Systemic herbicides are foliar-applied and are translocated through the plant and destroy a greater amount of the plant tissue. Modern herbicides such as glyphosate are designed to leave no harmful residue in the soil.
- Soil-borne herbicides are applied to the soil and are taken up by the roots of the target plant.
- Pre-emergent herbicides are applied to the soil and prevent germination or early growth of weed seeds.

In agriculture large scale and systematic weeding is usually required, often by machines, such as liquid herbicide sprayers, or even by helicopter (such as in the USA), to eliminate the massive amount of weeds present on farming lands. However there are a number of techniques that the organic farmer can employ such as mulching and carefully timed cutting of weeds before they are able to set seed.

ORGANIC METHODS

Typically a combination of methods are used in organic situations:

- *Drip Irrigation:* Rubber hoses and other methods are used to bring water directly to the roots of the desired plants. This limits weed access to water.
- *Manually Pulling Weeds:* Labourers are used to pull weeds at various points in the growing process.
- *Boiling Water:* Pour boiling water to weed, they will become more green and then die in few hours. Best for weed in cracks or other hard to reach locations.
- *Vinegar:* Vinegar kills the visible part of the weed. They will wrinkle and die next day, although the root will still be in place to continue growing.
- *Mechanically Tilling around Plants:* Tractors are used to carefully till weeds around the crop plants at various points in the growing process. Besides tilling, other mechanical weed control methods also exist
- *Ploughing:* Ploughing includes tilling of soil, intercultural ploughing and summer ploughing. Ploughing through tilling of soil uproots the weeds which causes them to die. In summer ploughing is done during deep summers. Summer ploughing also helps in killing pests.
- *Crop Rotation:* Rotating crops with ones that kill weeds by choking them out, such as hemp, *Mucuna pruriens,* and other crops, can be a very effective method of weed control. It is a way to avoid the use of herbicides, and to gain the benefits of crop rotation.

- *Weed Mat:* A weed mat is an artificial mulch, fibrous cloth material, bark or newspaper laid on top of the soil preventing weeds from growing to the surface.

THERMAL METHODS

There are several thermal methods known to control weed. Weed burners heat up soil quickly and destroy superficial parts of the plants. Weed seeds are often heat resistant and even react with an increase of growth on dry heat. Since the 19th century soil steam sterilization is used as a farming technique to clean soil completely from weeds. Several research results confirm the high effectivness of humid heat against weeds and its seeds.

UK LEGISLATION

The *Weeds Act, 1959* is described as *"Preventing the spread of harmful or injurious weeds"*, and is mainly relevant to farmers and other rural settings rather than the allotment or garden-scale grower. There are five 'injurious' (that is, likely to be harmful to agricultural production) weeds covered by the provisions of the Weeds Act.

These are:

- Spear thistle (*Cirsium vulgare*)
- Creeping or field thistle (*Cirsium arvense*)
- Curled Dock (*Rumex crispus*)
- Broad leaved dock (*Rumex obtusifolius*)
- Common ragwort (*Senecio jacobaea*) (n.b., this weed is poisonous to livestock. Livestock should not be allowed to graze where ragwort has grown until it is eradicated, and any traces have disintegrated. Ragwort should not be allowed to be harvested in hay or silage for feed).

The Department for Environment, Food and Rural Affairs (DEFRA) provides guidance for the treatment removal of these weeds from infested land. Much of this is oriented towards the use of herbicides, the majority of which may not be acceptable to the organic producer (apart from non-synthetic substances like sulphur, which in some circumstances are accepted within Soil Association standards) but in most cases there are manual techniques that can be used such as digging out the roots, mulching out or carefully timed cutting before seeds are able to spread.

Primary responsibility for weeds control rests with the occupier of the land on which the weeds are growing, therefore it is important to be alert to potential weed problems and to take prompt action. However, it should be remembered that most common farmland weeds are not "injurious" within the meaning of the Weeds Act, and many such plant species have conservation and environmental value. When dealing with complaints under the Weeds Act, DEFRA has a duty in law to try to achieve a reasonable balance among different interests.

These include agriculture, countryside conservation and the public in general. Constructive discussion about problems caused by weeds can often result in effective solutions and avoid the need for DEFRA to take official action. In addition to those weeds covered by the 1959 act, under section 14 of the *Wildlife and Countryside Act 1981*, it can be an offence to plant or grow certain specified plants in the wild, including Giant Hogweed and Japanese Knotweed. Problems involving these plants can be referred to the local authority for the area where those weeds are growing as some local authorities have bye-laws controlling these plants. There is no statutory requirement for landowners to remove these plants from their property.

7

Plant Biology of Fungi

INTRODUCTION TO FUNGI

The organisms of the fungal lineage include mushrooms, rusts, smuts, puffballs, truffles, morels, molds, and yeasts, as well as many less well-known organisms. More than 70,000 species of fungi have been described; however, some estimates of total numbers suggest that 1.5 million species may exist. As the sister group of animals and part of the eukaryotic crown group that radiated about a billion years ago, the fungi constitute an independent group equal in rank to that of plants and animals. They share with animals the ability to export hydrolytic enzymes that break down biopolymers, which can be absorbed for nutrition. Rather than requiring a stomach to accomplish digestion, fungi live in their own food supply and simply grow into new food as the local environment becomes nutrient depleted.

Most biologists have seen dense filamentous fungal colonies growing on rich nutrient agar plates, but in nature the filaments can be much longer and the colonies less dense. When one of the filaments contacts a food supply, the entire colony mobilizes and reallocates resources to exploit the new food. Should all food become depleted, sporulation is triggered. Although the fungal filaments and spores are microscopic, the colony can be very large with individuals of some species rivaling the mass of the largest animals or plants.

Prior to mating in sexual reproduction, individual fungi communicate with other individuals chemically via pheromones. In every phylum at least one pheromone has been characterized, and they range from sesquiterpines and derivatives of the carotenoid pathway in chytridiomycetes and zygomycetes to oligopeptides in ascomycetes and basidiomycetes.

Within their varied natural habitats fungi usually are the primary decomposer organisms present. Many species are free-living saprobes (users of carbon fixed by other organisms) in woody substrates, soils, leaf litter, dead animals, and animal exudates. The large cavities eaten out of living trees by wood-decaying fungi provide nest holes for a variety of animals, and extinction of the ivory billed woodpecker was due in large part to loss, through human activity, of nesting trees in bottom land hardwoods. In some low nitrogen environments several independent groups of fungi have adaptations such as nooses and sticky knobs with which to trap and degrade nematodes and other small animals. A number of references on fungal ecology are available.

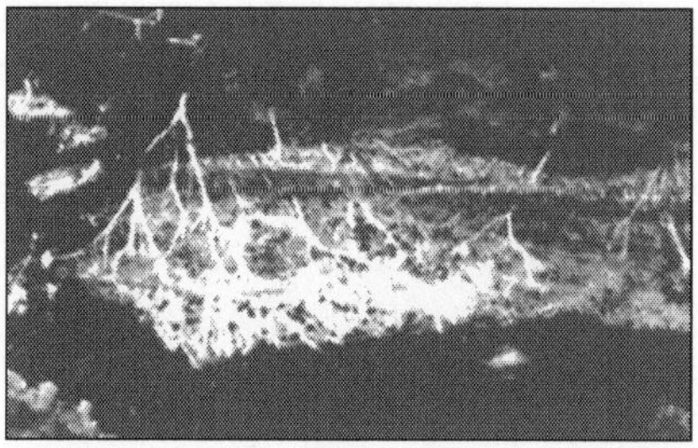

Fig. Hyphae of a Wood-decaying Fungus found Growing on the Underside of a Fallen log. The Metabolically Active Hyphae have Secreted Droplets on their Surfaces.

However, many other fungi are biotrophs, and in this role a number of successful groups form symbiotic associations with plants (including algae), animals (especially arthropods), and prokaryotes. Examples are lichens, mycorrhizae, and leaf and stem endophytes. Although lichens may seem infrequent in polluted cities, they can form the dominant vegetation in nordic environments, and there is a better than 80% chance that any plant you find is mycorrhizal.

Leaf and stem endophytes are a more recent discovery, and some of these fungi can protect the plants they inhabit from herbivory and even influence flowering and other aspects of plant reproductive biology. Fungi are our most important plant pathogens, and include rusts, smuts, and many ascomycetes such as the agents of Dutch elm disease and chestnut blight. Among the other well known associations are fungal parasites of animals. Humans, for example, may succumb to diseases caused by Pneumocystis (a type of pneumonia that affects individuals with suppressed immune systems), Coccidioides (valley fever), Ajellomyces (blastomycosis and histoplasmosis), and Cryptococcus (cryptococcosis).

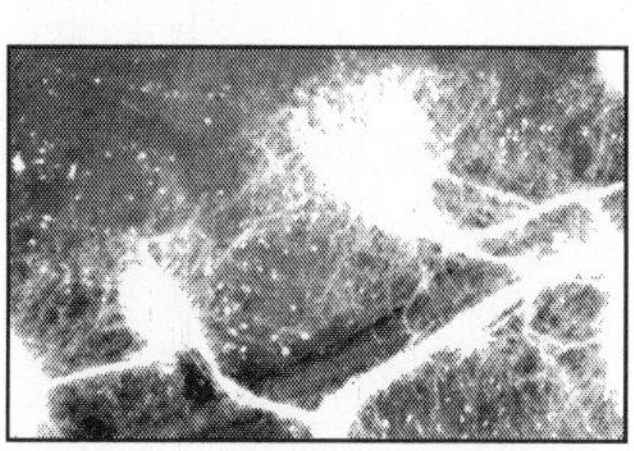

Fig. The Fluffy White Hyphae of the Mycorrhizal Fungus *Rhizopogon Rubescens* has Enveloped the Smaller Roots of a Virginia Pine Seedling. Note that Some of the Mycelium Extends out into the surrounding environment.

Fungal spores may be actively or passively released for dispersal by several effective methods. The air we breathe is filled with spores of species that are air dispersed. These usually are species that produce large numbers of spores, and examples include many species pathogenic on agricultural crops and trees. Other species are adapted for dispersal within or on the surfaces of animals (particularly arthropods). Some fungi are rain splash or flowing water dispersed. In a few cases the forcible release of spores is sufficient to serve as the dispersal method as well. The function of some spores is not primarily for dispersal, but to allow the organisms to survive as resistant cells during periods when the conditions of the environment are not conducive to growth.

Fig. *Entomophthora,* "Destroyer of Insects", is the Agent of a Fungal Infection that Kills Flies. After their Death the Fungal Growth Erupts through the fly Cuticle, and Dispersal by Forcible Spore Discharge is a Source of Inoculum for Infection of New Flies.

STRUCTURE OF FUNGI

The structures of fungi are microscopic and not visible to the naked eye. Some are unicellular like yeast, but most string their cells together in long, thread-like strands called hypha. Most fungi produce an extensive system of hyphae, which may be visible when growing thickly in a mass called mycelium (commonly referred to as mould). Mycelium can be of any size from tiny clusters to massive acre wide systems, which effectively form the feeding and growing body of the fungus.

HYPHAL STRUCTURE

A single fungus consists of a number of identifiable components. In most cases, the most common element is the hyphal network (mycelium) which ramifies through the substrate from which the fungus is gaining its organic energy. The fungus is likely to reproduce asexually and sexually at some period during its existence. The reproductive structures are usually morphologically distinct from the mycelium.

Asexual structures may simply be compartments which have separated from the thallus and fragmented, or identifiable structures where the spores are formed and released. The fungus may also form somatic structures which enable the fungus to survive a period of stress. Apart from spores, some fungi can form sclerotia, or stromata. The fungus may escape from reduced resources by aggregating hyphal structures called strands or rhizomorphs which may grow beyond the zone of depletion and in doing so, locate further resources. Each of these structures is constructed from modified hyphae.

Many fungi do not form hyphae. Yeasts are single celled fungi. Their life cycle in analogous to the filamentous lifecycle, but much simpler. In the Chytridiomycota, the thallus may be even simpler. Some fungi have only two forms: a sporangium and zoospore, and the sexual and asexual stages are encompassed within this simple staging.

Septum

In all fungi, septa form, either adventitiously in all filamentous fungi, or at regular intervals along the hypha in most members of the Ascomycota and Basidiomycota. Septa are barriers across the filament. In adventitious or secondary septa, the barriers are absolute, and no cytoplasm can pass from one compartment to the adjacent one. In the fungi primary septa, where the septa are regularly spaced, pores in the septum enable exchange of at least cytoplasm between adjacent compartments. Nuclei and most organelles within the cytoplasm can pass through pores.

In all regularly septate fungi, cytoplasmic continuity within the thallus is a feature regardless of the structure of the pore. This suggests that the linked compartments will have the potential to communicate rapidly. Closure of the pore enables differentiation of function within the thallus.

Septum formation is a simple process. Wall ingrowth towards the centre of the compartment results in a complete or incomplete blockage of the hypha. Inward growth may be followed by modification of the outer wall. The septum also increases rigidity of the hypha as it can function as a structural support to the turgor pressure within the compartment.

The septum in yeast cells has a function during reproduction. Replication of yeast cells differs from hyphal fungi. In the ascomycetous yeasts, where bud cells separate from the parent, a septum divides the two during growth of the bud. A wall is deposited within the septum and dissolution of the septum proceeds. The bud separates following septum dissolution. The parent cell is left with a scar at the point where the bud separates. This scar will not be able to bud again, and as a consequence, the number of buds formed by any one cell is limited.

Hyphae

Fungal hyphae are tubular structures. The cytoplasm is surrounded by a rigid wall which separates the fungus, physically and functionally, from the external environment.

The tube is divided into compartments in most cases. The dividing wall is called a septum, and the division is incomplete in the early stages of the life of the hypha. Elongation of the hypha takes place at the apex. The tip region extends as the cytoskeleton delivers materials for enzymic dissolution of the old and construction of the new wall at the apex. The precise mechanism whereby the tip extends is still unclear. However, turgor pressure is thought to drive the process which appears to be under the control of the cytoskeleton.

Closure of Pores

Closure of pores takes place by one of several processes. The pores may become blocked by a variety of materials, including electron dense materials, crystals or Woronin bodies. Blockage can be very rapid, in a few seconds in extreme cases.

The effect is to separate adjacent compartments, establishing two functional units on either side of the blocked septum. The separation enables development of the thallus to proceed differently on either side of the blockage. It also prevents or reduces loss of cytoplasm from the thallus should a hypha be damaged. Cytoplasmic continuity within the thallus can be re-established by growth of laterals followed by anastomosis of hyphae from either side of the closure. Closure of pores appears to have an important role in the life cycle of fungi enabling spatial separation of function.

Development of a variety of structures follows closure of pores. Asexual spores develop within a small number of cells above an adventitious septum. Sexual structures and stromata are more complex. These structures continue to grow as separate units after closure, before completion of the structure. The contribution of the thallus to the development of these complex and large structures is unclear. Even within the sexual structure, different elements of the structure appear to have different developmental pathways suggesting complex signalling and physiological processes.

Complex Hyphal Structures

Hyphae found in complex structures such as fruit bodies, rhizomorphs and sclerotia may differ. These differences were used taxonomically to differentiate various fruit bodies of some groups within the Basidiomycota. A system was developed where three different types of hyphae were used to distinguish polyporoid fungal taxa.

Generative hyphae were branched, thin walled, septate and commonly with clamp connections.

Skeletal hyphae were unbranched, thick walled hyphae lacking septa.

Binding hyphae were highly branched, thick walled, aseptate, found binding the other types.

Monomitic fungi contained only generative hyphae. Dimitic fungi contained both generative and skeletal hyphae. Trimitic fungi had all three types of hyphae. This system has been extended with the recognition that a variety of hyphal types exist, some are inflated, others of a specific shape and apparent function.

The point to understand here is that hyphae may develop a variety of forms which are under genetic and environmental control. Overall, fungi are able to radically alter their development pathway as the environment changes. These influences are expressed at the level of the hypha.

MACROSCOPIC FUNGI

Macroscopic fungi such as morels, mushrooms, puffballs, and the cultivated agarics available in grocery stores represent only a small fraction of the diversity in the kingdom Fungi. The molds, for example, are a large group of microscopic fungi that include many of the economically important plant parasites, allergenic species, and opportunistic pathogens of humans and other animals.

They are characterized by filamentous, vegetative cells called hyphae. A mass of hyphae forms the thallus (vegetative body) of the fungus, composed of mycelium. The more phylogenetically primitive molds (e.g., water molds, bread molds, and other sporangialsaclikeforms) produce cenocytic filaments (multinucleate cells without cross-walls), while the more advanced forms produce hyphae with cross-walls (septa) that subdivide the filament into uninucleate and multinucleate compartments.

The septum, however, still provides for cytoplasmic communication, including intercellular migration of nuclei. Many fungi occur not as hyphae but as unicellular forms called yeasts, which reproduce vegetatively by budding. Some of the opportunistic fungal pathogens of humans are dimorphic, growing as a mycelium in nature and as a vegetatively reproducing yeast in the body.

Candida is an example of such a dimorphic fungus. It can undergo rapid transformation from the yeast to the hyphal phase in vivo, which partly contributes to its success in invading host tissue.

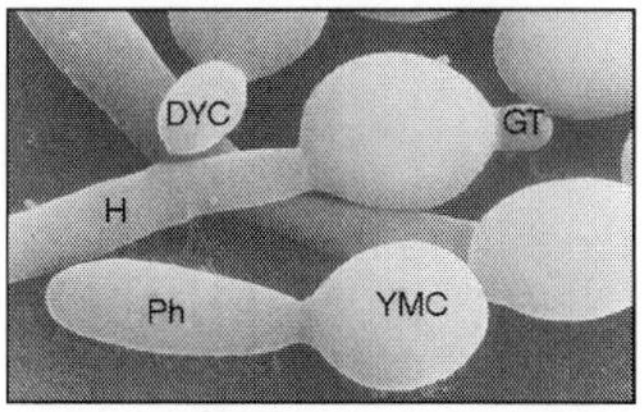

Fig. Dimorphism in C *Albicans*. DYC, Daughter Yeast Cell; GT, Germ Tube; H, Hypha; Ph, Pseudohypha; YMC, Yeast Mother cell.

The true fungi obtain their carbon compounds from nonliving organic substrates (saprophytes) or living organic material (parasites) by absorption of nutrients through their cell wall. Small molecules (e.g., simple sugars and amino acids) accumulate in a watery film surrounding the hyphae or yeast and simply diffuse through the cell wall.

Macromolecules and insoluble polymers (e.g., proteins, glycogen, starch, and cellulose), on the other hand, must undergo preliminary digestion before they can be absorbed by the fungal cell. This process involves release of specific proteolytic, glycolytic, or lipolytic enzymes from the hypha or yeast, extracellular breakdown of the substrate(s), and diffusion of the products of digestion through the fungal cell envelope. Fungal pathogens rely on these digestive enzymes to penetrate natural host barriers.

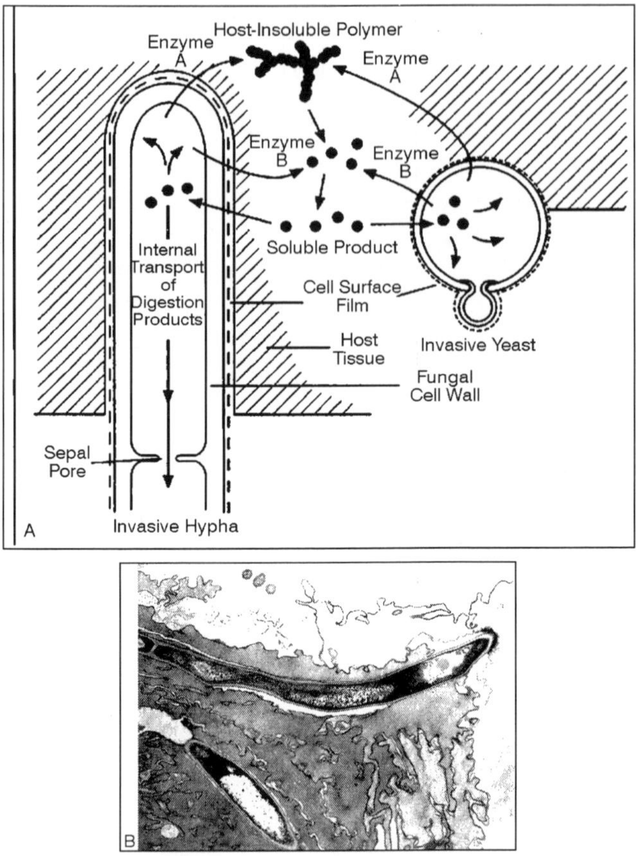

Fig. (A) Extracellular Digestion and Absorptive Nutrition in Fungi. (B) Invasive Hyphae of *C Albicans* in Stratified Epithelial Tissue of Mouse stomach..

CELL WALLS

Not all species of fungi have cell walls, but in those that do, cell wall synthesis is an important factor in determining the final morphology of fungal elements. Thus, our knowledge of fungal morphogenesis has evolved in parallel with our understanding of fungal cell wall biosynthesis. The fungal wall also protects cells against mechanical injury and blocks the ingress of toxic macromolecules.

This filtering effect may be especially important in protecting fungal pathogens against certain fungicidal products of the host. The fungal cell wall is also essential to prevent osmotic lysis. Even a small lesion in the cell wall can result in extrusion of cytoplasm as a result of the internal (turgor) pressure of the protoplast. The composition of fungal cell walls is relatively simple and includes substances not typically found in animal and plant hosts (e.g., chitin).

On this basis, it may be possible to identify pathogen-specific molecular targets from investigations of the biosynthesis of fungal wall components. Such targets may prove pivotal for the successful development of antifungal drugs that are not toxic to mammalian cells.

FILAMENTOUS FUNGI AND FILAMENTOUS BACTERIA

Fungi, like bacteria, are ecologically important as decomposers as well as parasites of plants and animals. Both groups of microbes often inhabit the same ecosystem and thus compete for the same food supply. Associated with this competition is the production by both the fungi and bacteria of secondary products that function as microbial growth inhibitors or toxins.

These compounds constitute a rich library of antimicrobial agents, many of which have been developed as pharmacologic antibiotics (e.g., penicillin from *Penicillium chrysogenum*, nystatin from *Streptomyces noursei*, amphotericin B from *S niveus*).

The superficial morphologic similarities between actinomycetes (filamentous bacteria) and molds suggest that the two groups have undergone parallel evolution. Despite the production of branching filaments and mold-like spores, the actinomycetes are clearly prokaryotes, whereas fungi are eukaryotes.

Moreover, the sexual reproduction of bacteria, which typically occurs by transverse binary fission, should not be confused with asexual processes of budding and fragmentation associated with mitotic nuclear division in fungi.

Most of the molds that produce septate vegetative hyphae reproduce exclusively by asexual means, giving rise to airborne propagules called conidia. On the other hand, elabourate mechanisms of sexual reproduction are also demonstrated by members of the Eumycota.

Four distinct kinds of meiospores (products of karyogamy-meiosis-cytokinesis) are recognized: oospores (Oomycetes), zygospores (Zygomycetes), ascospores (Ascomycetes), and basidiospores (Basidiomycetes). A summary of these and other diagnostic features of the fungi is presented.

Summary of Diagnostic Features of Fungi:

- Heterotrophic (no photosyuthetic nulrition).
- Thallus (vegetative body of fungus) on or in the substratum; may be plasmodial, ameboid, pseudoplasmodial, unlcellular (e.g., yeast), or filamentous (mycelial).
- Absorptive nutrition, typically by extracellular, enzymatic digestion (ingestion rare).
- Occurrence ubiquitous as saprobes, parasites, symbionts, or hyperparasites (fungi parasitic on another fungi).
- *Cell wall present in nonplasmodial forms*: Well defined and typically contains chitin (polymer of N-acetylglucosamine subunits joined by glycosidic bonds), β1-6-glucans; cellulose and b1-4-glucans present in the Oomycetes (water molds).
- *Eukaryotic (distinct from actinomycetes which are filamentous and prokaryotic unialike)*: heterokaryotic (genetically different unclei persent parent cells).

- *Asexual or sexual reproduction*: Spore-producing bodies (sporocarps) microscopic or macroscopic, showing limited tissue differentiation.

FUNGI AND THEIR HABITATS

Fungi can be found in just about any habitat you care to mention, from sea water through to freshwater, in soil, on plants and animals, on human skin and even growing on microscopic. Most of the fungi you would find in these places however, are very small and would need a microscope or magnifying lenses to see. The following discussion on fungal habitats concentrates on macrofungi, that is, those with fruiting bodies big enough to be easily seen.

Woods and meadows are the best places to go hunting for fungi. Of the two, woods are by far the best place to look, as over 80% of fungi are associated with trees. Some fungi, such as the Fairy Ring Toadstool, Marasmius oreades, and the edible field mushroom, Agaricus campestris, prefer open, grassy places.

Also watch out for a different group of fungi which grow on animal dung. These tend to have very specific habitat requirements and you will usually find that particular species will grow only on a certain kind of dung, at a particular stage of decay. In order to identify the fungus, you may first need to identify the type of dung! Fungi are common in woodlands because of all the rotting wood and leaf litter, which provides a wide range of the dead organic matter which most fungi feed on.

Many fungi are also associated with trees because they are linked into the tree roots. The association benefits both the fungi and the trees. This particular type of association between fungi and the roots of plants such as trees, is known as a mycorrhiza. Some fungi have very specific associations and will grow only with one kind of tree, for example, the bolete, Uloporus lividus, grows only under alders.

Other fungi may be found in association with several different trees. Chanterelles, for example can be found linked with birch, pine, oak and beech trees. Soil type is also important. Some fungi may be associated with a particular tree, but only where it is growing on suitable soil. The most important factor is usually whether the soil is acidic or calcareous (chalky). Soil fertility also plays a part. Fields which have been heavily fertilised with artificial nitrates are less likely to be good mushroom hunting territory than those which have been organically fertilised.

MYCORRHIZAE

Mycorrhizae are associations between fungal hyphae and the roots of plants. Almost all terrestrial plants, including wild plants, trees and commercial crops will form mycorrhizal associations with fungi in the soil. These mycorrhizae are vitally important for the growth and health of the plants. The fungi extract food (sugars) that they need from the plants, but in return they supply the plants with some of the nutrients and water which they may require. Thus both the fungi and the plants flourish because of the association. This is known as a symbiotic relationship. Certain plants, such as orchids, are totally dependent on a fungus associated with their roots in order to grow at all.

Through their enormous collecting network of hyphae in the soil, the fungi help to supply nutrients, especially phosphorus, to the plant roots they are associated with. This can have amazing effects on plant growth, particularly in soils which are not very fertile. In soils containing little phosphorus, plants with mycorrhizae have been shown to grow up to 20 times faster than those without.

The survival of plant seedlings may also be up to five times greater if they have mycorrhizae to help them collect nutrients and water from the soil. Helping plants to obtain phosphorus from the soil may also give them increased drought tolerance, as this is one of the effects of improved phosphorus nutrition.

Mycorrhizal fungi can often form associations with many different kinds of plants at the same time. The fungal hyphae of several different kinds of fungi may therefore be like a giant underground network connecting most of the plants in a habitat together. It is possible, although not proven, that this enables different plants to exchange nutrients between them via the fungal hyphae.

If this is so then this would hugely improve the chances of seedlings surviving, because they are not just dependent on their tiny root system, but have access to the great underground collecting network.

The presence of a good network of mycorrhizal fungi in the soil is therefore of vital importance for good plant growth in most habitats. Where the fungi are absent, or only present as isolated spores in the soil, plant growth will be reduced, apart from those species which do not require mycorrhizae. Most of these are what are commonly termed weed species.

Mycorrhizae also help to develop good soil structure through production of a protein which helps to stick small particles of soil together to form larger ones. This means that water can move more easily through the soil and provides more air spaces and thus air, for soil organisms and plant roots.

There are several different kinds of mycorrhizae. In some, the fungal hyphae actually enter the cells of the plant roots. These are called endomycorrhizae. There are several different kinds of these endomycorrhizae, of which a type abbreviated to AM (Arbuscular Mycorrhizae), is perhaps the best known. Eighty percent of the world's plant species, from grasses, to trees, most crop plants, shrubs and flowers, form AM associations.

Another type is called an Ectomycorrhiza (ECM). In this kind, the fungus forms associations with plant roots, but does not actually enter the root cells. This kind is mostly formed with different types of trees, such as pines, firs, spruces and oaks, amongst many others. ECM fungi may form thick hyphal strands known as rhizomorphs which can conduct water and nutrients relatively long distances. Many of the fungi common in woodland are ectomycorrhizal fungi, with mushroom or toadstool-like fruiting bodies.

HABITATS OF FUNGI

Fungi are found in a enormous diversity of habitats. The habitats provide a detailed understanding of how fungi organise their growth, sporulation, dispersal and colonisation. Some exquisitely complicated interactions have evolved indicating enormous periods of coevolution between fungi and their partners, be they living or dead. Fungi can be easily extracted from soil, the air, plants, animals, and less easily from water. These habitats have a number of characteristics that define the types of fungi found in them.

SOIL FUNGI

Soil is an oligotrophic medium for the growth of fungi. Readily available nutrients are present for short periods of time in a limited zone. For the remainder, fungi metabolise and grow very slowly utilising a range of recalcitrant organic molecules. In general, the concentration of microbes is greatest close to the surface of roots where exudates from the root supply a source of organic energy. Away from the root, the remains of plants and microbes are the primary source of energy. These have been composted and degraded, usually very rapidly, leaving modified waxes, lignin and other complex molecules behind.

Lifestyles

The lifestyles adopted by fungi in soil fall into three types: ruderals, mycorrhizal fungi and hyphal forms. Ruderals take advantage of the flushes of nutrients usually associated with rainfall. The water moving through soil carries with it dissolved organic molecules flushed from plant surfaces, surface litter and microbes.

The fungi respond immediately, and grow actively while soil remains moist. They usually sporulate rapidly, and exist through dry periods as asexual spores. Common genera include Absidia, Aspergillus, Chaetomium, Fusarium, Mortierella, Mucor, and Penicillium. These are the fungi that are commonly isolated from soil using soil dilution techniques.

Mycorrhizal fungi subsist almost entirely by tapping into the roots of plants for their organic energy . The fungi are extremely common in soil, though quantification of the fungi is extremely difficult. Up to 5m of living hyphae of arbuscular mycorrhizal fungi can be extracted from 1 g soil. The same quantity of soil may reveal 10,000 colonising propagules of the same fungi, and maybe the same number of spores of ruderals. As ruderals are defined as organisms that form huge numbers of survival units, one might argue that mycorrhizal fungi should be classified as ruderals . The lifestyle of mycorrhizal fungi is however, much more stable. While ever the host is alive, the fungi have access to organic carbon, and their proliferation is comparatively measured. Indeed, they probably exude simple organic molecules that are a suitable source of carbon for some microbes. AM fungi are ubiquitous yet lack clear methods of widespread dispersal found in ruderals. The life style is very different.

Finally, a group of fungi exist as hyphae in soil. Spores of these fungi are extremely difficult to find in soil, and some may never sporulate. While some of the fungi have the potential to cause disease, it is likely that for much of their lifecycle, these organisms metabolise and grow slowly. They are usually associated with organic fragments, which they slowly degrade. Isolation from soil requires special techniques. However, once in culture the fungi can grow rapidly utilising a wide range of sources of complex carbon. These fungi exhibit some characteristics of combative and stress tolerant fungi.

These fungi play an important role in soil, degrading the complex sources of organic carbon. While degradation is part of the normal cycle of carbon and other elements, loss of carbon from soil, especially agricultural soil, is responsible for a significant proportion of the increased CO2 in air, leading to the greenhouse condition of the world.

Fungi in soil demonstrate a variety of life styles, indicative of the variation in most habitats. Groups of different fungi utilise different carbon resources, with the result that very little organic carbon remains in soil, or passes through the soil in water.

THE PLANT AS A HOST FOR FUNGI

The most readily and widely available source of organic carbon is the plant. However, while all plants photosynthesise or have access to photosynthates, plants are not a homogenous or uniform group of organisms.

As well, plants have evolved a suite of defences against unwanted invaders. Thus, though very attractive to saprotrophs, plants are a highly diverse kingdom, with different approaches to their interactions with fungi.

In the first instance, plants have quite characteristic defences. The defences have been categorised as generalised, including tannins and lignin, and induced, including phytoalexins. Further, the nature of the defence appears to be related to the mineral nutrients available to the plant. Carbon-based defences are more common in poor soils while nitrogen-based defences are more frequently found in plants growing in rich soils. Thus we might ask why plants become colonised by fungi at all.

The fungus removes organic carbon from the plant. It might be argued that this loss is tolerated by the plant because the plant grows in sunlight and therefore can afford to pass some of its photosynthate to the second party. Others would argue that loss of organic carbon would only make sense if the plant benefited in some way from the association. The plant has three quite clear types of association with fungi: endophytes of the leaves, stems and maybe roots, mycorrhizal fungi of the roots, and plant pathogens *sensu lato*. Note that epiphytes are not considered as a separate group here.

Shoots

A clear relationship of benefit to both has been demonstrated by two groups of endophytes,vertically transmitted fungi such as Neotyphodium in grasses, and horizontally transmitted fungi such as Rhabdocline parkeri in Douglas fir. In general, the fungus has a complex interaction with the host. The fungus lives for some years within the leaf tissue and gains access to organic carbon of the host.

The host uses a diverse array of defences which appear to prevent the endophyte spreading through the tissue. Plant responses are apparent and in some cases, the fungus is constrained to one or two cells until the tissue senesces.

The fungus benefits the host in both the examples above. The host has reduced insect herbivory in the presence of the endophyte. The fungi produce secondary metabolites which can slow rates of metabolism of those insects that feed from the plant. Some further metabolites deter insects from landing and feeding. Overall, the colonised host plant gains a competitive advantage over uncolonised plants.

The fungus gains access to an almost unlimited supply of organic carbon. The host provides a protected and predictable environment. In the case of Neotyphodium, the host also provides the mechanism for initiation of colonisation of the next plant generation. Quite clearly, the association is mutually advantageous in these cases.

A wide range of fungi can be isolated from within healthy plant tissues. The cost to the host in these cases (endophytes sensu lato) is unknown. The benefit to the host is unknown. Benefit may be in the form of a reduction in herbivory. Alternatively, the endophyte may compete with potential pathogens, and other members of the general biota of plants to the benefit, or detriment, of the host plant.

Roots

A clear relationship of mutual benefit has been demonstrated in the case of mycorrhizal associations. In mycorrhizas, fungi colonise young roots and access organic carbon from the host. In all but orchid mycorrhizas the fungus supplies the host plant with increased mineral nutrients, commonly phosphate and/or nitrogen from soil. In the case of orchids, the precise physiological relationship is unclear because the fungus germinates orchid seed in the wild, supplying organic carbon to the plant. The fungus gains no apparent benefit, though density of mycorrhizal fungi declines dramatically away from the surface of the plant.

While many fungi can be isolated from leaves, fewer are present in the living part of the root. Many saprotrophic fungi exist outside the exodermis in young roots, and outside the endodermis once cortical cells have senesced. However, these fungi do not overcome host barriers, are living in dead or dieing tissue, and so are not equivalent to endophytes in leaves. On some occasions, the number of fungi inside the living zone of the root increases. Water and nutrient stress to the host plant and absence of mycorrhizal fungi appear to be associated with increased colonisation. These fungi are probably general pathogens which have entered because stress has reduced host defences.

The presence of competitors of pathogens has also been documented. Binucleate Rhizoctonia appears to competitively reduce colonisation of pathogenic Rhizoctonia solani. Similarly, pathogenic and saprotrophic species of *Fusarium* interact. However, apart from mycorrhizal fungi few root endophytes have been found to benefit the host plant.

Pathogens

Interestingly, most potential pathogens do not penetrate the barriers of the host plant. We see pathogens in plants, but only a few of the enormous array known to cause disease. In a few cases, the fungus may penetrate the host tissue, only to have it collapse around the penetration tube. Plant penetration is the exception, rather than the rule. Evolution of pathogen/plant interactions appears to have led to a situation where the plant is not seriously disadvantaged in natural ecosystems.

However, in cropping systems the interaction can be devastating. When the pathogenic fungus initiates infection, it spreads rapidly. This is because the crop is a genetically uniform target (monoculture), growing in uniform environments, at the same rate. The pathogen has the genetic constitution to avoid recognition by the host, and the physiology to parasitise the entire crop. Some of the most spectacular crop losses have been where a single crop has been almost wiped out by a single pathogen. While these incidents have historical importance, they also contain lessons for plant breeders and growers. The ramifications include appropriate targets for plant genetic manipulations, attitudes to sources and use of genetic diversity, and reliance on single genotypes.

Colonisation

In all these cases, the interaction between fungus and plant has involved a series of steps between plant and fungus . When the spore lands on the plant surface, it releases a chemical message which may be recognised by the plant. The spore attachs to the surface, germinates, and the emergent germ tube then forms an appressorium. A fine penetration peg grows through the plant surface, either directly through the epidermis, or via a stomatal opening. The hypha then ramifies through the plant tissue.

Initial energy for colonisation comes from the spore, perhaps with a small contribution from plant exudates on the surface. This leads to the collapse of the spore and the appressorium. Plant carbohydrates become increasingly important as the tissue is penetrated.

Presumably, those fungi that are recognised as pathogens will induce an immediate response to deter colonisation. The details of pathogenesis are now understood. The details of interaction between mutualistic symbionts are still being clarified, though may well be similar. It is clear that colonisation is an active process which involves a series of steps enabling colonisation and yet controlling spread. Specific fungi interact with each plant species. A small number of generalists also establish colonies in a range of plants. Presumably, the process of recognition and colonisation is under tight genetic control, and may be as complex as the immune system in mammals.

Much of the study of fungal colonisation of plants has concentrated on plant pathogens. However, a diversity of fungi interact with plants. Some of the fungi may cause disease, but many apparently lack detriment, and a few clearly benefit the host. The approach we take with fungi in plants needs to be modified by the knowledge that fungi may benefit the plant in ways we have yet to understand. These studies have the potential to reorganise the way we think about fungi, though they may also confirm the pathology- based thinking extant today.

FUNGI IN AIR - AEROMYCOLOGY, ASTHMA AND OTHER LUNG DISEASES

The air we breathe contains spores of many different fungi. Wind is commonly used as a mechanism to disperse fungi. Because of high irradiance and low water availability, air is a hostile environment for fungi. Spores of most fungi do not survive significant periods in air because of the absence of energy. Those that survive have quite specific mechanisms to prevent damage from desiccation and irradiation.

The Spores

The common air-borne fungal spores have thick walls and melanin deposited in the walls. The walls usually contain complex carbohydrates which are hydrophobic, and waxy. Hydrophobicity enables control of water loss, but the wall must also enable uptake of water prior to germination. Thus spore survival in air is a fine balance between water loss and retention of metabolic activity.

The cytoplasm of air-borne spores may be constituted of greater concentrations of compatible solutes such as glycerol and other polyols, which enable continued metabolism in conditions of reduced water availability., Use of glycerol in the cytoplasm enables loss of water from the cell without catastrophic consequences for water uptake in moist conditions.

Many molecules, including melanin reduce the penetration of radiation especially in the UV range. Melanin is contained in walls of spores, and so is the first barrier to UV. However, a range of other UV absorbing molecules are found in spores. The consequence is that energy is transformed usually to heat, which can be readily radiated from the spore in air.

Inhalation of Fungi

Spores are inhaled. In most cases, inhalation has no effect on humans. Most spores lodge on the surface of the lining of the airways, and are then expelled. In addition, the rest are usually neutralised because the immune response of the body is immediate and effective. The following segment is about those fungi that cause respiratory and other disease.

Asthma

Asthma is an allergic response in the lungs to allergens (Sinusitis is a similar response of the sinuses). The cells lining bronchi and alveoli become inflamed, severely reducing the capacity to inhale. Among common allergens are house dust mite, cat dander, rye grass pollen and some fungi including Alternaria, Cladosporium and Aspergillus. The fungus Alternaria is a common saprotroph and endophyte on many plants. It is commonly isolated from stems, and young and old leaves. It is particularly common on old leaves, and sporulates profusely especially as the leaf senesces.

Spores of the fungus are usually dispersed only short distances in high densities. However, mechanical manipulation of the plant debris during harvesting and handling of crops may result in the simultaneous release to the moving air of very high densities of spores. Once airborne, spores may disperse long distances in the air stream. In NSW Australia, the harvesting period for various crops coincides with the "spore season" and cases of asthma induced by Alternaria increase. Crop processing and transport are also likely to contribute to the spore load in air.

The fungi Alternaria and Cladosporium are similar in many ways. Cladosporium is a common saprotroph on plants, and can be isolated from soil. Spores are released to the air and dispersed significant distances in wind. Cladosporium is more common in the cooler areas of southern Australia; Alternaria is more common further north in the drier inland areas. Cladosporium causes a similar syndrome, but more commonly in southern Australia. Their distribution probably reflects their different biology. Cladosporium requires moist conditions to germinate and cool conditions to grow. Alternaria, on the other hand, can germinate at lower water potential than Cladosporium, and the fungus can grow in drier and hotter habitats.

Asthma in children is primarily caused by the fungus Alternaria in western NSW, especially in cropping areas. While the condition is associated with harvesting of crops, the fungus grows in dry conditions on all plants and can thus be expected across agricultural and non-agricultural areas. The fungus is uncommon on the coast, where many other fungi grow readily, thus competitively reducing populations of Alternaria in plants. Because plant growth will remain common in inland NSW, asthma caused by Alternaria will continue to be a major health problem.

Farmers Lung

This is a disease commonly found in people who handle hay and compost. In the strict sense, Farmers lung is caused by actinomycetes, not fungi, though the fungus Aspergillus fumigatus is often blamed. The fungus, however, can lodge and grow in the lungs where it may cause Aspergillosis, a serious condition when the person is immunocompromised (eg. the young and very old, pregnant, using drugs of recreation, immune suppressed by drugs). Most people with a functional immune system remain unaffected by Aspergillus. A. fumigatus is a thermotolerant and thermostable saprotrophic fungus commonly isolated from compost especially during the high temperature stage of composting. , The fungus is found world wide. It is more commonly associated with farming because of the readily available plant materials, but is also easily isolated from gardens and parks, bark, potting mixtures and mulches. Spore densities are normally low, maybe around 5 colony forming units per cubic metre air. In hay sheds, silos and other farm buildings, up to 20 million colony forming units per cubic metre have been collected. Thus Farmers lung is a syndrome associated with composting and plant materials, though the role of fungi in causing the disease is less than commonly indicated.

Other Medically Important Fungi

A number of fungi are becoming increasingly important because of the increased numbers of people whose immune system is compromised in some way. Traditional agents of fungal disease include Cryptococcus neoformans, Coccidioides immitis and Histoplasma capsulatum.

C. immitis

is a dimorphic fungus found in soil of the arid regions of the Americas. Spores become air-borne in dust and may become lodged in the lungs. Infection is established and the fungus spreads through the body. The fungus may lodge in bones, skin and meninges. The disease is most common in people who, never having been exposed to the fungus, suddenly become exposed to very high doses, such as army personnel, archaeologists and farm workers. The fungus is highly infectious, so can be readily transmitted from culture or tissue to laboratory workers. Immunisation is an effective preventative measure.

C. neoformans

is a yeast-like fungus, found naturally in bird droppings and the reproductive parts of some Eucalyptus trees. The spores are distributed in wind. When spores lodge in the lungs of susceptible individuals, development of pneumonia-like symptoms is followed by systemic infection, possibly involving the central nervous system. Normally, the immune system will remove the fungus and the symptoms will be transient.

H. capsulatum

is a dimorphic fungus found in soil. Spores become airborne in high winds. When spores lodge in the lungs of sensitive individuals, they may cause a mild, chronic or severe infection. Generally the bodies natural immune response clears the fungus from the lungs and the person shows antibody response to applied antigens.

With the spread of HIV, which suppresses the immune system, and use of immune-suppressant drugs in patients with transplant tissues, the range of fungi able to cause significant disease has increased . These fungi, have in the past, been unimportant because the human body has an effective immune response to the presence of the fungi. The response has been removed by agents of disease or medical intervention. With the increased medicalisation of human life, it is likely that diseases caused by fungi will increase.

AQUATIC FUNGI

Aquatic fungi, in the broadest sense, include fungi present transiently in water, terrestrial fungi that release spores which are dispersed in water, and species that function entirely within water. Water is found in transient pools after rain, swamps, estuaries, permanent rivers and oceans. Within a single environment, the water can have predictable characteristics. At the edge however, water bodies may contain soil and plant materials, and will be a changing enironment.

Ingoldian Fungi

The major input of plant materials in streams comes from leaf fall. These materials contain a range of endophytes of terrestrial fungi. However, the litter soon becomes colonised by a characteristic group of hyphomycetes with specialised spores. The fungi are known as Ingoldian fungi. They colonise the litter and replace the terrestrial fungi. Further, a typical succession takes place, analogous to succession found in compost and terrestrial leaf litter .Like the common terrestrial fungi, the Ingoldian fungi include many species that appear to lack specificity to the host plant; they are also found around the world.

Other types

Ingoldian fungi are not the only fungal inhabitants of litter. Another group form large coiled or rounded conidia which trap air, and these are known as aero-aquatic hyphomycetes. The advantage of their spores is that when the water dries, spores can lodge in the mud. On wetting again, the conidia float away on the water, being available to colonise leaf material when they make contact. Colonisation is rapid, and sporulation may take place within 7 days.

The hydrophobic conidia are formed on aerial conidiophores, which on release, float on the surface of the water.

The Chytridiomycota can only disperse in water . The zoospore requires at least a film of water to move away from the sporangium. They are readily trapped from bodies of water, and are also found in soil, or on surfaces where a water film forms transiently, on a regular basis.

Distribution of Spores

Ingoldian fungi have a number of interesting features. While the spores are characterised by a set of similar morphological features, it turns out that the teleomorphs are in a diverse range of both Ascomycota and Basidiomycota. The fungi can enzymically degrade the litter and are probably far more common on the leaves than bacteria, at least in the early stages of decomposition. The fungi form conidia while ever leaf litter is available. In Australian streams, seasonal peaks of spores in late summer will be found in the south, and in the late winter in the north, coinciding with the main season of leaf fall. However, the spore season is much longer and less pronounced in Australia where the leaf fall is less seasonal than in northern Europe and America where most leaves of deciduous trees fall in autumn.

Many of the fungi also colonise roots of trees on the stream banks. The fungi form sexual spores that lack any features enabling dispersal in water. In fact, the sexual stages may enable dispersal from stream to stream, with the asexual stages increasing colonisation within a stream. After all, release of conidia into a stream inevitably leads to dispersal downstream, not upstream. Further, as the source of leaf litter is highly variable, a source of propagules unrelated to the litter would enable faster build-up of inoculum in the stream.

Interactions with Invertebrates

The size of the population of Ingoldian fungi is also controlled by invertebrates. These animals graze on leaves, preferring colonised leaves, presumably because they have higher protein, lipid and mineral phosphate content due to the fungi, and because the leaf has been markedly predigested. Invertebrate competition markedly increases the rate of litter breakdown, and litter usually disappears rapidly from streams. However, it is argued that the presence of fungi is essential to the invertebrates because the fungi increase the nutritive value, providing essential nutrients not available in leaf material.

Fungi found in aquatic habitats have to gain access to organic materials to maintain themselves. Ingoldian and other aquatic fungi appear to have evolved unique life history strategies. The fungi are essentially biphasic, with one life component dispersing the fungi over land, and the second taking advantage of the transient flushes of nutrients following leaf fall to water. It is interesting that the asexual stage with its potential to rapidly increase numbers of propagules, utilises the aquatic leaf fall, and the sexual stage which disperses overland.

CHARACTERISTICS OF FUNGI

Fungi are characterized by non-motile bodies (thalli) constructed of apically elongating walled filaments (hyphae), a life cycle with sexual and asexual reproduction, usually from a common thallus, haploid thalli resulting from zygotic meiosis, and heterotrophic nutrition. Spindle pole bodies, not centrioles, usually are associated with the nuclear envelope during cell division. The characteristic wall components are chitin (beta-1,4-linked homopolymers of N-acetylglucosamine in microcrystalline state) and glucans primarily alpha-glucans (alpha-1,3- and alpha-1,6- linkages).

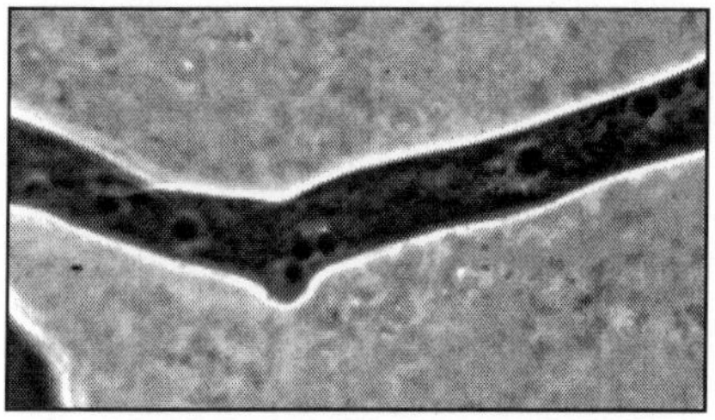

Fig. Portion of a Hypha of a Zygomycete Stained with a Blue dye to show the many Nuclei Present. Many other Fungi have Septations that Divide the Hyphae into Compartments that Usually Contain one to Several Nuclei Per Compartment.

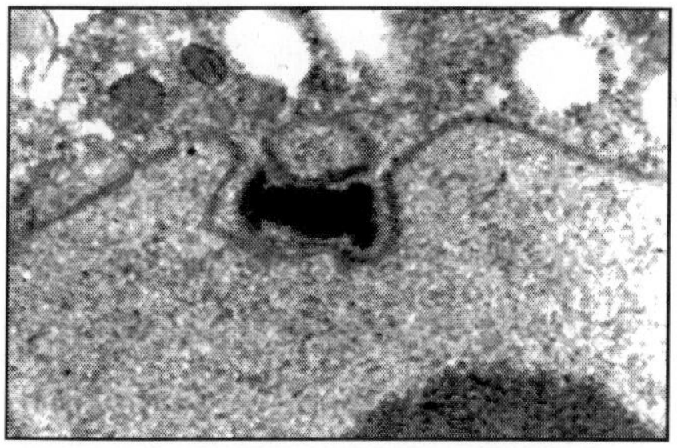

Fig. Transmission Electron Micrograph Showing Duplicated Spindle Pole Body of a Prophase I Meiotic Nucleus of a Basidiomycete *Exobasidium*. Only Chytrids among Fungi have Centrioles and Lack Spindle Pole Bodies.

Exceptions to this characterization of fungi are well known, and include the following: Most species of chytrids have cells with a single, smooth, posteriorly inserted flagellum at some stage in the life cycle, and centrioles are associated with nuclear division. The life cycles of most chytrids are poorly studied, but some (Blastocladiomycota) are known to have zygotic meiosis (therefore, alternation between haploid and diploid generations). Certain members of Mucoromycotina, Ascomycota, and Basidiomycota may lack hyphal growth during part or all of their life cycles, and, instead, produce budding yeast cells.

Most fungal species with yeast growth forms contain only minute amounts of chitin in the walls of the yeast cells. A few species of Ascomycota (Ophiostomataceae) have cellulose in their walls, and certain members of Blastocladiomycota and Entomophthoromycotina lack walls during part of their life cycle.

FOSSIL RECORD

Based on the available fossil record, fungi are presumed to have been present in Late Proterozoic (900-570 mya). Terrestrial forms of purported ascomycetes are reported in associations with microarthropods in the Silurian Period (438-408 mya).

Fossil hyphae in association with wood decay and fossil chytrids and Glomales-Endogonales representatives associated with plants of the Rhynie Chert are reported from the Devonian Period (408-360 mya). Fungal fossil diversity increased throughout the Paleozoic Era with all modern classes reported in the Pennsylvanian Epoch (320-286 mya).

A first attempt to match molecular data on fungal phylogeny to the geological record shows general agreement, but does point out some conflicts between the two types of data.

BIOGEOGRAPHY

Wherever adequate moisture, temperature, and organic substrates are available, fungi are present. Although we normally think of fungi as growing in warm, moist forests, many species occur in habitats that are cold, periodically arid, or otherwise seemingly inhospitable. It is important to recognize that optimum conditions for growth and reproduction vary widely with fungal species. Diversity of most groups of fungi tends to increase in tropical regions, but detailed studies are only in their infancy.

Although many saprobic and plant pathogenic species with low substrate specificity and effective dispersal systems have broad distributions, gene flow appears to be restricted in many fungi. For these species large bodies of water such as the Atlantic and Pacific Oceans create barriers to gene exchange. Some distributions are limited by substrate availability, and dramatic examples come from parasites of Gondowanan plants; one of these is the Southern Hemisphere distribution of the ascomycete Cyttaria, corresponding with part of the distribution of its host plant Nothofagus.

The fossil record shows that fungi were present in Antarctica, as is the case for other organisms with Gondwanan distributions. Arthropod associates also may show distributions throughout part or all of a host range, and some fungal species (ex. wood wasp associates) occur outside the range of the associated arthropod.

Notable Fungi

The largest known basidioma (mushroom or fruiting body) was that of a

Rigidioporus ulmarius (Agaricomycetes), hidden-away in a shady corner of the Royal Botanic Gardens, Kew, Richmond, Surrey, England.

This fruiting body was mentioned in the Guinness Book of World Records. At the beginning of every New Year the Annual Mensuration Ceremony of the fruiting body took place, and on 19 January 1996 it had increased to 170 cm maximum length (up from 159 in 1995) and 146 cm maximum width (up from 140 in 1995).

It also grew 4 cm taller from the soil level, measuring 54 cm. The weight of the fruiting body was estimated to be 284 kg (625 pounds)! Amid rumors of its destruction, Dr. Brian M. Spooner, Head of Mycology, Royal Botanic Gardens, has brought us up to date on the fate of the record specimen. Unfortunately, the basidioma began to rot at the edges a few years ago, likely because the hyphal body of the fungus digested away its elm root substrate, reminding us that a fungus needs a good dispersal system to escape the substrate that eventually inevitably is destroyed.

In the life of the fruiting body many trillions of spores must have been produced, and some of these surely fell on an appropiate substrate to establish a new infection. The final insult to the fruiting body came from a fox that burrowed under one side and caused it to collapse.

Fig. The Largest Basidioma World Record Holder Rigidioporus Ulmarius at Kew when it was Still Intact. The Mushroom is shown in its Largest Dimension (170 cm or over 5 1/2 feet).

Other large basidiomata are those of a Canadian puffball almost 9 feet in circumference (over 48 pounds) and a basidiocarp of the sulfur mushroom in England (100 pounds). A previous Guinness Book of World Records record-sized fruiting body of Bridgeoporus nobilissimus, an endangered species of the Pacific Northwest of the United States, is over 160 kg (300 pounds) and may have regained the title of"largest" with the demise of the Kew specimen. This polypore also may do itself in because its great weight is likely to eventually cause it to fall as the mycelium depletes its food source, often the noble fir tree.

"Reproductive structures clearly can be very large, but what about the body of the fungus, which often is hidden from view within the substrate? One fungus body constructed of tubular filaments (hyphae) was brought to our attention when molecular techniques were used to show that it was extensive (37 acres and an estimated blue whale equivalent size of 110 tons). The Michigan fungus clone (Armillaria bulbosa, Agaricomycetes) grew in tree roots and soil.

This report drew attention to an even larger fungal clone of Armillaria ostoyae, reported earlier in the state of Washington, which covered over 1,500 acres. Each clone began from the germination of a single spore over a thousand years ago. Although they probably have fragmented and are no longer continuous bodies, such organisms give us cause to think about what constitutes an individual.

"Penicillium chrysogenum (Ascomycota) is known for its production of the antibiotic penicillin. Although other antibiotics are produced by a variety of organisms, penicillin was the first to be developed. In the spring of 1996 a long dried out culture of the original isolate prepared by its discoverer, Sir Alexander Fleming in the late 1920s, was auctioned by Sotheby's of London and sold to a pharmaceutical company for 23,000 pounds.

This price is insignificant when one considers the worth of this fungus, not only in sales of penicillin, but in forms of illnesses cured and lives saved. In the past a simple scratch or blister sometimes could result in a fatal infection such as the blister that resulted in the death of John Calvin Coolidge (1918-1924), the son of a U. S. president.

However, misuse of penicillin and other antibiotics has resulted in selection of resistant microorganisms, and the threat of untreatable bacterial infections and diseases (for example Staphylococcus aureus and Escherichia coli and tuberculosis and syphilis) is still present in our homes and recreation areas.

"Fungal spores fill the air we breathe. On many days in some localities the number of fungal spores in the air far exceeds the pollen grains. Fungal spores also cause allergies; however, unlike seasonal pollen production, some fungi can produce spores all year long. The largest number of fungal spores ever sampled was over 5.5 million per cubic foot in Wales.

"Basidiomycetes have always attracted a lot of attention because some of them have large basidiocarps, but the realization that all fungi are important in ecosystem function has drawn more attention to microscopic forms as well. For example a report on the secret sex life of a yeast-like ascomycete human pathogen, Coccidioides immitis, made a headline of the New York Times. This fungus causes Valley Fever and is endemic in parts of the southwestern United States. Although no one has been able to observe sexual reproduction in this species, molecular studies show genetic diversity that is best explained by occurrence of sexual reproduction in the life cycle.

"Another yeast-like ascomycete reported in the Dallas Morning News lives in the gut of cigar beetles and is essential to the beetle's health. Without the gut fungi to detoxify the plant material of toxins, the beetles would be poisoned. Keep on the lookout for other reports of fascinating fungal feats.

Discussion of Phylogenetic Relationships

The kingdom Fungi is a diverse clade of heterotrophic organisms that shares some characters with animals such as chitinous structures, storage of glycogen, and mitochondrial codon UGA encoding tryptophan. Both animals and fungi have spores or gametes with a single smooth, posteriorly inserted flagellum, but only species of the basal chytrid phyla have retained this primitive character. Fungi, animals, and other heterotrophic protist-like organisms such as choanoflagellates and Mesomycetozoea are now considered part of the larger group termed opisthokonts in reference to the posterior flagellum.

The branch uniting the fungi and animals is well-supported based on a number of molecular phylogenetic datasets, including the nuclear small subunit ribosomal RNA gene, unique and shared sequence insertions in proteins such as elongation factor 1á, entire mitochondrial genomes and concatenated protein-coding genes.

Prior classification systems of Fungi based primarily on morphology are in need of updating to more accurately reflect phylogenetic relationships as determined by molecular systematics. Molecular characters have been essential for phylogenetic analysis in cases when morphological characters are convergent, reduced, or missing among the taxa considered. This is especially true of species that never reproduce sexually, because characters of sexual reproduction traditionally have been the basis for classification of Fungi. Use of molecular characters allows asexual fungi to be placed among their closest relatives.

Previous classifications placed early-diverging fungal groups (non-Ascomycota or Basidiomycota) into two phyla: Chytridiomycota and Zygomycota. Numerous phylogenetic studies now suggest that neither is monophyletic, and the latest classification scheme includes six phyla and an additional four unplaced subphyla.

At present, because of the ancient divergence times between the fungal phyla, the exact phylogenetic relationships are ambiguous. Chytrids appear to be a paraphyletic group at the base of the fungal phylogeny and merely fungal lineages which have retained the character of flagellated spores.

Three phyla of flagellated fungi are proposed and two chytrid genera *Olpidium* and *Rozella,* are of uncertain phylogenetic position. These genera are interesting because they are both highly reduced endoparasites (living inside the host cell) whose entire thallus consists of only a spherical body absorbing nutrients from the host material that surrounds it. *Rozella* appears in an isolated position in the fungal phylogeny as the very earliest lineage to diverge from the rest of the fungi. In contrast, *Olpidium brassicae* appears to have diverged after the majority of chytrids and is more closely related to some zygomycete fungi.

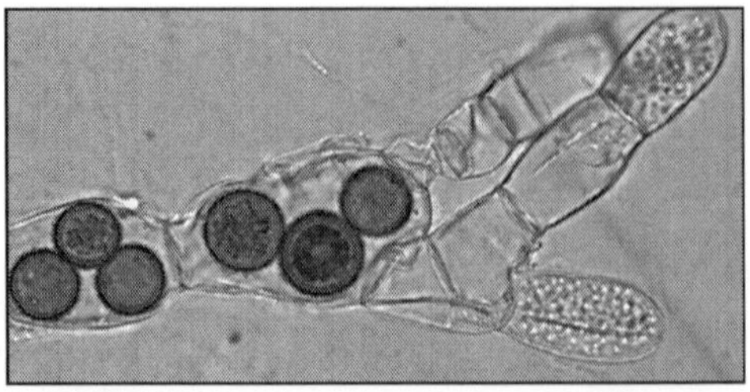

Fig. The Endoparasitic Chytrid *Rozella Allomycis* Inside the Hyphae of Another Chytrid *Allomyces*. Thick Spiny Spores of the Parasite are seen Inside some cells while Zoospores are Produced in other Cells.

Fungi with non-septate or irregularly septate hyphae and thick-walled spores were traditionally placed in the phylum Zygomycota. However, evidence for a monophyletic Zygomycota is lacking, and the deconstruction of the Zygomycota into four unordered subphyla (Entomophthoro-mycotina, Kickxellomycotina, Mucoromycotina, Zoopagomycotina) has been proposed. The separation of the superficially similar arbuscular mycorrhizal fungi (that lack septa in hyphae but also lack zygospores) into the phylum Glomeromycota has been previously proposed. Whether this phylum is more closely related to the Ascomycota and Basidiomycota lineage or to other zygomycete lineages is controversial.

Evidence from shared morphological characters such as regularly septate hyphae and a dikaryotic stage (two separate and different nuclei in a single hyphal segment) in the life cycle usually has been interpreted as support for a close relationship between Basidiomycota and Ascomycota. Numerous phylogenetic studies such as SSU rDNA, RNA polymerase genes, and mitochondrial genome sequencing provide strong support for this relationship. A subkingdom termed Dikarya is proposed, creating a division between a highly speciose subkingdom (Dikarya) and the remaining early diverging lineages whose relationships are not precisely known.

Fungal classification is far from static, and even which organisms are actually members of Fungi is changing. For example, the group trichomycetes describes gut inhabitants of arthropods that share similarities with zygomycetes. Molecular phylogenetic studies have demonstrated that two of the four orders of trichomycetes are actually members of the Mesomycetozoea protist group.

Other organisms that were previously considered to be Fungi because of their heterotrophic, mold-like growth forms are now classified as stramenopiles (Oomycota, Hyphochytriomycota, and Labyrinthulomycota) or slime molds (Myxomycota, Plasmodiomycota, Dictyosteliomycota, Acrasiomycota). More interesting for mycologists are the findings that some species previously considered protozoa are actually Fungi. For example, the species *Hyaloraphidium curvatum* was assumed to be a green alga that had adopted a heterotrophic lifecycle concomitantly with losing its chloroplast. It is now known to be a chytrid fungus related to Monoblephariomycetes but lacking a flagellated stage. Other examples include the parasitic organisms presumed to be protozoa, such as the cockroach parasite *Nepridiophaga* and the *Daphnia* parasite *Polycarum* recently demonstrated to be members of the fungal kingdom based on SSU rDNA phylogenies.

The most revolutionary addition to the fungal lineage has occurred with phylogenetic evidence indicating the protist group microsporidia is closely related to Fungi–possibly derived from zygomycetes or sister to the genus *Rozella* on the earliest branch in the fungal kingdom. Microsporidia are highly specialized intracellular parasites (primarily of animals) that lack mitochondria but have chitin and trehalose in their spores (similar to Fungi). All molecular studies have shown that microsporidia evolve at an extremely accelerated rate of evolution, making their placement in the Tree of Life difficult. The relationship with fungi is supported by many single and multiple gene phylogenies, but an exact placement within the fungi has not received strong support.

More recently the nuclearid amoebae have been demonstrated to be a sister group to the Fungi with strong support. This finding is significant because *Nuclearia* lacks a cell wall and has phagotrophic nutrition in which the food source (such as a bacterium or algal cell) is engulfed wholly, unlike fungi and microsporidia which utilize absorptive nutrition. Further sampling of basal fungal lineages will be needed to determine whether a *Nuclearia*-like organism was the cenancestor (most recent common ancestor) of Fungi.

MUTATION IN FUNGI

Mutation in fungi can occur for most physiological and morphological characters. The change may be in one or more characters and in many degrees of magnitude. Mutation is common for cultural characters, including colour, topography, consistency of colonies, direction of growth, nature of margin, zonation, rate of growth, type of growth, and amount of sporulation. There is mutation for physiological characters (such as enzyme production), reaction to known chemicals and to toxic substances, in temperature requirements, and in tendency to mutate. There often is mutation for morphological characters, principally size, shape, and colour of spores, fruiting bodies, and resting bodies. Mutation in sexual vigour and in pathogenicity also is common.

Mutation in Ustilago zeae, Helminthosporium sativum, and many other fungi has been studied at the University of Minnesota continuously for more than 30 years, and many thousands of mutants have been isolated and studied. The number and kinds of mutants produced by U. zeae and H. sativum are indefinite. Mutants may differ widely or very slightly from their parents. As one example, a series of mutants from a single brown monosporidial line of U. zeae ranged from near black through a score of tints to colourless. From another line, a white mutant was isolated from a pigmented line of U. zeae, and this white mutant in turn produced white mutants, which in turn produced additional ones.

The original mutant was then crossed with certain pigmented lines to find out its breeding behaviour, and all of the white segregates from the cross were isolated and kept. In this way, 98 clearly distinct white lines, mutants and segregates, were obtained, which comprised a number of sex groups. Some of the numerous white- x white crosses produced large galls, but none produced chlamydospores.

When some of the white lines were crossed with pigmented lines, however, both galls and chlamydospores were produced. Evidently, therefore, the original white mutant had lost factors for colour production and for nuclear fusion and consequent production of chlamydospores, although the nuclei of opposite sex in some of its progeny had the necessary factors for pairing. Certain of the white x white crosses therefore had the necessary factors for pairing Of nuclei and for pathogenicity, but not for the final stage in the sexual process; hence there had been a loss of some but not all of the factors for sex. Although mutation in factors for sex and pathogenicity are common in smuts, complete sex reversal has not been reported.

It is harder to observe mutation for pathogenicity than for cultural characters. When it does occur, however, there usually is partial or complete loss in pathogenicity and only occasionally is there a gain. This is conspicuously true of H. sativum and of U. zeae, probably the most thoroughly studied in this respect. When successive mutants of U. zeae are isolated (that is, mutant from mutant from mutant and so on), a clearly descending series from strongly pathogenic to non-pathogenic is likely to result. Occasional mutants, however, are decidedly and consistently more pathogenic than the parental line. This is true also of other fungi.

Colour mutations are not uncommon in rusts, and at least two such recorded cases also involved changes in pathogenicity. There is evidence that uredospores of orange and white races of Puccinia graminis are destroyed more easily by ultraviolet light than those with the normal darker colour. This may partly account for the relatively poor survival ability and infrequent occurrence of the abnormally light-coloured races in nature.

Mutation in pathogenicity occurs occasionally in rusts and it probably is fairly common, although minor mutations are hard to detect. In cooperative investigations between the Department of Agriculture and the University of Minnesota, E. C. Stakman, M. N. Levine, and R. U. Cotter studied four distinct mutations for pathogenicity in race 1 of Puccinia graminis tritici, and two of the mutants differed so much from anything previously described that they were designated as new races, numbers 60 and 68. Margaret Newton and Dr. T. Johnson, at the Dominion Labouratory of Plant Pathology, reported two separate mutations for pathogenicity in Puccinia graminis, one in the variety tritici and the other in the variety avenae. Mutation for pathogenicity also has been reported for Puccinia hordei, leaf rust of barley; P. glumarum, stripe rust of wheat and other cereals; and P. rubigo-vera tritici, orange leaf rust of wheat. Some of these rust mutants were less virulent than their parents on some varieties, but some were more virulent than their parents on some varieties of host plants.

In some fungi the number of visible mutants produced differs on different media. Whether this is always due to different rates of mutation or to differences in the conspicuousness of the mutants on different media is not known in all instances. Certain sugars, potassium salts, uranium and polonium salts, salts of some of the heavy metals, and some other chemicals generally increase the number of visible mutants. Either high or low temperature may also increase the number, depending on the species or biotype of the fungus. Ultraviolet light, radiations, and some of the bacterial products also are mutagenic.

J. J. Christensen observed that mercury bichloride, ethyl mercury phosphate, and malachite green in nutrient media were decidedly mutagenic to Gibberella zeae, Helminthosporium sativum, H. carbonum, Fusarium moniliforme, and Colletotrichum lini. The most significant fact derived from these experiments has been discussed in the section on adaptation, where it was pointed out that some of the mutants grew from 5 to 15 times as fast as the parental lines on these chemicals. Christensen and F. R. Davies also found that bacteria-staled media caused very abundant mutation in monosporous lines of H. sativum. Among the mutants there were zero, plus, and minus deviations from the original lines in tolerance for the bacterial products, in pathogenicity, in rate of growth, and in many cultural characters.

M. L. Gattani, E. C. Stakman, J. M. Daley, Shih I. Lu, and J. B. Rowell, at Minnesota, increased the rate of mutation in haploid and diploid lines of Ustilago zeae by adding Uranyl nitrate, at the rate of about one gram per liter, to potato-dextrose agar.

Most mutants grew as well as or better than the parental lines on the mutagenic medium. Lu studied 13 characters in 198 uranium-induced mutants from a monosporidial haploid line of U. zeae and summated algebraically the zero, plus, and minus deviations for each of the 13 characters for each mutant. About 20 per cent had a net minus deviation, 20 per cent a plus deviation, and 60 per cent a zero deviation, although all were, of course, different from the parent in one or more characters.

Thus, deviation would be zero if a mutant exceeded the parent in size of colonies and in two other characters, had lost some factors for colour and for two other characters, but was like the parent in the other seven characters.

There would be a plus deviation if the mutant exceeded the parent in two characters, had lost in one, and remained the same in 10, and there would be a minus deviation if the plus and minus numbers were reversed.

In similar studies on a solopathogenic (diploid) line, 10 per cent of the mutants had a net minus deviation, 30 per cent plus, and 60 per cent were zero. In similar experiments I. Wahl, also at Minnesota, produced mutants of the common mushroom, Agaricus campestris, that were much more vigourous than the original line and that produced mushrooms of more desirable colour.

Uranyl nitrate is not equally mutagenic to all fungi, being effective in about 10 per cent of 194 species and lines studied by E. C. Stakman, J. B. Rowell, Howard Ehrlich, and others in an investigation made cooperatively at the University of Minnesota by the United States Atomic Energy Commission and the University. Polonium salts have been found more effective than uranium salts as mutagenic agents in the investigations.

H. sativum and U. zeae have been discussed so fully merely because they have been studied extensively and continuously by many investigators at Minnesota, under the general supervision of the writers, for about a quarter of a century, and because evidence is increasing that the principles derived from them apply in varying degrees to many plant pathogens. Above all, it is noteworthy that in U. zeae where single, haploid sporidia can be isolated and propagated asexually, thus theoretically establishing single-biotype cultures, it is extremely difficult, because of visible or invisible mutations, to maintain the genetic purity even of the apparently least-mutable monosporidial lines. At best, purity appears to be relative only.

Stability among mutants differs greatly: Many mutants continue mutating indefinitely. Others are relatively stable. There are many intermediate degrees of mutability. Many mutants of Helminthosporium sativum, H. oryzae, H. carbonum, Gibberella zeae, and Fusarium lini have been grown side by side with their parental lines for many years and have retained their distinctive characters.

Some have been grown on many kinds of media several years, and have retained their distinctive cultural characters. Many mutant lines of Ustilago zeae have been grown under a wide range of conditions for many years without perceptible change, but others continue to mutate abundantly. C. S. Holton, in longtime experiments started at the University of Minnesota and continued under State and Federal cooperation at Washington State College, maintained an albino mutant of covered smut of oats, Ustilago kolleri, through many asexual and sexual generations during the past 18 years. G. W. Keitt and M. H. Langford, at the University of Wisconsin, passed three haploid lines of Venturia inaequalis through the leaves of the McIntosh apples four successive times without perceptible change in cultural characters. J. J. Christensen obtained similar results by passing mutants of Diplopia zeae, Gibberella zeae, Helminthosporium sativum, and H. oryzae through their respective hosts, although in one case the cultural characters of a mutant of H. sativum were changed slightly. Mutant characters of Ustilago zeae have persisted through the sexual stage on corn plants, even though they sometimes were recombined with other characters.

INHERITANCE OF FACTORS

Faithful inheritance of mitochondria is essential for growth and development. Uniparental inheritance of mitochondria is a common phenomenon in sexual eukaryotes and has been reported for numerous fungal species. Uniparental inheritance is a genetically regulated process, aimed to gain a homoplasmic state within cells, and this is often associated with selective elimination of one parental mitochondria population. This review will focus on recent developments in our understanding of common and specified regulatory circuits of selective mitochondrial inheritance during sexual development.

It further refers to the influence of mitochondrial fusion on generation of recombinant mitochondrial DNA molecules. The latter aspect appears rather exciting in the context of intron homing and could bring a new twist to the debate on the significance of uniparental inheritance. The emergence of genome-wide studies offers new perspectives to address potential relationships between uniparental inheritance, vegetative inheritance and last but not least cellular scavenging systems to dispose of disintegrated organelles.

The inheritance of factors for spore-wall markings of hybrid chlamydospores of smuts usually is simple. When one with smooth-walled spores is crossed with one that has spiny-walled spores (echinulate), factors for spiny wall tend to dominate over those for smooth wall, but in crosses between species of Tilletia with smooth and reticulate-walled spores (those with net-like ridges) more than one set of factors apparently is involved. In some crosses smooth completely dominates over reticulate, and in others reticulate is partly dominant over smooth, and intergrading types of reticulation may occur in the F, and F2 chlamydospore generations.

The range of variation in types of smut is well illustrated in crosses between the loose and covered smuts of oats. When varieties of oats are inoculated with different combinations of monosporidial lines of Ustilago avenae, loose smut, and U. kolleri, covered smut, many kinds of smutted heads (panicles) may be produced, ranging from loose to covered, through many intermediate types.

One combination may produce loose smut on one variety and covered smut on another, and another combination may produce the same type of smut on both varieties, which indicates that the type of smutted panicles is determined by both the fungus and the variety of oats.

In interspecific crosses between the barley smuts Ustilago hordei (covered type) and U. nigra (false loose type), made by C. C. Allison at Minnesota, the F_1 head types were intermediate, but tended towards the loose type. The progeny from the F_1 chlamydospores produced not only the parental types of sori, compact heads with smooth spores and loose heads with echinulate spores, but several new types, including intermediates with smooth or echinulate spores, compact heads with echinulate spores, and loose heads with smooth spores.

L. J. Tyler and C. P. Shumway, at Minnesota, made crosses between the sorghum smuts Sphacelotheca sorghi (covered kernel smut) and Sorosporium reilianum (head smut). The characters of the F, sori and of the chlamydospores tended to be intermediate between those of the parents. Syed Vaheeduddin crossed Sorosporium reilianum (head smut of sorghum) with Sphacelotheca cruenta (loose kernel smut of sorghum).

Different combinations Of monosporidial lines produced sori differing in shape and size. Some of the F_1 sori resembled one or the other parent, while others were intermediate. Similar types were obtained again in F_2 from inoculating with different combinations of f_1 (gametic) segregates. Naturally, there was segregation of factors for many other characters also such as size and echinulation of chlamydospores, cultural characters, sex, and pathogenicity.

Some of the hybrids between the two smuts had pronounced hybrid vigour. They caused extreme elongation of the ovaries of sorghum, thus producing an effect similar to that caused by the long smut of sorghum, Tolyposporium filiferum. Moreover, the chlamydospores, although intermediate in size between the sizes of the parental spores, germinated over an exceptionally wide range of temperature and produced promycelia two and a half to three times as long as those of either parent, and the sporidia and hyphal branches were correspondingly large. In addition, some haploid segregates had extraordinary tolerance for certain chemicals. Since sporidia and promycelia) cells are haploid, not diploid as in higher plants, a study of hybrid vigour in smut fungi may aid in interpreting that in higher plants.

INTERSPECIFIC HYBRIDIZATION

Fungi (kingdom Mycota) and oomycetes (kingdom Stramenopila, phylum Oomycota) are crucially important in the nutrient cycles of the world. Their interactions with plants sometimes benefit and sometimes act to the detriment of humans. Many fungi establish ecologically vital mutualisms, such as in mycorrhizal fungi that enhance nutrient acquisition, and endophytes that combat insects and other herbivores. Other fungi and many oomycetes are plant pathogens that devastate natural and agricultural populations of plant species. Studies of fungal and oomycete evolution were extraordinarily difficult until the advent of molecular phylogenetics.

Over the past decade, researchers applying these new tools to fungi and oomycetes have made astounding new discoveries, among which is the potential for interspecific hybridization. Consequences of hybridization among pathogens include adaptation to new niches such as new host species, and increased or decreased virulence. Hybrid mutualists may also be better adapted to new hosts and can provide greater or more diverse benefits to host plants.

Interspecific hybridization apparently produces new parasitic races of some smuts in nature. New virulent races of buff smuts of oats have been produced artificially by crossing Ustilago avenae with U. kolleri. Some of the hybrids had a wider host range than either parent. Some of the hybrid races of the buff smut attacked both the susceptible variety Monarch, and the variety Gothland, which was immune to the parental buff race. The inter-specific hybrids combined some of the factors for pathogenicity of both parents. It appears that new races of buff smut can be produced readily by crossing any buff race with normal black races of U. avenae or of U. kolleri.

Many new parasitic races of Tilletia that cause bunt or stinking smut of wheat have been produced by hybridization between T. caries and T. foetida. Although some hybrid races are less virulent than either parent, others combine factors for pathogenicity of both parents. Crosses between T. foetida and T. caries can result in new morphologic types also. As an example, the hybrid sori and chlamydospores were smaller in one cross than those of either parent. Hybrids also may have varying degrees of spore-wall reticulations, and size of chlamydospores can vary considerably. Such hybrids occur in nature also, and in some cases have been given taxonomic rank, such as T. caries intermedia.

That there are innumerable species, parasitic races, and biotypes of plant disease fungi is a banal truism. That innumerable new biotypes and races can be produced by mutation, hybridization, and heterocaryosis has been shown by extensive experimentation. That new biotypes and parasitic races often appear in nature is known from long-continued observation. The new biotypes and races, even though they were produced infrequently, could still become widely prevalent in a short time because many fungi are prodigiously prolific and can be disseminated widely and quickly by the wind. A single kernel of smutted wheat contains between 2 million and 12 million spores; a single pustule of wheat stem rust may contain a quarter of a million spores; stem rust may produce 70 billion spores on a single barberry bush; there are about 50,000 billion rust spores on one acre of fairly heavily rusted wheat. Countless billions of spores are literally carried on the wings of the wind. Many of the new biotypes that are produced in these enormous populations are not dangerous, but some are. The potentially rapid multiplication of new biotypes and parasitic races of the fungi and the potentially rapid spread make the more virulent ones potentially dangerous, and too often potentiality has become reality.

The practical implications of the vast variation in plant pathogenic fungi are manifold, but the most important is the menace to food supplies. Some of the most devastating epidemic diseases of basic food and feed crops can be controlled economically only by the development of disease-resistant varieties. New parasitic races have repeatedly appeared to attack these varieties in the past and the menace still exists for the future. The genetic diversity and phenotypic variability of many of the most destructive pathogens are so great as to create extremely complex problems of disease control in the present and to raise the question as to how complex they can become in the future. How much virulence can nature put into plant pathogens and how much resistance can man put into crop plants? This is one of the most important questions for present and future agriculture. It can be answered only by basic studies to determine the limits of genic combinations for virulence in pathogens and for productivity and disease resistance in crop plants.

TERM ADAPTATION OF FUNGI

The term adaptation, as used here, means the capacity of a single biotype to acquire and transmit the ability to do something that it either could not do originally or could not do well. Thus, race 19 of wheat stem rust can produce only small pustules on Marquis wheat. It would be a case of adaptation if genetically pure race 19 were grown on Marquis several successive rust generations and produced successively larger pustules until it acquired the ability to grow well on this variety, or at least better than it did originally. Likewise it would be adaptation if a single biotype of the corn smut fungus (Ustilago zeae) were grown for several successive generations on a medium containing arsenic and thus acquired the ability to tolerate several times as much arsenic as it did originally.

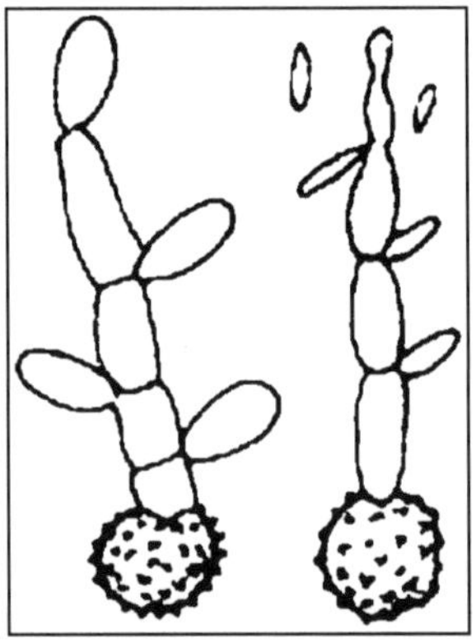

Fig. Ustilago Zeae

Between 1900 and 1910 the question of adaptation was studied by several investigators, who concluded that certain rust fungi and powdery mildews could adapt themselves quickly to resistant varieties.

Marshall Ward made many experiments with rust of bromegrasses (Puccinia dispersa) at Cambridge University, England, and published a paper on adaptive parasitism in 1903. He concluded that the brome rust could develop the ability to grow well on a resistant species of Bromus if grown on the resistant species for one or more successive rust generations.

He also concluded that there were "bridging species" of Bromus that enabled the rust to attack very resistant varieties; for example, the rust could grow well on variety A but not on variety C.

If, then, a variety B could be found that was intermediate taxonomically between A and C, the rust could be grown on B and there it acquired the ability to infect C. Variety B, therefore, was considered a bridge or "bridging host" between the susceptible A and the resistant C.

Fig. Puccinia Dispersa

In cooperative investigations between the United States Department of Agriculture and the University of Minnesota, E. C. Stakman, F. J. Piemeisel, and M. N. Levine studied the possible adaptation of stem rust of wheat and other cereals and grasses and of powdery mildew of wheat and barley, but obtained no evidence that these fungi could adapt themselves to resistant varieties by any of the methods previously tried nor by any new methods that could be devised.

They also pointed out that differential hosts must be used to separate different biologic forms from mixtures before making experiments on adaptation; otherwise what appears to be adaptation may be merely the result of the selective effect of host plants on a mixed population of the pathogen. At that time, the now obsolete term "biologic form" was used to designate what are now called varieties of stem rust; races were just being discovered within the tritici and other "biologic forms"; and eventually new facts led to new concepts and to changed terminology.

These results seemed to prove that stem rust and powdery mildew of cereal grains did not change by adaptation. But the rusts and powdery mildews are obligate parasites; it has not been possible to grow them on anything except living plants. Accordingly it was desirable to study possible adaptation or changes in virulence in fungi that can grow both in living host plants and on nutrient media that is, both as parasites and as saprophytes.

The results of extensive experiments made by J. J. Christensen and C. L. Schneider at the University of Minnesota from 1946 to 1950 with Helm inthosporiun? sativum, which causes leaf spots, foot rots, and root rots of barley, wheat, and other cereals and grasses, support the view that genetically pure lines of fungi do not change their virulence easily. The isolate of H. sativum they studied had been grown on artificial nutrient media for 28 years and had been purified by making many successive single-spore isolations. Wheat plants were inoculated with single spores; when new spores were produced, single ones were again picked and inoculated into plants with a fine needle. This process was repeated ten times in succession, and there never was any evidence of change in virulence. This extreme refinement of method was necessary because H. sativum mutates rather freely, producing many mutants that are less pathogenic and a few that are more pathogenic than the parental line.

Some of the less pathogenic lines grow very rapidly and therefore tend to overgrow the original parental line. Unless the fungus is grown under conditions that permit the recognition of mutants and their separation from the parental line, an initially pure isolate may soon comprise a diverse mixture of mutant lines, and the original line may even have been lost in the process of making periodic transfers to new tubes of nutrient media.

Obviously, then, an isolate of H. sativum derived from a single spore may change in virulence, but the change is due to the production of new biotypes resulting from mutation and not to a change in virulence of the original biotype. It would be easy to conclude that some of the smut fungi increased in virulence for resistant varieties as a result of successive passages through those varieties if the intraspecific complexity of smut species were not known.

Many investigators, notably W. A. R. Dillon-Weston in England, tested the resistance of varieties of wheat, barley, oats, and other crop plants to various smuts. When first inoculated with a collection of smut spores (chlamydospores), very little smut develops on some varieties.

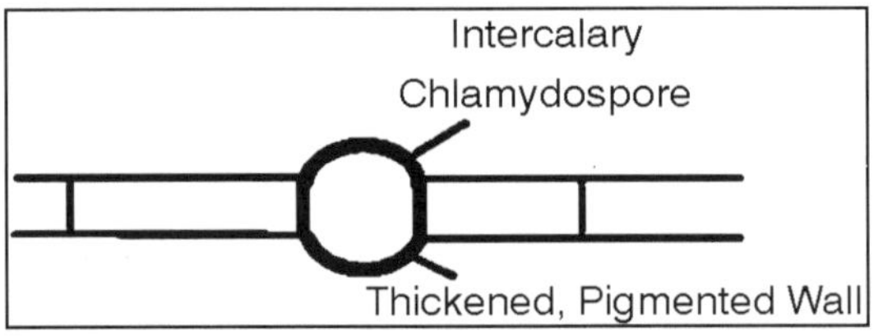

Fig. Chlamydospores

If the smut from a resistant variety is used to inoculate plants of the same variety, heavy infection may result, because nonpurified collections of smut spores are likely to comprise many parasitic races that can be separated from each other by certain varieties. Assuming that three wheat varieties, A, B, and C, are inoculated with a smut collection containing races 1, 2, and 3 in the ratio of 90:9:1 and assuming that all three races infect variety A normally, that race 2 infects B but not C, and that race 3 infects C and not B, the percentages of infection theoretically should be the following:

Obviously, if plants of variety A are then inoculated with the smut produced on A, spores of all three races will again be produced. But if variety B is inoculated with spores from B, 100 per cent of them will be of race 2; and if variety C is inoculated with spores from C, 100 per cent of the spores will be of race 3, the only one in the original mixture that can attack C. Thus, varieties B and C are biologic indicators, or differential varieties, that make it possible to find out not only which races were present in the original smut collection but also the relative percentages of each. Moreover, the apparent adaptation of the smut to varieties B and C is not adaptation at all but is due to selection of the races that can attack those varieties out of a mixture in which they were present in small amounts only.

Even after races of smuts have been isolated and purified as we have described, the degree and permanence of purity are only relative because the spores are not always exactly alike. And, even if they were, the chlamydospores of smuts are the result of sexual fusions. When the spores germinate, segregation and recombination usually occur, so that the progeny of a single spore might be diverse genetically. For practical purposes, physiologic races of smuts are usually considered to be collections of chlamydospores that behave with relative constancy in successive generations. Because of sexual recombinations and mutations within the races, however, they often comprise many biotypes, usually closely enough related so that the race behaves fairly consistently.

The degree of refinement commonly practiced in classifying smut races usually suffices for determining the relative resistance of crop varieties, but certainly not for studies on adaptation or other physiologic and genetic phenomena. It suffices for practical procedures but not for scientific understanding.

There are several other reasons why races or biotypes of fungi may appear to have changed, when in reality there has been a change in the kinds of biotypes. It has been pointed out in the discussion of Helminthosporium sadvum that apparent changes in a monosporous line may be due to the production of new biotypes by mutation, so that the culture soon comprises not only the original biotype but several others in addition; it becomes a mixed or heterogeneous population instead of a pure or homogeneous one. That the original biotype may even be lost has been shown by many experiments. And even if it is not, the behaviour of mixtures of biotypes cannot be predicted.

Different races or biotypes of pathogenic fungi may not survive equally well in mixtures. This is to be expected if the environmental conditions favour one race more than the others. But it is sometimes true when conditions, as far as can be determined, are equally favourable to the different races. Extensive studies of the relative survival of races of wheat stem rust have been made at the University of Minnesota and at Sydney University, and typical results were published by I. W. Watson in 1942 and by W. Q. Loegering in 1951.

As the principle is more important than the details, which must be ascertained for each individual mixture of each fungus, a few examples are given to illustrate differential survival of crop plants and of fungi.

H. V. Harlan and Mary L. Martini of the Department of Agriculture published in 1938 the results of experiments on the relative survival of barley varieties in a mixture grown in successive years. H. H. Lande and A. F. Swanson made similar experiments on winter wheats in Kansas and published their results in 1942. The experiments with barley and winter wheats established the important principle that certain varieties persisted better than others in mixtures in which the relative percentage of each variety was determined in the first year and in each successive year, when seed was taken each year and used for the next year's planting.

The Darwinian principles of competition, natural selection, and survival of the fittest applied to these mixtures of varieties of barley and of wheat. Watson and Loegering tested the validity of the same principle on mixtures of races of wheat stem rust.

Mixtures of other fungi may behave like those of wheat stem rust, which have been used as an example because of the relative ease and precision of identification. Paul E. Hoppe at the University of Wisconsin inoculated corn with equally pathogenic strains Of Diplodia zeae, which causes ear rot and stalk rot, and the antagonism between strains was so great that only One survived. Similarly V. F. Tapke made experiments with different races of covered smut of barley and observed that there was differential survival of races.

Fig. Diplodia Zeae

It appeared for a number of years that Phytophthora infestans, the fungus that causes late blight of potatoes, could adapt itself to resistant varieties, as Donald Reddick and associates, in their attempts to develop blight-resistant varieties in New York, observed that resistant varieties might be slightly infected when first inoculated with certain races of the blight fungus, but became more severely infected when successive inoculations were made with the blight fungus from the same variety or certain others. Helena L. G. DeBruyn made similar experiments in the Netherlands and confirmed the results of Reddick and others. H. D. Thurston and C. J. Eide, at the University of Minnesota, have obtained results that indicate that the apparent adaptation of the potato blight fungus is due, in some cases at least, to the selective effect of varieties on a mixed population of biotypes.

As an example, two races of the blight fungus infect Irish Cobbler equally well, but one attacks Cherokee and the other does not. When these races were grown in mixture on Irish Cobbler, the Cherokee race decreased rapidly in prevalence but did not disappear entirely; consequently, when the Cherokee variety is inoculated with the blight fungus from the Cobbler variety the infection is very light, because so small a proportion of the racial mixture can attack it. But the blight that does develop is caused by the Cherokee race, and when this is used to inoculate Cherokee again, there is abundant infection because all of the inoculum is potentially effective.

The present authors and some of their associates began a comprehensive investigation of adaptation in 1935.As they had not been able to obtain evidence for adaptive changes in parasitism, they investigated the possibility of adaptation to chemicals in nutrient media. Certain protozoa and bacteria were reputed to develop the ability to adapt themselves to deleterious chemicals; therefore studies were made independently by three groups of investigators to find out if this were true also of fungi.

The results were not uniform. In some instances mutation clearly accounted for what could have been considered adaptation. J. J. Christensen and several associates grew the asexual (Fusarium) stage of Gibberella zeae, which parasitizes corn and causes head blight of wheat and barley, on nutrient media containing, respectively, ethyl mercury phosphate, a widely used seed disinfectant; mercury bichloride; and malachite green in concentrations that dwarfed the growth of the fungus.

Fig. Fusarium

The fungus produced numerous clearly visible mutants on each medium. Some grew more poorly than the parent on the special media. Some grew about equally well. Some made from 5 to 15 times as much growth as the parent. Some of the mutants, in fact, grew so much faster on the mercury and malachite green media that they soon overgrew the original line completely. The relative ability of the original line and its mutants to tolerate these chemicals persisted through many generations. In the experiments there was no increase in ability to tolerate the chemicals tried, except as a result of visible mutations, which might easily have been overlooked if special methods had not been used to detect them.

E. O. Mader and C. L. Schneider grew the asexual, or conidial, stage (Monilia) of Sclerotinia fructicola, the cause of brown rot of peaches, plums, and similar fruits, on nutrient media containing injurious amounts of copper sulfate. Some of their results were comparable with those of Christensen: Mutants were produced that grew much better than the parent on the copper media and they differed also in appearance. The increased tolerance for copper was persistent and constant in the mutant lines.

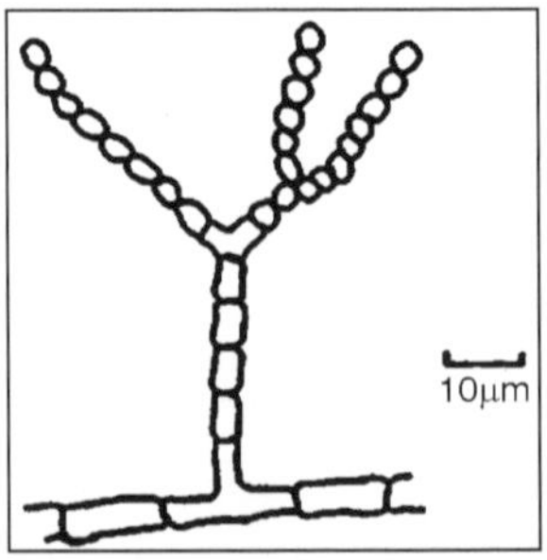

Fig. Monilia

But the fungus also acquired increasing tolerance for copper sulfate, without visible mutation, when grown in successive transfer generations on copper-containing media. When the fungus was grown for successive generations on copper-free media, however, it lost the acquired ability.

The third series of Minnesota experiments were made by Coyt Wilson, Frank Stevenson, Donald Munnecke, J. M. Daly, Elisa Hirschhorn, and E. C. Stakman. This group studied the adaptation of corn smut (Ustilago zeae) to arsenic. They grew mutable and relatively non-mutable monosporidial lines on nutrient media containing 2,400 parts per million of calcium arsenite. All lines grew poorly at first, but after having been transferred successively for about 10 generations to increasing concentrations of arsenic, all lines eventually grew as well on media containing 12,000 parts per million of calcium arsenite as they had originally grown on 2,400 parts per million, and some lines even grew on 14,000 parts per million. When grown for five successive generations or more on arsenic-free media, the arsenic-adapted lines lost their acquired ability. Numerous mutants appeared in the mutable line growing on arsenic media.

Some appeared in the relatively stable line, but only two grew slightly better than the parents on arsenic. It was the consensus of this group of investigators, some of whom worked independently at different times with different lines and with different methods, that perceptible mutations did not account for the results. All smut lines that were tried, including the least mutable ones among thousands that were studied by Stakman and many associates for more than 20 years, increased their ability to grow on arsenic. It is possible that there had been unseen mutation in physiologic characters, but, if so, these mutants escaped detection.

From extensive experiments with several fungi, at Michigan State College about 25 years ago, P. D. Caldis and G. H. Coons concluded that white variants that appeared in single-sporecultures of several fungi were "semi- permanent) variations, which differed from the parent form somatically rather than genetically and were corn parable with the "dauermodifications" described for paramecia by Victor Jollos. The genetic explanation for this type of semi-permanent change is still lacking. It is commonly observed that there may be what is sometimes called a "hangover" effect when fungi are grown for some time on one kind of nutrient medium and then transferred to another kind.

This may be accounted for partly by the production of adaptive enzymes or other chemical substances, but the exact explanation cannot be given for many of these temporary changes.

Although it cannot be asserted categorically that parasitic fungi never adapt themselves to resistant varieties of plants, there is convincing evidence that apparent adaptation often is due to the selective effect of crop varieties on a mixed population of biotypes, either because the isolate or line of the fungus was genetically diverse at the beginning, even though it appeared pure by all tests then available, or because it became mixed as a result of inconspicuous mutations. The apparent loss of virulence in a single biotype frequently unquestionably is due to the production of mutants that look like the original biotype but have lost genetic factors for virulence. The apparent acquisition of ability to tolerate injurious chemicals certainly is due to mutation in some instances but appears to occur independently of mutation in others. This is still an unsolved problem and one that requires intensive research. The difficulty of obtaining and maintaining genetically pure cultures of pathogenic fungi cannot be overemphasized in research of this kind.

Index